SPICE

The Oxford Series in Electrical and Computer Engineering

M. E. Van Valkenburg, Senior Consulting Editor
Adel S. Sedra, Series Editor, Electrical Engineering
Michael R. Lightner, Series Editor, Computer Engineering

SPICE

Second Edition

Gordon W. Roberts
McGill University

Adel S. Sedra
University of Toronto

New York   Oxford
Oxford University Press
1997

Oxford University Press

Oxford New York

Athens Auckland Bangkok Bogota
Bombay Calcutta Cape Town Dar es Salaam
Delhi Florence Hong Kong Istanbul
Karachi Kuala Lumpur Madras Madrid
Melbourne Mexico City Nairobi Paris
Singapore Taipei Tokyo Toronto

and associated companies in

Berlin Ibadan

Published by Oxford University Press, Inc.,
198 Madison Avenue, New York, New York, 10016-4314

Library of Congress Cataloging-in-Publication Data

Roberts, Gordon W., 1959-
Spice / Gordon W. Roberts, Adel S. Sedra. – 2nd ed.
p.   cm. – (The Oxford series in electrical and computer
engineering)
Includes bibliographical references and index.
ISBN 0-19-510842-6
1. Semiconductors–Design and construction–Data processing.
2. SPICE (Computer file) I. Sedra, Adel. S. II. Title.
III. Series.
TK7871.85.R52 1996   96-33744
621.3815'01'1353–dc20   CIP

Printing (last digit): 9 8 7

Printed in the United States of America
on acid-free paper

Contents

Chapter 3: Diodes   75

Chapter 4: Bipolar Junction
Transistors (BJTs)   105

Chapter 5: Field-Effect Transistors (FETs)   138

Chapter 6: Differential and Multistage Amplifiers   201

Chapter 7: Frequency Response   233

Chapter 8: Feedback   256

Chapter 9: Output Stages and Power
Amplifiers   287

Preface

Today most, if not all, microelectronic circuit design is carried out with the aid of a computer-aided circuit analysis program such as Spice. Spice, an acronym for *Simulation Program with Integrated-Circuit Emphasis*, is considered by many to be the de facto industrial standard for computer-aided circuit analysis for microelectronic circuits, mainly because it is used by the majority of IC designers in North America today. It is reasonable to say that to master electronic circuit design, one must also develop a fair amount of expertise in a circuit analysis program such as Spice. It is therefore our aim in this book to describe how Spice is used to analyze microelectronic circuits and, more important, to outline how Spice is used in the process of design itself.

It is our view that electronic circuit design begins with the assembling of various known subcircuits in a systematic manner, assuming rather simple mathematical models of transistor behavior. Keeping the mathematical model of the transistor simple enables the designer to quickly configure an electronic network and to determine through hand analysis whether the resulting circuit has the potential to meet required specifications. Once satisfied, the designer can use a more complex model for the transistors with Spice to better judge the behavior of the overall circuit as it will appear in integrated form. If the circuit fails to meet specifications, the designer can revert to a simpler computer model, preferably the same one used during the initial design, and identify the reason for the discrepancy. In this way, the designer is in a position to decide where the shortfall lies, in the designer's own understanding of circuit operation or in inherent problems caused by the nonidealities of the devices that require additional circuitry to circumvent. Examples throughout the text will emphasize the importance of this approach.

There is a tendency for new designers of electronic circuits to be awed by the analysis capability of a circuit analysis program such as Spice and to ignore the thought process provided by a hand analysis using simple models for the transistors. They usually begin their designs directly with complex transistor models, falsely believing that the results generated by the computer will provide the necessary insight into circuit operation if the circuit fails to perform as required. Experience has shown that this generally leads to poor designs, because most of the design effort is spent blindly searching for ways to improve the design using a brute-force hit-and-miss approach. It is our intention in this book to help the reader avoid this pitfall by teaching what *not* to do with Spice. This is accomplished by relating examples included here to those presented in *Microelectronic Circuits* (3rd ed.) by A. S. Sedra and K. C. Smith, where a complete hand analysis is provided. In this way, the insight provided by a hand analysis is readily available to the reader. To allow the reader to quickly locate the hand analysis in Sedra and Smith, each example of this text that has a corresponding hand analysis will be denoted by the appropriate example number in a bold box located in one of the corners of the schematic that illustrates the example.

Spice, developed in the early 1970s on mainframe computers, is now being used by undergraduates in engineering schools all across North America. Although other programs for computer-aided circuit analysis exist and are being used by various groups, none is as widely used as Spice. This is largely a result of the generous distribution policies of the Electronics Research Laboratory of the University of California, Berkeley, during the early stages of the program's development. Until recently, Spice was largely limited to mainframe computers on a time-sharing basis. However, today one can find versions of Spice for personal computers (PCs).

There are many Spice-like simulators for the PC; however, a version of PSpice, developed and distributed by the MicroSim Corporation, is available free of charge to students and their instructors and runs on IBM PCs or compatibles with at least 512 kilobytes of resident memory. Although limited to circuits containing no more than 10 transistors or 20 electrical nodes (whichever takes precedence), this simplified version is usually more than adequate for the types of circuit problems facing students at the undergraduate level. Hence, PSpice enables the integration of computer-aided circuit analysis into the undergraduate curriculum at a reasonable cost. In this text all circuit examples will be simulated using the student version of PSpice unless they exceed the circuit size limit. In these few cases, we will resort to the professional version of PSpice, which may be purchased from the MicroSim Corporation, or Spice version 2G6, distributed by the University of California.

The student version of PSpice can be obtained from the MicroSim Corporation by writing directly to them at the address

MicroSim Corporation
20 Fairbanks
Irvine, CA 92718
USA

or by accessing their home page via the World Wide Web using URL

http://www.microsim.com/.

Although we make direct reference to the text by Sedra and Smith, the material has been presented in such a way that this book can be used as a stand-alone text. It is intended for undergraduate students learning microelectronics for the first time but can also serve as a tutorial to many industry professionals on computer-aided circuit analysis using Spice.

Based on feedback received from our readers regarding the first edition, we have streamlined this edition by reducing the number of examples to give the overall size of the text a more reasonable length; it now consists of 400 pages instead of 620. For those who wish to continue using the examples that were removed from this edition, they can be accessed via the World Wide Web using URL http://www.macs.ee.mcgill.ca/~roberts/SpiceBook/. Spice decks for the second edition are also available from the same location and can be downloaded using a Web browser such as Netscape.

The organization of this book is as follows.

Chapter 1 provides an introduction to electronic circuit simulation using Spice. A brief description of the capabilities of Spice and the computer concept of electrical and electronic elements are outlined. Moreover, this chapter illustrates the role that computer-aided circuit simulation plays in the process of circuit design.

Chapter 2 demonstrates how Spice can be used to simulate the ideal and nonideal behavior of op amp circuits. Various models of op amps are introduced to assist the user in investigating the effect of op amp behavior on closed-loop circuit operation. Additional Spice commands are also introduced.

Chapters 3 to 5 present simulation details for circuits containing semiconductor diodes, zener diodes, bipolar junction transistors (BJTs), metal-oxide-semiconductor field-effect transistors (MOSFETs), junction field-effect transistors (JFETs) and metal-semiconductor field-effect transistors (MESFETs). The main objective of these chapters is to demonstrate how to simulate circuits containing active devices and how to calculate the quiescent point of each circuit from which the small-signal model of the circuit can be determined. Most Spice results are compared with those computed by hand analysis.

Chapter 6 investigates both the large- and small-signal operation of differential and multistage amplifiers using Spice. Various attributes of a current-source circuit are also investigated using Spice.

Chapter 7 investigates the frequency response behavior of various amplifier circuits using Spice. Spice is ideally suited to frequency response calculations. The accuracy of the method of short- and open-circuit time constants for estimating the 3 dB bandwidth of wideband amplifiers is investigated with Spice.

Chapter 8 deals with the topic of feedback. Stability issues are also investigated with Spice.

Chapter 9 investigates the DC and transient behavior of various types of output stages.

Chapter 10 presents several circuit simulation studies of analog integrated circuits. This includes a detailed investigation of the 741 bipolar op amp and a two-stage CMOS op amp.

Chapter 11 investigates the frequency response behavior of various types of active-RC filter circuits and LC tuned amplifiers. In addition, the reader is exposed to the use of computer-aided circuit design to fine-tune the behavior of a circuit.

Chapter 12 investigates the nonideal behavior of various types of signal-generator and waveform-shaping circuits. Many of the analyses involve the use of an op amp macro-model of a commercial op amp circuit.

Chapters 13 and 14 deal with bipolar and MOS digital circuits.

All of the chapters have in-depth problem sets that are intended to be solved using Spice. In most cases, the student version of PSpice is sufficient to solve these exercises.

We owe a debt of gratitude to a number of our friends and colleagues who assisted us by reading and commenting on selected chapters. For the first edition, these include Philip Crawley, Michael Toner, Andrew Bishop, Jean-Charles Maillet, Xavier Haurie, Stuart Banks, and Antoine Chemali, as well as the students in the Analog Microelectronics course at McGill University conducted in the Fall of 1991. The help of Pierre Parent, Mei Sum Kwan, Mark Moraes, and Jacek Slaboszewicz in preparing the manuscript was much appreciated. For the second edition, we would like to thank Alice Lium, who devoted many hours to proofreading the manuscript. Furthermore, we would like to acknowledge the assistance of a number of individuals who made this book possible. We are grateful to our developmental editor, Kysia Bebick, and to our editor, Bill Zobrist. Finally, we wish to thank our families for much encouragement and support.

We hope the book is readable and useful. As always, we appreciate comments and suggestions from the readers. They can be sent directly to G. Roberts or A. Sedra at roberts@macs.ee.mcgill.ca.

Gordon W. Roberts
McGill University

Adel S. Sedra
University of Toronto

April 28, 1996

1

Introduction to Spice

This chapter introduces electronic circuit simulation using Spice,[†] outlines its basic philosophy of circuit simulation, and explains why it has become so important for today's electronic circuit design. It then summarizes the capabilities of Spice and the computer conception of electrical and electronic elements, and gives examples.

Although this chapter is an **Introduction to Spice,** it could just as easily be an **Introduction to PSpice.** When we make specific reference to Spice by name, our discussion applies equally well to PSpice. However, statements made about PSpice do not, in general, apply to Spice.

1.1

Computer Simulation of Electronic Circuits

Traditionally, electronic circuit design was verified by building prototypes, subjecting the circuit to various stimuli (such as input signals, temperature changes, and power supply variations), and then measuring its response using appropriate laboratory equipment. Prototype building is somewhat time-consuming, but it produces practical experience from which to judge the manufacturability of the design.

The design of an integrated circuit (IC) requires a different approach. Due to the minute dimensions associated with the IC, a breadboarded version of the intended circuit will bear little resemblance to its final form. The parasitic components that are present in an IC are entirely different from the parasitic components present in the breadboard, and signal measurements obtained from the breadboard usually do not provide an accurate representation of the signals that appear on the IC.

Measuring the appropriate signals directly on the IC itself requires extreme mechanical and electrical measurement precision and is limited to specific types

---

[†]Spice is an acronym for *S*imulation *P*rogram with *I*ntegrated *C*ircuit *E*mphasis.

of measurements (e.g., it is very difficult to measure currents). Furthermore, an IC implementation does not lend itself easily to circuit modifications, which must be made at the IC mask level prior to circuit fabrication. Because of processing time, weeks may elapse between executing the modification and observing its effect.

Computer programs that simulate the performance of an electronic circuit provide a simple, cost-effective means of confirming the intended operation prior to circuit construction and of verifying new ideas that could lead to improved circuit performance. Such computer programs have revolutionized the electronics industry, leading to the development of today's high-density monolithic circuit schemes such as VLSI. Spice, the de facto industrial standard for computer-aided circuit analysis, was developed in the early 1970s at the University of California, Berkeley. Although other programs for computer-aided circuit analysis exist and are used by many different electronic design groups, Spice is the most widespread. Until recently, it was largely limited to mainframe computers on a time-sharing basis, but today various versions of Spice are available for personal computers (PCs). In general, these programs use algorithms slightly different from Spice's for performing the circuit simulations, but many of them adhere to the same input format description, elevating the Spice input syntax to a programming language.

Commercially supported versions of Spice can be considered to be divided into two types: mainframe versions and PC-based versions. Generally, mainframe versions of Spice are intended to be used by sophisticated integrated-circuit designers who require large amounts of computer power to simulate complex circuits. Commercial versions of Spice include *HSpice* from Meta-Software and *IG-Spice* from A. B. Associates. PC-based versions of Spice allow circuit simulation to be performed on a low-cost computer system. The version of interest to us in this text is *PSpice,* from MicroSim Corporation.

Although Spice was originally intended for analyzing integrated circuits, its underlying concepts are general and can apply to any type of network that can be described in terms of a basic set of electrical elements (i.e., resistors, capacitors, inductors, and dependent and independent sources). Today, Spice is often used for such applications as the analysis of high-voltage electrical networks, feedback control systems, and the effect of thermal gradients on electronic networks.

Integrated circuit design usually begins with a set of specifications (e.g., frequency response, step response, etc.). The designer's objective is to configure an electronic circuit that satisfies these specifications. This task intrigues most circuit designers, because computers so far have not yet acquired enough intelligence for it; the designer must rely on his or her knowledge of electronic circuit design. By using approximate methods of analysis, designs can be configured and quickly analyzed by hand to determine whether they have the potential for meeting the proposed specifications.

Once a design is found that might meet the specifications, the designer applies more complex models of device behavior (such as those in Spice). The behavior of the design, as simulated by Spice, is checked against the required specifications. If the circuit fails to meet the specs, the designer can return to a simpler computer model

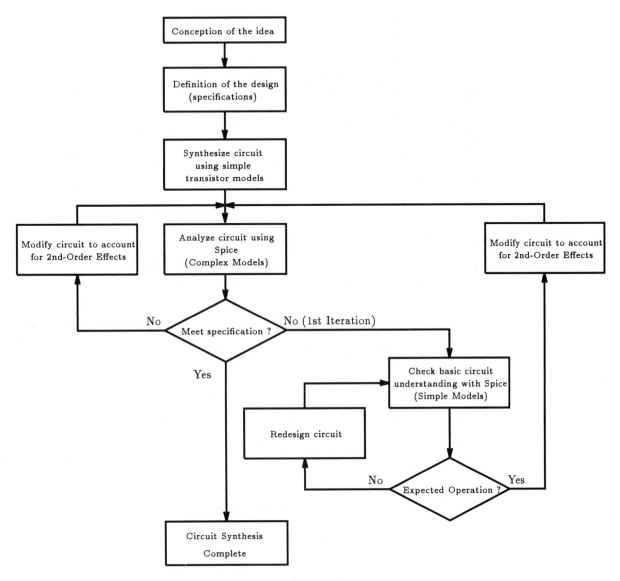

Figure 1.1 Illustrating the role of circuit simulation in the process of circuit design.

(preferably the one used during the initial design) and identify the reason for the discrepancy. When computer simulation shows that the performance of a circuit is not adequate, the designer who understands the components of the design can system-atically alter them to improve performance. (Otherwise, the designer must rely on a brute force hit-or-miss approach—which usually results in a lot of wasted effort and probably no improvement to the circuit.) The design process is depicted in the flow-chart in Fig. 1.1.

1.2

An Outline of Spice

Spice simulates the behavior of electronic circuits on a digital computer and tries to emulate both the signal generators and measurement equipment such as multimeters, oscilloscopes, curve tracers, and frequency spectrum analyzers. This section briefly outlines the analysis available in Spice and the proper way to describe a circuit to Spice. The following description is meant as only an introduction. Later chapters provide more detailed examples.

1.2.1 Types of Analysis Performed by Spice

Spice is a general-purpose circuit simulator capable of performing three main types of analysis: nonlinear DC, nonlinear transient, and linear small-signal AC circuit analysis.

Nonlinear DC analysis, or simply **DC analysis,** calculates the behavior of the circuit when a DC voltage or current is applied to it. In most cases, this analysis is performed first. The result of this analysis is commonly referred to as the *DC bias* or *operating-point* characteristic.

The **transient analysis,** probably the most important analysis type, computes the voltages and currents in the circuit with respect to time. This analysis is most meaning-ful when time-varying input signals are applied to the circuit; otherwise this analysis generates results identical to the DC analysis. The third type of analysis is a small-signal **AC analysis.** It linearizes the circuit around the DC operating point and then calculates the network variables as functions of frequency. This, of course, is equiv-alent to calculating the sinusoidal steady-state behavior of the circuit, assuming that the signals applied to the network have infinitesimally small amplitudes.

Spice is capable of performing the following other types of analysis, which are generally viewed as special cases of the three main types.

DC sweep allows a series of DC operating points to be calculated while sweeping or incrementally changing the value of an independent current or voltage source. This analysis is used largely to determine the DC large-signal transfer characteristic. A related analysis is the **transfer function** analysis. It computes the small-signal DC gain from a specified input to a specified output (i.e., voltage gain, transconductance, transresistance, or current gain) and the corresponding input and output resistance.

In a manner similar to DC sweep, **temperature analysis** allows a series of analy-ses to be performed while varying the temperature of the circuit. Because the charac-teristics of many devices depend on temperature, this facility provides a useful tool for investigating the effect of temperature variation on circuit operation. Any of the fore-going main analysis types can be performed in conjunction with temperature analysis, thus providing insight into temperature dependencies.

Sensitivity analysis indicates which components affect circuit performance most critically (critical components may require tighter manufacturing tolerances). There are usually two sensitivity analyses available. The first, **DC sensitivity,** is used to compute changes in the DC operation of the circuit that result from infinitesimally small changes in the values of various circuit components and is available in most, if not all, versions of Spice. The second, **Monte Carlo analysis,** performs multiple runs of selected analysis types (DC, AC, and transient) using a predetermined statistical distribution for the values of various components. It is rather complicated and is beyond the scope of this text.

Finally, **noise** and **Fourier** analysis procedures calculate the dynamic range of a circuit. Noise analysis calculates the noise contribution of each element, injects its effect back into the circuit, and calculates its total effect on the output node in a mean-square sense. The Fourier analysis computes the Fourier series coefficients of the circuit's voltages or currents with respect to the period of the input excitation(s).

1.2.2 Input to Spice

A circuit to be simulated must be described to Spice in a sequence of lines entered via computer terminal into a computer file commonly referred to as the **Spice input deck** or **file.** Each line is either a statement, which describes a single circuit element, or a control line, which sets model parameters, measurement nodes, or analysis types. (Simplified versions will be given later.) The first line in the Spice input deck *must* be a **title** to identify the output generated by Spice, and the last line *must* be an **.END** statement to indicate an end to the Spice input file. The sequence of the remaining lines is arbitrary. Based on the authors' experience, the format shown in Fig. 1.2 is recommended for the Spice input file's layout, but this arrangement is arbitrary, and even the authors will sometimes deviate from it in examples. **Comments** sprinkled throughout the file improve readability, identify components of the design, and explain the rationale for the simulation. They are designated by inserting an asterisk (*) as the first character of the comment line.

Title Statement

    Circuit Description

        Power Supplies / Signal Sources
        Element Descriptions
        Model Statements

    Analysis Requests

    Output Requests

.END

Figure 1.2 Suggested format for a Spice input file.

Each statement is of the free-format type; that is, the words used in each statement can be separated by either arbitrary-sized spaces (limited, of course by the line length), commas, or both. Lines longer than 80 characters (i.e., the screen width) can be continued on the next line by entering a + (plus sign) in the first column of the new line. (The original version of Spice was all uppercase, but more recent versions make no distinction between upper and lowercase. In the examples in this book the cases will be mixed for easy reading.) A number can be represented either as an integer or as floating-point using either decimal, scientific notation, or engineering scale factors. The recognized scale factors are listed in Table 1.1. Not included in this table, but recognized by Spice, is the suffix MIL, which is equivalent to $\frac{1}{1000}$ of an inch. In addition, the dimensions or units of a given value can also be appended to any element value to clarify its context. The allowed suffix types are listed in Table 1.2.

One word of caution about attaching Farads to a capacitor value. Spice, unfortunately, uses the same letter (F) to denote a scale factor of 10^{-15} (femto)—see Table 1.1. One must therefore be careful not to confuse these two suffixes in a Spice input file. Placing a single suffix F on the value of a capacitor indicates that the value

Table 1.1   Scale-factor abbreviations recognized by Spice.

Power-of-Ten Suffix Letter	Metric Prefix	Multiplying Factor
T	tera	10^{+12}
G	giga	10^{+9}
Meg	mega	10^{+6}
K	kilo	10^{+3}
M	milli	10^{-3}
U	micro	10^{-6}
N	nano	10^{-9}
P	pico	10^{-12}
F	femto	10^{-15}

Table   1.2   Element dimensions.

Spice Suffix	Units
V	volts
A	amps
Hz	hertz
Ohm	ohm (Ω)
H	henry
F	farad
Degree	degree

of the capacitor is to be expressed in femtofarads, not in farads. Thus, 1 F is 10^{-15} farads, and 1 is one farad.

The Spice simulation process begins when we draw a clearly labeled circuit diagram in which all nodes are distinctly numbered with nonnegative integers between 0—for the ground node—and 9999. All other components also must be uniquely labeled.

Figure 1.3(a) shows a linear network consisting of various resistors, capacitors, and sources (both dependent and independent) with the values indicated. In Fig. 1.3(b) each element is assigned a unique name, in which the beginning letter (e.g., R, C, E, V) indicates the element type (e.g., R for resistor, D for diode). Table 1.3 lists such key letters available in Spice. For example, the 1 Ω resistor is assigned the name R_1, the load resistor of 10 Ω is R_{load}, the 2.65 mF capacitor is C_1, the voltage-controlled voltage source is E_1, and the input sinusoidal voltage source is v_i. The ground node is labeled 0, and nongrounded nodes are assigned the numbers 1, 2, and 3.

The Spice input deck is made up of three major components: a detailed circuit description, analysis requests, and output requests. Following is an outline of the basic syntax of the various commands in these three Spice components.

Circuit Description

Each circuit element is described to Spice by an **element statement,** which contains the element name, the circuit nodes to which it is connected, and its value. Spice

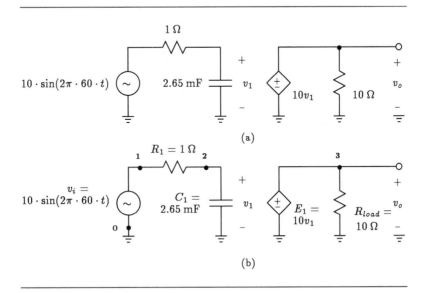

(a)

(b)

Figure 1.3 Preparing a network for Spice simulation. (a) Schematic drawing of a linear network. (b) Each element is uniquely labeled, and each node is assigned a positive number; the ground reference point is assigned the number 0.

Table 1.3   Basic element types in Spice.

First Letter Representation	Element
B	GaAs field-effect transistor (MESFET)
C	Capacitor
D	Diode
E	Voltage-controlled voltage source (VCVS)
F	Current-controlled current source (CCCS)
G	Voltage-controlled current source (VCCS)
H	Current-controlled voltage source (CCVS)
I	Independent current source
J	Junction field-effect transistor (JFET)
K	Coupled inductors
L	Inductor
M	MOS field-effect transistor (MOSFET)
Q	Bipolar transistor (BJT)
R	Resistor
V	Independent voltage source

recognizes four general classes of network elements: passive elements, independent sources, dependent sources, and active devices (e.g., diodes and transistors).

The general Spice syntax for an element description is that the first letter indicates the element type, followed by an alphanumeric name, limited to seven characters, which uniquely identifies that element. The information following the element type and name depends on the nature of the element.

Passive elements: Figure 1.4 illustrates the Spice single-line descriptor statements for an arbitrary resistor, capacitor, and inductor. The first field (or set of characters separated by blank spaces) of each statement describes its type and provides a unique name for each element. The next two fields indicate the numbers of the connecting nodes. Although these elements are bilateral, each element is assigned a positive and a negative terminal. This convention assigns direction to the current flowing through each device, as shown in Fig. 1.4, but more importantly, it specifies the polarity of the initial conditions as associated with the energy storage devices. The fourth field is used to specify the value of the passive element. Resistance is in ohms (Ohm), capacitance in farads (F), and inductance in henries (H). These values are usually positive, but the elements can also be assigned a negative value (in which case they are not passive). For either the capacitor or inductor, an initial (time zero) voltage or current condition can be specified in its fifth field (see Fig. 1.4).

For the circuit in Fig. 1.3 the element statements for passive elements R_1, C_1, and R_{load} would appear in the Spice deck:

```
R1      1 2 1Ohm
C1      2 0 2.65mF
Rload 3 0 10Ohm
```

An Outline of Spice

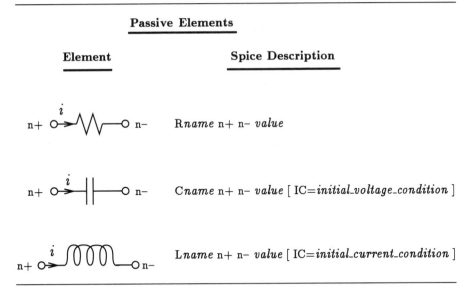

Passive Elements

Element	Spice Description

Rname n+ n− *value*

Cname n+ n− *value* [IC=*initial_voltage_condition*]

Lname n+ n− *value* [IC=*initial_current_condition*]

Figure 1.4 Spice descriptors for passive elements. Fields surrounded by [ ] are optional.

For easy reading, the dimensions of each element have been attached on the end of each parameter value.

Independent sources: Three types of independent sources can be described to Spice: a DC source, a frequency-swept AC generator, and various types of time-varying signal generators. The independent signal associated with any one source can be either voltage or current. Figure 1.5 gives a shortened summary of the description of these various sources with the kind of analysis that would be most appropriate for the source type.

The first field begins with the letter V or I, indicating a voltage or a current source, followed immediately by a unique seven-character name. The next two fields describe the nodes that connect the source to the rest of the network. The order of the nodes is important because of the signal polarity associated with the source. For example, for a voltage source, the first node is connected to the positive side and the second node to the negative side.

The Spice convention concerning current polarity (sign) is that current flowing into the positive terminal of the source is taken as positive. For a current source, positive current is pulled from the positive node (n+) and returned to the negative node (n−).

The next field indicates whether the signal source is DC, AC, or time-varying.

The remaining fields are then used to specify the characteristics of the source's signal waveform. The parameters should be obvious from the Spice syntax descriptions in Fig. 1.5. The signal level of the DC source is specified by the field labeled *value*. The peak amplitude and the phase (in degrees) of the AC source are specified in *magnitude* and *phase_degrees,* respectively. If *phase_degrees* is left blank, Spice assumes that the phase is zero.

Independent Source Representation In Spice

voltage source   current source

Spice Description		**Type Of Analysis**

$\left\{\begin{array}{c} V\,name \\ I\,name \end{array}\right\}$ n+ n− DC *value* — All Types

$\left\{\begin{array}{c} V\,name \\ I\,name \end{array}\right\}$ n+ n− AC *magnitude phase_degrees* — AC Frequency Response

$\left\{\begin{array}{c} V\,name \\ I\,name \end{array}\right\}$ n+ n− SIN (V_o V_a *freq* t_d *damp*) — Transient

$\left\{\begin{array}{c} V\,name \\ I\,name \end{array}\right\}$ n+ n− PULSE (V_1 V_2 t_d t_r t_f *PW T*) — Transient

$\left\{\begin{array}{c} V\,name \\ I\,name \end{array}\right\}$ n+ n− PWL (t_1,v_1 t_2,v_2 ... t_n,v_n) — Transient

Figure 1.5 Independent sources and their Spice descriptions. Also shown is the analysis type for which each type of source is normally used. One exception is for DC sources, which are commonly used to set bias conditions in all types of circuits.

Some of the time-varying signal sources available for transient analysis in Spice are listed in Fig. 1.5; they include a sinusoidal signal (denoted by the SIN flag), a periodic pulse signal (PULSE), and an arbitrary waveform consisting of piecewise linear segments (PWL). Figure 1.6 shows the waveform with the appropriate signal-determining parameters superimposed on each one, and the Spice description. We express these waveforms in terms of voltage, but similar waveforms can be described for current sources.

For the circuit example of Fig. 1.3, the input sinusoidal voltage source v_I described by $10 \cdot \sin(2\pi \cdot 60 \cdot t)$ would have the following Spice description:

```
Vi 1 0 SIN ( 0V 10V 60Hz  0 0 )
```

Time-Varying Signal   **Spice Description**

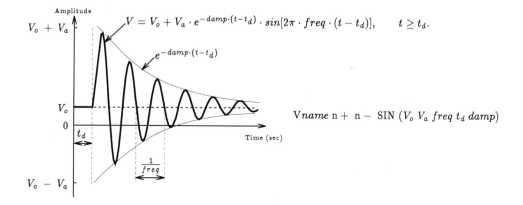

$$V = V_o + V_a \cdot e^{-damp \cdot (t - t_d)} \cdot sin[2\pi \cdot freq \cdot (t - t_d)], \qquad t \geq t_d.$$

V*name* n+ n− SIN (V_o V_a *freq* t_d *damp*)

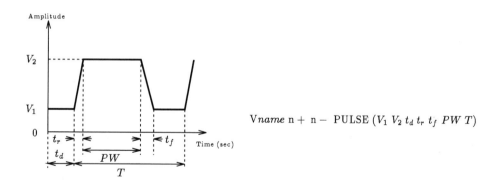

V*name* n+ n− PULSE (V_1 V_2 t_d t_r t_f *PW* *T*)

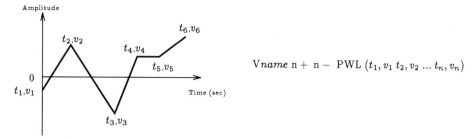

V*name* n+ n− PWL (t_1, v_1 t_2, v_2 ... t_n, v_n)

Figure 1.6 Various time-varying signals available in Spice and the corresponding element statements. Top curve: damped sinusoid; middle curve: periodic pulse waveform; bottom curve: piecewise linear waveform.

In many cases the delay time t_d and the damping factor *damp* are both zero, so we commonly shorten the above Spice statement to

```
Vi 1 0 SIN ( 0V 10V 60Hz )
```

This is acceptable to Spice.

Linear dependent sources: Spice knows about four dependent sources: voltage-controlled voltage source (VCVS), voltage-controlled current source (VCCS), current-controlled voltage source (CCVS), and current-controlled current source (CCCS). These can be either linear or nonlinear, but here we are concerned only with the linear ones. Figure 1.7 illustrates all four sources with the relationship between their input and output variables made clear and also shows the statement used to describe the element to Spice. The name of each dependent source begins with a unique letter (i.e., E, G, H, F) followed by a unique seven-character name exactly like the passive elements described earlier.

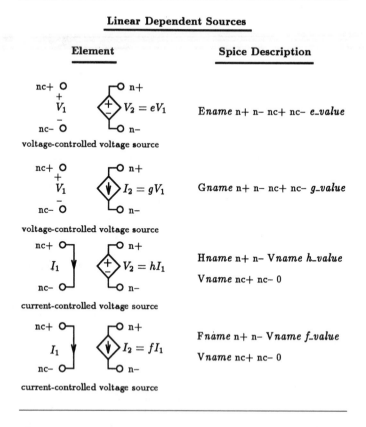

Linear Dependent Sources

Element	Spice Description

$V_2 = eV_1$ — voltage-controlled voltage source

$E name$ n+ n− nc+ nc− *e_value*

$I_2 = gV_1$ — voltage-controlled voltage source

$G name$ n+ n− nc+ nc− *g_value*

$V_2 = hI_1$ — current-controlled voltage source

$H name$ n+ n− V*name h_value*
V*name* nc+ nc− 0

$I_2 = fI_1$ — current-controlled voltage source

$F name$ n+ n− V*name f_value*
V*name* nc+ nc− 0

Figure 1.7 Linear dependent sources. Notice that the CCVS and the CCCS are both specified using two Spice statements, unlike the other two dependent sources.

Each dependent source is a two-port network, with the voltage or current at one port (terminals denoted as n+ and n−) controlled by the voltage or current at the other port (terminals denoted as nc+ and nc−). For the voltage-controlled dependent source the controlling voltage is derived directly from the network node voltages. A current-controlled source, however, must sense a current through a short circuit that is described to Spice using a zero-valued voltage source (i.e., V*name* nc+ nc− 0). This means a current-controlled source requires two statements where a voltage-controlled source needs only one—an aspect to keep in mind when working with current-controlled dependent sources.

The gain factor associated with the input and output variables is specified in the field labeled *value,* and its dimensions will depend on the type of the dependent source. For example, the voltage-controlled voltage source in the circuit of Fig. 1.3 can be described to Spice as follows:

```
E1  3  0  2  0  10
```

Active devices: The real computational strength of Spice lies in its ability to simulate the behavior of various types of active or electronic devices such as diodes, bipolar transistors, and field effect transistors. More recent versions of Spice have been extended to include gallium arsenide transistors.

Active devices are described to Spice in much the same manner as electrical elements: a statement indicating the device type and name, followed by the nodes by which it is connected to the rest of the network. The subsequent fields refer to a specific model statement found on another line of the Spice input deck. The model then contains the parameters of the device and the nature of the device model (e.g., *npn* bipolar transistor). Most active-device models are quite sophisticated and consist of many parameters, so this approach has the advantage that more than one device can reference the same model, simplifying data entry to the Spice input file. Detailed discussion on active devices will be deferred until Chapter 3 on diode circuits.

Analysis Requests

Once a circuit is described to Spice via an input file, we must specify the analyses required for our simulation. We have three main choices: DC operating-point, AC frequency response, and transient response. Table 1.4 shows their syntax plus that of the DC sweep command. Notice that each of these commands begins with a dot ".", which tells Spice that the line is a command line requesting action, not part of the circuit description.

The command requesting a DC operating-point calculation is .OP, and it comprises finding all the DC node voltages and currents and the power dissipation of all voltage sources (both dependent and independent). The .OP command automatically prints the calculation results in the output file.

In general, to determine a circuit's DC transfer characteristic, we need to vary the level of some DC source. We could run repeated .OP commands, but Spice provides a DC sweep command (.DC) that performs this calculation automatically. The syntax of this command includes the name of the DC source to be varied (*source_name*) beginning at the value marked by *start_value* and increased or decreased in steps of

Table 1.4   Main analysis commands.

Analysis Requests	Spice Command
Operating-point	.OP
DC sweep	.DC *source_name start_value stop_value step_value*
AC frequency response	.AC DEC *points_per_decade freq_start freq_stop* .AC OCT *points_per_octave freq_start freq_stop* .AC LIN *total_points freq_start freq_stop*
Transient response	.TRAN *time_step time_stop* [*no_print_time max_step_size*] [UIC] .IC V(*node₁*) = *value* V(*node₂*) = *value* ....

step_value until the value *stop_value* is reached. We can also vary the temperature of the circuit by replacing the name of the source in the field labeled *source_name* by TEMP.

With the AC frequency response command (.AC), Spice performs a linear small-signal frequency response analysis. It automatically calculates the DC operating point of the circuit, thereby establishing the small-signal equivalent circuit of all nonlinear elements. The linear small-signal equivalent circuit is then analyzed at frequencies beginning at *freq_start* and ending at *freq_stop*. Points in between are spaced logarithmically, either by decade (DEC) or octave (OCT). The number of points in a given frequency interval is specified by *points_per_decade* or *points_per_octave*. We can specify a linear frequency sweep (LIN) and the total number of points in it by *total_points*. We usually use a linear frequency sweep when the bandwidth of interest is narrow and a logarithmic sweep when the bandwidth is large.

Finally, with the transient response command (.TRAN), Spice computes the network variables as a function of time over a specified time interval. The time interval begins at time $t = 0$ and proceeds in linear steps of *time_step* seconds until *time_stop* seconds is reached. Although all transient analysis must begin at $t = 0$, we have the option of delaying the printing or plotting of the output results by specifying the *no_print_time* in the third field enclosed by the square brackets. This is a convenient way of skipping over the transient response of a network and viewing only its steady-state response. In order to have Spice avoid skipping over important waveform details within the time interval specified by *time_step*, the field designated by *max_step_size* should be chosen to be less than or equal to the *time_step*. The origins of *max_step_size* are rather involved, and interested readers should consult the *PSpice Users' Manual*. For most, if not all, examples of this text we chose the *max_step_size* equal to the *time_step*.

Before the start of any transient analysis, Spice must determine the initial values of the circuit variables, usually from a DC analysis of the circuit. If the optional UIC (use initial conditions) parameter is specified on the .TRAN statement, Spice will skip the DC bias calculation and instead use only the IC= information supplied on each capacitor or inductor statement (see Fig. 1.4). All elements without an IC= specification are assumed to have an initial condition of zero.

Initial conditions can also be set using an **.IC** command, which clamps specific nodes of the circuit at the user-specified voltage levels during the DC bias calculation.

This DC solution is then used as the initial conditions for the transient analysis. The syntax of the .IC statement is listed under the .TRAN command in Table 1.4. Note that this command is not used with the UIC flag of the transient analysis command.

Spice can perform many variations of these analyses, and they will be discussed in later chapters.

Output Requests

Circuit simulation produces a lot of data, and it would be impractical to pass all of it on to the user. Instead, Spice provides display features that enable us to specify which network variables we want to see and the best format for them. This is much like placing a measurement probe at the node of interest. Table 1.5 lists the syntax of print and plot formats.

The .PRINT command prints out variables in tabular form as a function of the independent variable associated with the analysis. With it, we must also specify the analysis (i.e., DC, AC, or TRAN) for which the specified outputs are desired. Next, we specify a list of voltage or current variables (denoted as *output_variables*). Generally, a voltage variable is specified as the voltage difference between two nodes, say *node*$_1$ and *node*$_2$, as V(*node*$_1$, *node*$_2$). When one of the nodes is omitted, it is assumed to be the ground node (0).

Spice allows only those currents flowing through independent voltage sources to be observed. Such a current would be specified by I(V*name*) where V*name* is the name of the independent voltage source through which the current is flowing. If we wish to observe a particular branch current without a voltage source, then we add a zero-valued voltage source in series with this branch and request that the current flowing through this source be printed or plotted.

Table 1.5   Spice output requests.

Output Requests	Spice Command
Print data points	.PRINT DC *output_variables*
	.PRINT AC *output_variables*
	.PRINT TRAN *output_variables*
Plot data points	.PLOT DC *output_variables* [(*lower_plot_limit, upper_plot_limit*)]
	.PLOT AC *output_variables* [(*lower_plot_limit, upper_plot_limit*)]
	.PLOT TRAN *output_variables* [(*lower_plot_limit, upper_plot_limit*)]

Notes:

1. Spice *output_variables* can be a voltage at any node V(*node*), the voltage difference between two nodes V(*node*$_1$, *node*$_2$), or the current through a voltage source I(V*name*).

2. AC *output_variables* can also be

   Vr, Ir: real part

   Vi, Ii: imaginary part

   Vm, Im: magnitude

   Vp, Ip: phase

   Vdb, Idb: decibels

3. PSpice provides a greater flexibility for specifying *output_variables*.

For a DC analysis the variables printed are the network node voltages or branch currents computed as a function of a particular DC source in the network.

For an AC analysis the output variables are sinusoidal or phasor quantities as a function of frequency and are represented by complex numbers. Spice accesses these results in the form of real and imaginary numbers or in magnitude and phase form. Magnitude can also be expressed in terms of dBs when convenient. To access a specific variable type, Table 1.5 shows how a suffix is appended to the letter V or I.

The results of a TRAN analysis are the network node voltages or branch currents computed as a function of time.

Spice's graphical feature generates a simple line plot from the list of output variables as a function of the independent variable. The syntax for the plot command is identical to that of the print command, and the .PRINT keyword is replaced by .PLOT. The range of the *y* axis given by (*lower_plot_limit*, *upper_plot_limit*) can be specified as an optional field on the .PLOT command line. (See Table 1.5.)

There are no restrictions on the number of .PRINT or .PLOT commands that can be specified in the Spice input file. This is a convenient way of controlling the number of data columns appearing in the output file.

A Simple Example

For the simple circuit of Fig. 1.3 let us create a Spice input file that would be used to compute the transient response of this circuit for three periods of a 10 V, 60 Hz sinusoidal input signal. The Spice input file for this circuit would appear:

```
Transient Response of a Linear Network

** Circuit Description **

* input signal source
Vi 1 0 SIN ( 0V 10V 60Hz )
* linear network
R1     1 2 1Ohm
C1     2 0 2.65mF
Rload 3 0 10Ohm
E1 3 0 2 0 10

** Analysis Request **
* compute transient response of circuit over three full
* periods (50 ms) of the 60 Hz sine-wave input with a 1 ms
* sampling interval
.TRAN 1ms 50ms 0ms 1ms

** Output Request **
* print the output and input time-varying waveforms
.PRINT TRAN V(3) V(1)
* plot the output and input time-varying waveforms
* set the range of the y-axis between -100 and +100 V
.PLOT   TRAN V(3) V(1) (-100,+100)

* indicate end of Spice deck
.end
```

The first line begins with the title, "Transient Response of a Linear Network," followed by a circuit description, an analysis request, and several output request statements. The final statement is an .END statement. Comments are sprinkled throughout this file to improve its readability. The transient analysis statement,

```
.TRAN 1ms 50ms 0ms 1ms
```

is a request to compute the transient behavior of this circuit over a 50 ms interval using a 1 ms time step. The results of the analysis are to be stored in resident memory beginning at time $t = 0$ and will later be available for printing or plotting. The last field specifies that the maximum step size is to be limited to 1 ms, the same value as the time step. Finally, to observe the output response behavior, we request that the voltage at the output (node 3) and the voltage appearing across the input terminal (node 1) be both printed and plotted and that the two node voltages be plotted on the same graph with the y-axis having a range of -100 to $+100$ V.

1.2.3   Output from Spice

Once the Spice input file is complete, the Spice computer program is executed with reference to this file, and the results will be found in the **Spice output file.** We can examine the contents of this file for the results of the different analyses requested in the input file. If other analysis is needed, we must alter the element statements or add additional analysis commands and then re-execute the Spice program until we have all the information we require.

For our example, the results found in Spice output file appear:

```
******* 11/19/91 ******* Student PSpice (Dec. 1987) ******* 10:24:00 *******

Transient Response of a Linear Network

****      CIRCUIT DESCRIPTION

*****************************************************************************

** Circuit Description **
* input signal source
Vi 1 0 SIN ( 0V 10V 60Hz )
* linear network
R1     1 2 1Ohm
C1     2 0 2.65mF
Rload 3 0 10Ohm
E1     3 0 2 0 10

** Analysis Request **
* compute transient response of circuit over three full
* periods (50 ms) of the 60 Hz sine-wave input with a 1 ms
* sampling interval
.TRAN 1ms 50ms 0ms 1ms

** Output Request **
* print the output and input time-varying waveforms
```

```
.PRINT TRAN V(3) V(1)
* plot the output and input time-varying waveforms
* set the range of the y-axis between -100 and +100 V
.PLOT  TRAN V(3) V(1) (-100,+100)

* indicate end of Spice deck
.end
```

******* 11/19/91 ******* Student PSpice (Dec. 1987) ******* 10:24:00 *******

Transient Response of a Linear Network

****   INITIAL TRANSIENT SOLUTION   TEMPERATURE =   27.000 DEG C

****

  NODE   VOLTAGE   NODE   VOLTAGE   NODE   VOLTAGE   NODE   VOLTAGE

(1)   0.0000 (2)   0.0000 (3)   0.0000

    VOLTAGE SOURCE CURRENTS
    NAME   CURRENT

    Vi   0.000E+00

    TOTAL POWER DISSIPATION   0.00E+00   WATTS

******* 11/19/91 ******* Student PSpice (Dec. 1987) ******* 10:24:00 *******

      Transient Response of a Linear Network

      ****   TRANSIENT ANALYSIS   TEMPERATURE =   27.000 DEG C

      ****

        TIME   V(3)   V(1)

        0.000E+00   0.000E+00   0.000E+00
        1.000E-03   6.504E+00   3.652E+00
        2.000E-03   2.120E+01   6.745E+00
        3.000E-03   3.923E+01   8.920E+00
        4.000E-03   5.645E+01   9.842E+00
        5.000E-03   6.896E+01   9.382E+00
        6.000E-03   7.398E+01   7.604E+00
        7.000E-03   7.010E+01   4.758E+00
        8.000E-03   5.739E+01   1.244E+00
        9.000E-03   3.731E+01 -2.445E+00
        1.000E-02   1.247E+01 -5.790E+00
        1.100E-02 -1.381E+01 -8.322E+00
        1.200E-02 -3.792E+01 -9.686E+00
```
```

```
1.300E-02   -5.657E+01   -9.689E+00
1.400E-02   -6.716E+01   -8.331E+00
1.500E-02   -6.825E+01   -5.803E+00
1.600E-02   -5.971E+01   -2.460E+00
1.700E-02   -4.276E+01    1.228E+00
1.800E-02   -1.977E+01    4.744E+00
1.900E-02    6.007E+00    7.594E+00
2.000E-02    3.095E+01    9.377E+00
2.100E-02    5.155E+01    9.843E+00
2.200E-02    6.492E+01    8.927E+00
2.300E-02    6.917E+01    6.757E+00
2.400E-02    6.371E+01    3.638E+00
2.500E-02    4.931E+01    8.124E-03
2.600E-02    2.798E+01   -3.623E+00
2.700E-02    2.716E+00   -6.745E+00
2.800E-02   -2.292E+01   -8.920E+00
2.900E-02   -4.535E+01   -9.842E+00
3.000E-02   -6.140E+01   -9.382E+00
3.100E-02   -6.883E+01   -7.604E+00
3.200E-02   -6.659E+01   -4.758E+00
3.300E-02   -5.500E+01   -1.244E+00
3.400E-02   -3.568E+01    2.445E+00
3.500E-02   -1.136E+01    5.790E+00
3.600E-02    1.456E+01    8.322E+00
3.700E-02    3.844E+01    9.686E+00
3.800E-02    5.692E+01    9.689E+00
3.900E-02    6.740E+01    8.331E+00
4.000E-02    6.842E+01    5.803E+00
4.100E-02    5.983E+01    2.460E+00
4.200E-02    4.283E+01   -1.228E+00
4.300E-02    1.982E+01   -4.744E+00
4.400E-02   -5.972E+00   -7.594E+00
4.500E-02   -3.093E+01   -9.377E+00
4.600E-02   -5.154E+01   -9.843E+00
4.700E-02   -6.491E+01   -8.927E+00
4.800E-02   -6.917E+01   -6.757E+00
4.900E-02   -6.371E+01   -3.638E+00
5.000E-02   -4.965E+01   -2.800E-06
```

******* 11/19/91 ******* Student PSpice (Dec. 1987) ******* 10:24:00 *******

Transient Response of a Linear Network

****   TRANSIENT ANALYSIS   TEMPERATURE =   27.000 DEG C

****

LEGEND:

*: V(3)
+: V(1)

```
     TIME        V(3)
     (*+)---------   -1.0000E+02  -5.0000E+01   0.0000E+00   5.0000E+01   1.0000E

     - - - - - - - - - - - - - - - - - - - - - - - -
0.000E+00   0.000E+00 .              .            X         .           .
1.000E-03   6.504E+00 .              .            .+*       .           .
2.000E-03   2.120E+01 .              .            . +    *  .           .
3.000E-03   3.923E+01 .              .            . +       *           .
4.000E-03   5.645E+01 .              .            .  +      .  *        .
5.000E-03   6.896E+01 .              .            . +       .    *      .
6.000E-03   7.398E+01 .              .            . +       .     *     .
7.000E-03   7.010E+01 .              .            .+        .    *      .
8.000E-03   5.739E+01 .              .            +        .  *        .
9.000E-03   3.731E+01 .              .            +.       *  .        .
1.000E-02   1.247E+01 .              .           +  . *     .           .
1.100E-02  -1.381E+01 .              .        *  + .       .           .
1.200E-02  -3.792E+01 .              .     *      +  .      .           .
1.300E-02  -5.657E+01 .            *  .           +  .      .           .
1.400E-02  -6.716E+01 .          *    .           +  .      .           .
1.500E-02  -6.825E+01 .         *     .           +  .      .           .
1.600E-02  -5.971E+01 .          *    .           +.       .           .
1.700E-02  -4.276E+01 .            . *            +        .           .
1.800E-02  -1.977E+01 .              .      *     .+        .           .
1.900E-02   6.007E+00 .              .           . X       .           .
2.000E-02   3.095E+01 .              .            . +    *  .           .
2.100E-02   5.155E+01 .              .            .  +      *           .
2.200E-02   6.492E+01 .              .            . +       .  *        .
2.300E-02   6.917E+01 .              .            . +       .    *      .
2.400E-02   6.371E+01 .              .            .+        .    *      .
2.500E-02   4.931E+01 .              .            +        .  *        .
2.600E-02   2.798E+01 .              .           +.       *  .           .
2.700E-02   2.716E+00 .              .          +  . *     .           .
2.800E-02  -2.292E+01 .              .      *     +  .      .           .
2.900E-02  -4.535E+01 .            . *            +  .      .           .
3.000E-02  -6.140E+01 .          *    .           +  .      .           .
3.100E-02  -6.883E+01 .         *     .           +  .      .           .
3.200E-02  -6.659E+01 .         *     .           +.       .           .
3.300E-02  -5.500E+01 .           *.              +        .           .
3.400E-02  -3.568E+01 .            . *            .+        .           .
3.500E-02  -1.136E+01 .              .        *   . +       .           .
3.600E-02   1.456E+01 .              .           . + *     .           .
3.700E-02   3.844E+01 .              .            . +    *  .           .
3.800E-02   5.692E+01 .              .            . +      . *         .
3.900E-02   6.740E+01 .              .            . +      .    *      .
4.000E-02   6.842E+01 .              .            . +      .    *      .
4.100E-02   5.983E+01 .              .            .+       .   *       .
4.200E-02   4.283E+01 .              .            +        .  *        .
4.300E-02   1.982E+01 .              .           +.       *  .           .
4.400E-02  -5.972E+00 .              .            X .      .           .
4.500E-02  -3.093E+01 .              .        *   +  .      .           .
4.600E-02  -5.154E+01 .            *  .           +  .      .           .
4.700E-02  -6.491E+01 .          *    .           +  .      .           .
4.800E-02  -6.917E+01 .         *     .           +  .      .           .
4.900E-02  -6.371E+01 .          *    .           +.       .           .
5.000E-02  -4.965E+01 .            *  .           +        .           .
     - - - - - - - - - - - - - - - - - - - - - - - -
```

```
JOB CONCLUDED

TOTAL JOB TIME            5.82
```

The output file from Spice contains (1) a replica of the Spice input file or circuit description, (2) the initial conditions for the transient analysis, (3) the results of the transient analysis in tabular form generated by the .PRINT command, and (4) a graphical plot of the transient results produced by the .PLOT command. The output waveform (V(3)), represented by asterisks (*), almost completes three periods of oscillation and lags behind the input voltage waveform (V(1)), represented by plus signs (+), by about 2 ms or 45 degrees. The transient portion of the output waveform is less than one complete cycle of the input signal of 60 Hz. The amplitude of the output voltage is about 70 V. Exact values corresponding to points on the waveform can be read from the two number columns on the left of the graphical plot; the first column is the time axis, and the second column is the output voltage. For example, at time $t = 4$ ms the output voltage is 56.45 V. To find the input voltage level at this time, look in the table from the .PRINT command, where we find that at $t = 4$ ms the input voltage is 9.842 V.

1.3

Output Postprocessing Using Probe

To improve the accessibility of Spice's information, commercial vendors are making available postprocessing facilities that display results graphically on a computer monitor. This provides easier access, and it generates a higher-quality graph with more detail than the line plot produced by Spice. Furthermore, cursor facilities are available that enable us to determine the numerical value of any point on the graph. This eliminates searches of long numerical tables for specific values.

In this text we use the *Probe* facility available in PSpice. Probe is designed to function like a software version of an oscilloscope. It enables us to use an interactive graphic process to look at results. In addition, Probe has many built-in computational capabilities that allow an interactive investigation of circuit behavior after a completed PSpice simulation. For example, Probe can compute and graphically display the instantaneous power dissipated by a transistor by multiplying its collector current as a function of time by the corresponding collector–emitter voltage.

Table 1.6 lists the most important of the powerful mathematical commands available in Probe, including such functions as integration and differentiation. The variable **x** used in the argument of each function represents any output variable recognized by Spice as well as network variables generated by PSpice. Table 1.7 lists the variables in PSpice and recognized by Probe.

In order to use Probe, the input file must contain a **.Probe** statement, which causes PSpice to create the necessary data file for Probe's later use. The data file will contain **all** the network variables associated with the simulation (e.g., the results of DC sweep, AC frequency response, and transient response).

Table 1.6 Probe mathematical functions.

Probe Command	Available Functions
abs(x)	$\|x\|$
sgn(x)	$+1$ (if $x > 0$), 0 (if $x = 0$), -1 (if $x < 0$)
sqrt(x)	$x^{1/2}$
exp(x)	e^x
log(x)	$\ln(x)$ (log base e)
log10(x)	$\log(x)$ (log base 10)
db(x)	$20\log(\|x\|)$ (log base 10)
pwr(x,y)	$\|x\|^y$
sin(x)	$\sin(x)$ (x in radians)
cos(x)	$\cos(x)$ (x in radians)
tan(x)	$\tan(x)$ (x in radians)
atan(x)	$\tan^{-1}(x)$ (result in radians)
arctan(x)	$\tan^{-1}(x)$ (result in radians)
d(x)	derivative of x with respect to the x-axis variable
s(x)	integral of x over the range of the x-axis variable
avg(x)	running average of x the range of the x-axis variable
rms(x)	running RMS average of x over the range of the x-axis variable
min(x)	minimum of x
max(x)	maximum of x

Table 1.7 Variables generated by PSpice and recognized by Probe.

Voltage Variables		
Node Voltage	**Voltage across a Two-Terminal Element**	**Voltage at Transistor Terminal x†**
V(*node*)		
	V(*element_name*)	Vx(*trans_name*)
V(*node₁, node₂*)		

Current Variables	
Current through a Two-Terminal Element	**Current into Transistor Terminal x‡**
I(*element_name*)	Ix(*trans_name*)

† x can be any one of the following transistor terminals:

    BJT (Q): C (collector) B (base) E (emitter) S (substrate)

    FET (B, J, M): D (drain) G (gate) S (source) B (bulk, substrate)

‡ AC suffixes can also be appended—see Table 1.5.

Hard copies of any graphical result easily can be created by Probe for future reference. In fact, these plots have been used to illustrate the results of most of the simulations in this text.

Add a .Probe statement to the foregoing Spice deck and then rerun it. On completion, the results of the PSpice simulation are stored in a special file. Invoking Probe enables us to view the simulation results directly on the monitor screen. Figure 1.8 displays the actual view of the computer screen that appeared when Probe was used to plot the voltage waveform at node 3 of the circuit in Fig. 1.3. In the lower right corner of the screen are the *x* and *y* axis values that correspond to the positions of the two cursors on the voltage waveform. The first cursor, C1, is on the second negative peak of the waveform, corresponding to a time of 31.734 ms and a voltage of -69.079 V. The second cursor, C2, is on the second positive peak of the waveform corresponding to a time of 22.734 ms and a voltage of 69.961 V. Below the coordinates is the distance between the two cursors. From this difference, we see, for instance, that the peak-to-peak amplitude of this waveform is 139.040 V.

Two or more traces can be added to the same graph, as shown in Fig. 1.9, and the two cursors can be used to access information on either graph. See the *PSpice Users' Manual* for more details on other variants of these graphical features.

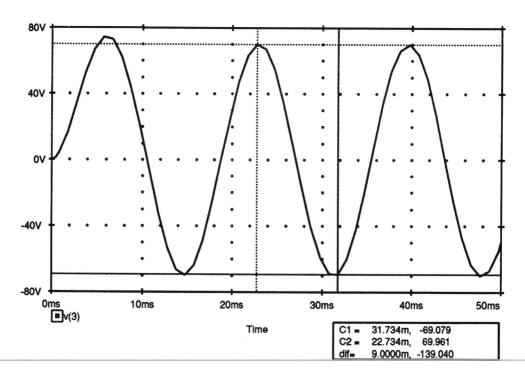

Figure 1.8 Screen display seen by user of the Probe facility of PSpice. Two cursors are superimposed on the waveform in order to read values directly off the waveform.

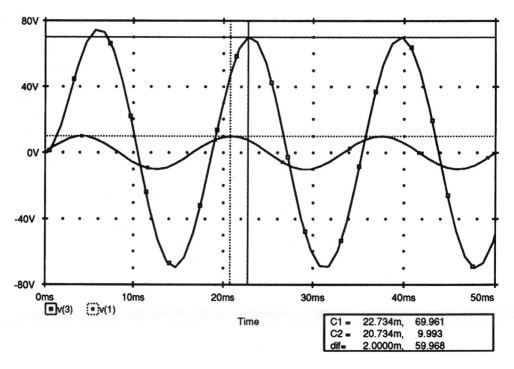

Figure 1.9 Probe screen display showing two waveforms on the same graph with a separate cursor placed on each.

1.4

Examples

The first example involves calculating the DC node voltages of a linear network; the second explores the transient behavior of a three-stage linear amplifier subject to a sine-wave input; the third illustrates how circuit initial conditions are established during a transient analysis; and the last example computes the frequency behavior of a linear amplifier.

Example 1: DC Node Voltages of a Linear Network

Figure 1.10(a) shows our first example: a rather complicated network of resistors and sources. Sources V_{S1}, V_{S2}, and I_{S1} are independent DC sources with values given on the circuit schematic. Here v_{c1} and i_{c1} are dependent sources, where v_{c1} is a current-controlled voltage source (CCVS) and i_{c1} is a voltage-controlled current source (VCCS). The voltage generated by v_{c1} is proportional to the current flowing through the 1 Ω resistor, designated by i_1. The current generated by i_{c1} is proportional to the voltage appearing across the 3 Ω resistor.

To do a Spice simulation on the circuit in Fig. 1.10(a), we must identify each element with a unique name. Then we label the ground node 0 and each other node of the circuit with a unique nonnegative integer. Fig. 1.10(b) displays the circuit diagram with labels. Further, recall from Table 1.7 that, for a CCVS, Spice must have a line specifying a zero-valued voltage source. We add, then, zero-valued voltage source V_{meter1} in series with the 1 Ω resistor (R_1) in order to sense the current through it.

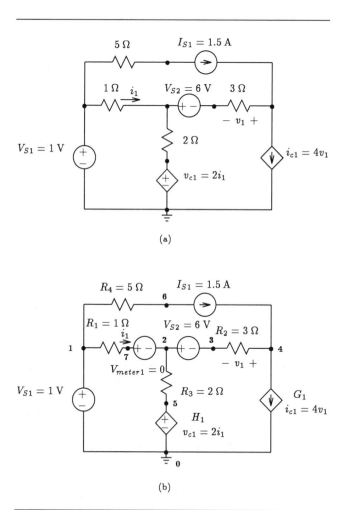

Figure 1.10 (a) Resistive network with dependent sources. (b) Each node is assigned a nonnegative integer number, and each element is assigned a unique name. Also, a zero-valued voltage source is placed in series with R_1 to monitor the current denoted by i_1.

```
Resistive Network with Dependent Sources

** Circuit Description **
* signal sources
Vs1 1 0 dc 1V
Vs2 2 3 dc 6V
Is1 6 4 dc 1.5A
* resistors
R1 1 7 1ohm
R2 3 4 3ohm
R3 2 5 2ohm
R4 1 6 5ohm
* CCVS with ammeter
H1 5 0 Vmeter1 2
Vmeter1 7 2 0
*VCCS
G1 4 0 4 3 4

** Analysis Requests **
* compute DC solution
.OP

** Output Requests **
* by default the ".OP" command prints all node voltages
.end
```

Figure 1.11   Spice input deck for the circuit shown in Fig. 1.10(b).

Next we title the Spice input deck for this circuit—in this case, "Resistive Network with Dependent Sources," as shown in Fig. 1.11. Next, under "Circuit Description," we describe each element of the circuit in one line, using the syntax discussed in previous sections. Each line in the Spice input deck corresponds directly to an element of the circuit in Fig. 1.10(b). Then, under "Analysis Requests," we request a DC operating-point analysis (.OP). This tells Spice to compute the DC node voltages of the circuit. Normally we would follow this command, and any other analysis request command, by a series of output requests, but the .OP command prints all the node voltages into the Spice output file. Finally, the ".end" statement signifies the end of the Spice input file.

We submit this input file to Spice for execution, and get the Spice output file shown in part in Fig. 1.12. (The only part not shown is a description of the input circuit, already given in Fig. 1.11.) The results here consist of two parts: a small-signal bias solution and operating-point information. The small-signal bias solution refers to the voltages on each node of the circuit relative to node 0, and the current flowing through each independent voltage source. Notice that the current

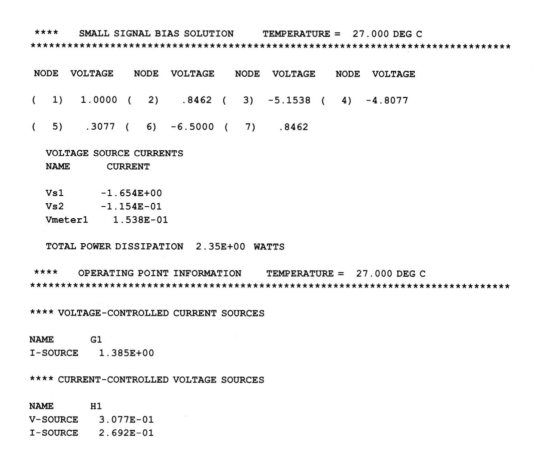

```
****    SMALL SIGNAL BIAS SOLUTION      TEMPERATURE =  27.000 DEG C
***************************************************************************

NODE   VOLTAGE     NODE   VOLTAGE     NODE   VOLTAGE     NODE   VOLTAGE

(   1)   1.0000  (   2)    .8462  (   3)  -5.1538  (   4)  -4.8077

(   5)    .3077  (   6)  -6.5000  (   7)    .8462

    VOLTAGE SOURCE CURRENTS
    NAME          CURRENT

    Vs1          -1.654E+00
    Vs2          -1.154E-01
    Vmeter1       1.538E-01

    TOTAL POWER DISSIPATION  2.35E+00  WATTS

****    OPERATING POINT INFORMATION    TEMPERATURE =  27.000 DEG C
***************************************************************************

**** VOLTAGE-CONTROLLED CURRENT SOURCES

NAME        G1
I-SOURCE    1.385E+00

**** CURRENT-CONTROLLED VOLTAGE SOURCES

NAME        H1
V-SOURCE    3.077E-01
I-SOURCE    2.692E-01
```

Figure 1.12 DC node voltages of the circuit shown in Fig. 1.10(b). Also shown are the voltages and currents associated with the independent and dependent sources.

supplied by V_{S1} and V_{S2} is negative. According to the Spice convention (positive current flows from the positive terminal of the voltage source to its negative terminal), both currents are actually flowing away from the positive terminal of each source. The total power dissipated by the circuit is also in the output file.

The second part of output file refers to the DC operating-point information for the dependent sources. Specifically, the controlled signal (not the controlling signal) of each dependent source is listed. For dependent voltage sources, the output file contains both the controlled voltage and the current it supplies to the circuit. The current that controls this source is the current that flows through V_{meter1} (1.538E-01), seen listed in the small-signal bias solution.

Example 2: Transient Response of a Three-Stage Linear Amplifier

Figure 1.13 shows our next example: a three-stage linear amplifier fed by a signal source with a source resistance of 100 kΩ. According to the hand analysis provided by Sedra and Smith in Section 1.5 of their textbook *Microelectronic Circuits* (3rd ed.), the overall voltage gain, $A_v = v_L/v_s$, was found to be 743.6 V/V; the current gain, $A_i = i_o/i_i$, to be 8.18×10^6 A/A; and the power gain, $A_p = A_v \cdot A_i$, to be 98.3 dB. Here we compute the same gains by using the transient analysis capability of Spice and the graphical postprocessing features of Probe; then we compare our results with those found by hand.

We begin our analysis by creating the Spice circuit description shown in Fig. 1.14 for the circuit of Fig. 1.13. All nodes have been prelabeled except for the ground node, which is assumed to be node 0. The input generator is a 1 volt time-varying sinusoidal voltage source of 1 Hz frequency with zero voltage offset. Using the .TRAN statement we instruct PSpice to calculate the time response of the circuit from $t = 0$ to $t = 5$ s in time steps of 10 ms. Instead of specifying an output request, we use PSpice's graphical postprocessor facility, Probe, to compute the signal gains. This requires that we enter the .Probe command into the Spice input file.

On completion of Spice, we use Probe to view the results. In Fig. 1.15(a) we plot both the input voltage (v(1)) and output voltage (v(8)) as a function of time. The range of the y-axis is set equal to the peak-to-peak value of the signal shown. Because the two signals are in phase, the voltage gain V(8)/V(1) is simply the ratio of the peak values, which in this case is 743.8 V/V.

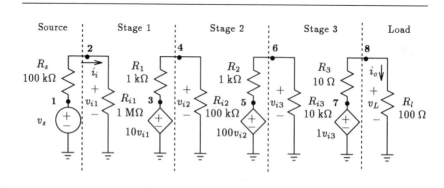

Figure 1.13   Three-stage amplifier with input signal and load.

```
Transient Response of a 3-Stage Linear Amplifier

** Circuit Description **
* signal source
Vs 1 0 sin (0V 1V 1Hz)
Rs 1 2 100k
* stage 1
Ri1 2 0 1Meg
E1 3 0 2 0 10
R1 3 4 1k
* stage 2
Ri2 4 0 100k
E2 5 0 4 0 100
R2 5 6 1k
* stage 3
Ri3 6 0 10k
E3 7 0 6 0 1
R3 7 8 10
* output load
Rl 8 0 100

** Analysis Requests **
* compute transient response from t=0 to 5s in time steps of
* 10ms with an internal time-step no greater than 10ms.
.TRAN 10ms 5s 0s 10ms

** Output Requests **
* graphical post-processor
.PROBE
.end
```

Figure 1.14   Spice input deck for circuit shown in Fig. 1.13.

Figure 1.15(b) shows the waveforms of the input current (i_i = i(Ri1)) and the load current (i_o = i(Rl)). The resulting current gain is then calculated to be 8.18 MA/A.

The waveforms in Fig. 1.15(c) are the instantaneous power delivered to the amplifier stage (v(2)*i(Ri1)) and to its load (v(8)*i(Rl)). The power gain is found to be 6.69 GW/W or 98.3 dB.

We find that we are in exact agreement with Sedra and Smith's hand analysis, as expected.

Example 3: Setting Circuit Initial Conditions during a Transient Analysis

There are three ways to set the initial conditions of a circuit at the start of a transient analysis. We demonstrate these on the simple *RC* circuit in Fig. 1.16.

For our first case, we begin our transient analysis with the initial conditions established by the DC operating point, as shown in the input file in Fig. 1.17. The transient analysis request commands Spice to compute the behavior of the circuit over a 10 ms interval using a 500 μs step interval. Spice is requested to plot the

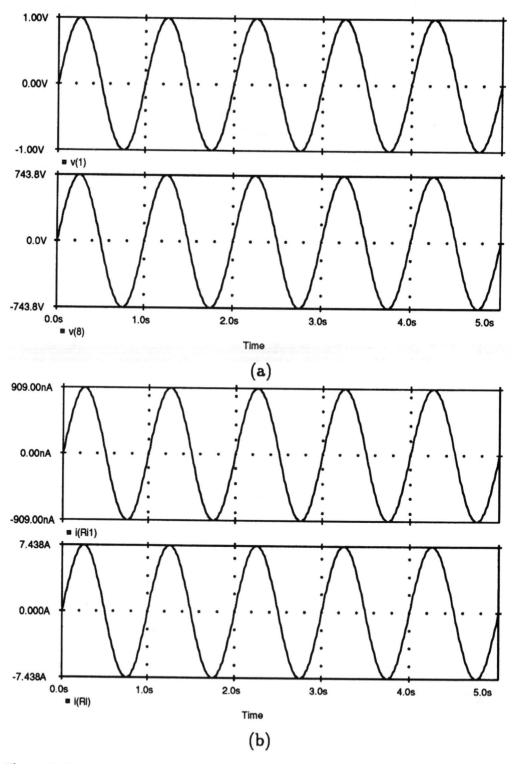

Figure 1.15 Various transient results for the circuit of Fig. 1.13: (a) input and output voltage signals; (b) input and output current signals; *continued*

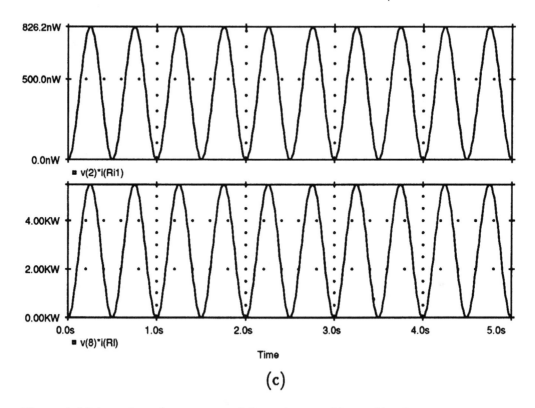

(c)

Figure 1.15 (*continued*)   (c) power delivered to amplifier and load.

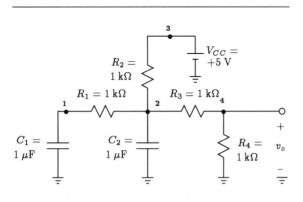

Figure 1.16   *RC* network for investigating the different ways in which Spice sets the initial conditions prior to the start of a transient analysis.

```
Investigating Initial Conditions Established by Spice

** Circuit Description **
Vcc 3 0 DC +5V
R1 1 2 1k
R2 3 2 1k
R3 2 4 1k
R4 4 0 1k
C1 1 0 1uF
C2 2 0 1uF

** Analysis Requests **
.Tran 500us 10ms 0ms 500us

** Output Requests **
.PLOT TRAN V(1) V(2) V(4)
.probe
.end
```

Figure 1.17 Spice input deck for circuit shown in Fig. 1.16. No explicit initial conditions are indicated.

voltages across each capacitor (node voltages 1 and 2), as well as the voltage that appears at the output terminals (node voltage 4) so that we can observe the initial conditions established by Spice at the start of the transient analysis and the effect that they have on the output.

The results of Spice's analysis are plotted in Fig. 1.18, with the top graph displaying the voltage across capacitor C_1, the middle graph displaying the voltage across capacitor C_2, and the bottom graph displaying the voltage at the output. Obviously, no change in the output voltage is taking place, and the node voltages remain steady with node 1 at 3.33 V, node 2 at 3.33 V, and output at 1.667 V, which suggests that the initial conditions found by Spice were also the final time values. To illustrate that these initial conditions correspond to the DC operating-point solution, we list below the results of an .OP analysis:

```
****     SMALL SIGNAL BIAS SOLUTION      TEMPERATURE =   27.000 DEG C
******************************************************************************

NODE   VOLTAGE     NODE   VOLTAGE     NODE   VOLTAGE     NODE  VOLTAGE

( 1)    3.3333 ( 2)     3.3333 ( 3)     5.0000 ( 4)    1.6667
```

A second method is to set initial conditions in the element description lines. To set the initial voltage across capacitor C_1 at +1 V, we modify the element statement:

```
C1 1 0 1uF IC=+1V
```

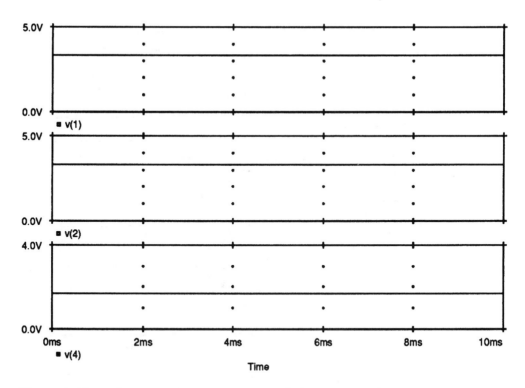

Figure 1.18   Voltage waveforms associated with the *RC* circuit shown in Fig. 1.16. No initial conditions were explicitly given; instead, the DC operating-point solution is used as the circuit initial conditions. The top graph displays the voltage appearing across capacitor C_1, the middle graph displays the voltage appearing across capacitor C_2, and the bottom graph displays the voltage appearing at the output.

and specify that Spice use this initial condition by attaching the flag "UIC" on the end of the .TRAN statement:

```
.TRAN 500us 10ms 0ms 500us UIC
```

Figure 1.19 shows the results of this analysis. Now the voltages in the circuit are changing with time. At time $t = 0$ we see that the voltage across capacitor C_1 is $+1$ V, as expected. The voltage across C_2 at this time is 0 V by default (since it was not specified in the Spice deck), and as a result, the output voltage is initially zero. With time, we see that these three voltages converge to values identical to those found in the previous case (i.e., 3.33 V, 3.33 V, and 1.667 V, respectively).

Initial conditions also can be set with the .IC command line, which is a combination of the two previous methods. The specific node voltages can be explicitly set, and the remaining nodes will take on values that result from the DC operating-point analysis (with the initial value of appropriate nodes taken into account) instead of defaulting to zero.

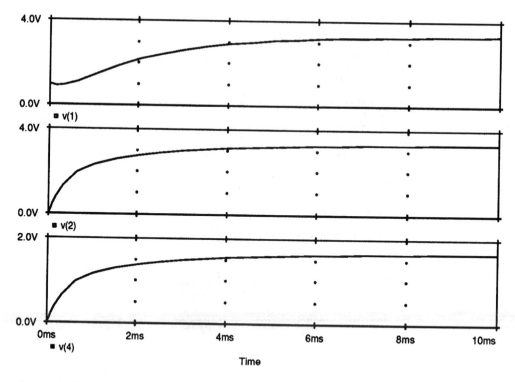

Figure 1.19 Voltage waveforms associated with the *RC* circuit shown in Fig. 1.16 when the voltage across capacitor C_1 is initially set to +1 V using IC = +1V on the element statement of this capacitor.

For example, let us set the voltage at node 1 by using the following .IC command line:

    .IC V(1)=+1V

We change the element statement for C_1 back to its original form,

    C1 1 0 1uF

and remove the UIC flag on the .TRAN statement:

    .TRAN 500us 10ms 0ms 500us

Figure 1.20 shows the results of this analysis. At time $t = 0$ the voltage at node 1 begins at +1 V, as specified. The voltage at the second node does not begin at 0 V but at 2.4 V. The output voltage is no longer zero but instead begins at 1.20 V. The final values settle to the same steady-state values found in the other two cases.

Example 4: Frequency Response of a Linear Amplifier

The final example of this chapter demonstrates how Spice is used to compute the frequency response of a linear amplifier. Figure 1.21 shows the small-signal

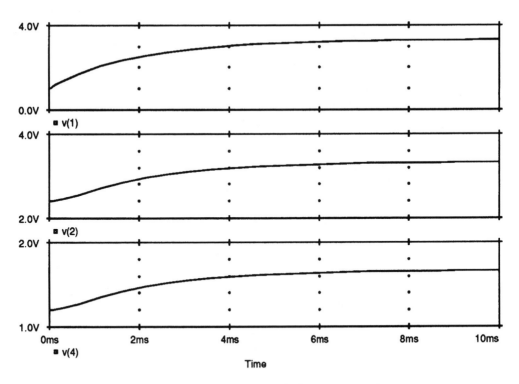

Figure 1.20   Voltage waveforms associated with the *RC* circuit shown in Fig. 1.16 when the voltage at node 1 is initially set to +1 V using an .IC command.

equivalent circuit of a one-stage amplifier, and Fig. 1.22 is the corresponding Spice input file used to describe it. The input to the circuit is a 1 V AC voltage source whose frequency will be varied between 1 Hz and 100 MHz logarithmically with 5 points per decade, as is indicated by the .AC analysis command. By selecting a 1 V input level, the output voltage level will also correspond to the voltage transfer function V_o/V_s, since $V_s = 1$.

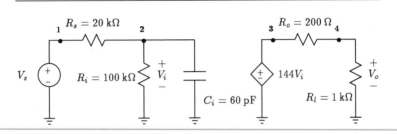

Figure 1.21   Frequency-dependent voltage amplifier with signal input and load.

```
Frequency Response Behavior of a Voltage Amplifier

** Circuit Description **
* signal source
Vs 1 0 AC 1V
Rs 1 2 20k
* frequency-dependent amplifier
Ri 2 0 100k
Ci 2 0 60p
Eamp 3 0 2 0 144
Ro 3 4 200
* load
Rl 4 0 1k

** Analysis Requests **
* compute AC frequency response from 1 Hz to 100 MHz
*   using 5 frequency steps per decade.
.AC DEC 5 1 100Meg

** Output Requests **
* print the magnitude  and phase of the output voltage
*   as a function of frequency
.PRINT AC Vm(4) Vp(4)
.PROBE
.end
```

Figure 1.22   Spice input deck for circuit shown in Fig. 1.21.

The frequency response behavior of this amplifier was calculated by Spice, and the magnitude and phase of the output voltage V_o were plotted using Probe. Figure 1.23 shows the plot on the screen of the computer monitor. It consists of two graphs: the top one for the magnitude response and the bottom one for the phase response of the amplifier. We can make a rough estimate that the 3 dB bandwidth ranges somewhere around 100 kHz. The cursor facility of Probe provides a better estimate, 158.5 kHz.

1.5

Spice Tips

- Spice is an acronym for *Simulation Program with Integrated-Circuit Emphasis*. It was originally developed for large mainframe computers.
- PSpice is a PC version of Spice, and a student version is distributed freely by MicroSim Corporation.
- Circuits are designed by people, not computers; Spice can only verify operations of human-designed circuits.

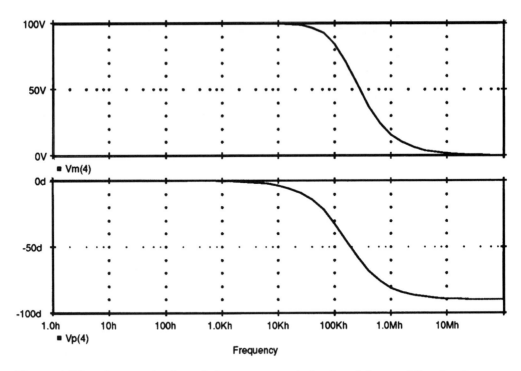

Figure 1.23 The magnitude and phase response behavior of the amplifier circuit shown in Fig. 1.21.

- A Spice input file consists of three main parts: circuit description, analysis requests, and output requests.
- The first line in a Spice input file must be a title statement, and the last line must be an .END statement.
- A circuit is described to Spice by a sequence of element statements that describe how each element is connected to the rest of the circuit and specifying its value.
- Each element type has a unique first-letter representation, (e.g., R for resistor, C for capacitor).
- Spice performs three main analyses: nonlinear DC analysis, transient analysis, and small-signal AC analysis.
- Spice can perform other analyses as special cases of the three main analysis types. One example is the DC sweep command, used to compute DC transfer characteristics.
- The results of circuit simulation are placed in an output file in either tabular or graphical form.
- PSpice is equipped with a postprocessor called Probe, which allows for interactive graphical display of simulation results.
- Probe can perform many powerful mathematical functions, such as differentiation and integration, on any network variable created during circuit simulation.

1.6

Bibliography

PSpice Users' Manual, MicroSim Corporation, Irvine, CA, Jan. 1991.

A. S. Sedra and K. C. Smith, *Microelectronic Circuits,* 3rd ed., Saunders College Publishing, Philadelphia, 1991.

A. Vladimirescu, K. Zhang, A. R. Newton, D. O. Pederson, and A. Sangiovanni-Vincentelli, *SPICE Version 2G6 User's Guide,* Dept. of Electrical Engineering and Computer Sciences, University of California, Berkeley, CA, 1981.

1.7

Problems

1.1 For each of the circuits shown in Fig. P1.1, compute the corresponding node voltages using Spice.

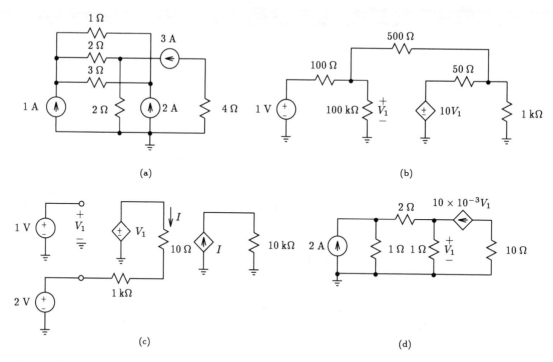

Figure P1.1

1.2 For each of the circuits shown in Fig. P1.2, compute the corresponding node voltages and branch currents indicated using Spice.

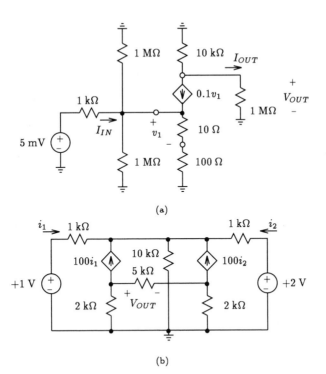

(a)

(b)

Figure P1.2

1.3 Using the simple circuit arrangement shown in Fig. P1.3, generate a voltage waveform across the 1 Ω resistor having the following described form:
(a) $10 \cdot \sin (2\pi \cdot 60 \cdot t)$
(b) $1 + 0.5 \cdot \sin (2\pi \cdot 1000 \cdot t)$
(c) $1 + 1 \cdot e^{-0.5t} \sin (2\pi \cdot 1000 \cdot t)$
(d) $v_I = \begin{cases} 1 & \text{for } t < 1\text{ms} \\ 1 + 5 \cdot \sin (2\pi \cdot 60 \cdot t) & \text{for } t \geq 1\text{ ms} \end{cases}$

Verify your results using Spice by plotting the voltage waveform that appears across the 1 Ω resistor for at least six cycles of its waveform. Use a time step that samples at least 20 points on one cycle of the waveform.

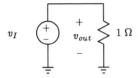

Figure P1.3

1.4 Using the simple circuit arrangement shown in Fig. P1.3, generate a voltage waveform across the 1 Ω resistor having the following described form:
(a) 10 V peak-to-peak symmetrical square wave at a frequency of 1 kHz. Let the rise and fall times be 0.1% of the total period of the waveform.
(b) 8 V peak-to-peak asymmetrical square wave at a frequency of 100 kHz having a DC offset of 2 V. Assign the rise and fall times of this waveform to be 0.1% of the total period of this waveform.

(c) 10 V peak-to-peak asymmetrical square wave at a frequency of 5 kHz having a DC offset of 2 V. Let the rise and fall times be 0.1% of the total period of the waveform. Verify your results using Spice by plotting the voltage waveform that appears across the 1 Ω resistor for at least six cycles of its waveform. Use a time step that samples at least 20 points on one cycle of the waveform.

1.5 Replace the voltage source seen in Fig. P1.3 by a current source. Generate a current into the 1 Ω resistor using the PULSE source statement of Spice such that it has a triangular shape with an amplitude of 2 mA and a period of 2 ms. The average value of the waveform is zero. Plot this current for at least six cycles of its waveform using 10 points per period. *Hint: Spice will not accept a pulse width of zero, so use a value that is at most 0.1% of the pulse period.*

1.6 Using the PULSE source statement of Spice, together with the circuit setup shown in Fig. P1.3, generate the saw-tooth voltage waveform shown in Fig. P1.6. Verify your results by plotting the voltage across the 1 Ω resistor for at least six cycles of its waveform.

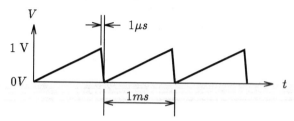

Figure P1.6

1.7 Using the PWL source statement of Spice, together with the circuit setup shown in Fig. P1.3, generate the voltage waveform shown in Fig. P1.7. Verify your results by plotting the voltage across the 1 Ω resistor for the full duration of this waveform. What voltage appears across the 1 Ω resistor if the simulation time extends beyond 50 ms?

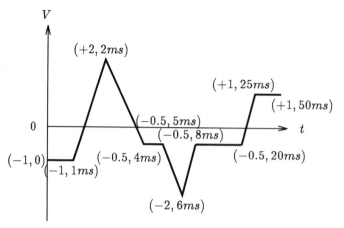

Figure P1.7

1.8 Using two voltage sources with appropriate loads, generate two signals that are nonoverlapping complementary square waves, such as those shown in Fig. P1.8. Verify your results by plotting the voltage across each load resistor for at least 10 cycles of each waveform.

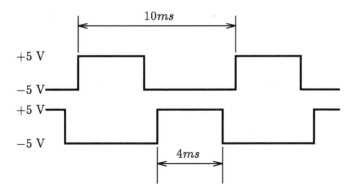

Figure P1.8

1.9 Using the simple circuit arrangement shown in Fig. P1.3, generate a 0 to 1 V step signal across the 1 Ω resistor having a rise time of no more than 1 μs. Verify your results using Spice by plotting the voltage waveform that appears across the 1 Ω resistor for at least 1 ms using a 50 μs time step.

1.10 For the first-order RC circuit shown in Fig. P1.10, simulate the behavior of this circuit with Spice subject to a 0 to 1 V step input having a rise time of no more than 10 ns. Plot the voltage waveform that appears across the 1 kΩ resistor and the 1 μF capacitor. Verify that the voltage across the capacitor changes by 63% of its final value in a time of one time constant.

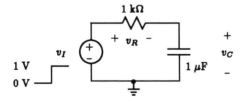

Figure P1.10

1.11 For the first-order RL circuit shown in Fig. P1.11, simulate the behavior of this circuit with Spice subject to a 0 to 1 A step input having a rise time of no more than 10 ns. Plot the current that flows in both the 1 kΩ resistor and the 1 mH inductor. Verify that the current that flows in the inductor changes by 63% of its initial value in one time constant. *Hint: One way of monitoring the current through a resistor or inductor is to connect a zero-valued voltage source in series with that element.*

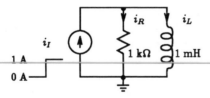

Figure P1.11

1.12 Repeat Problem 1.10 with the 1 μF capacitor initially charged to 0.5 V.

1.13 Repeat Problem 1.11 with the 1 mH inductor initially conducting a current of 2 A.

1.14 For the first-order RC circuit shown in Fig. P1.10 subject to a 1 V peak sine wave input signal of 1 kHz frequency, simulate the behavior of this circuit with Spice. Plot the voltage waveform that appears across the 1 μF capacitor for at least six cycles of the input signal. Use a time step that acquires at least 20 points per period.

1.15 Repeat Problem 1.14 with a 1 V peak symmetrical square wave-input of 10 kHz frequency. How would you describe the voltage waveform that appears across the capacitor?

1.16 For the first-order RL circuit shown in Fig. P1.11 subject to a 1 A peak-to-peak triangular input signal of 1 kHz frequency, simulate the behavior of this circuit with Spice. Plot the current waveform that flows through the 1 kΩ resistor for at least six cycles of the input signal. Use a time step that acquires at least 20 points per period. How would you describe this current waveform?

1.17 For the second-order RLC circuit shown in Fig. P1.17 subject to a 1 V step input, simulate the transient behavior of the circuit and plot the voltage waveform that appears across each element for about 40 ms. Use a time step of no more than 100 μs.

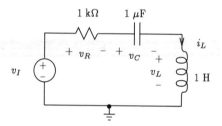

Figure P1.17

1.18 Repeat Problem 1.17 with value of the resistor decreased by a factor of 10. How do the waveforms compare with that in Problem 1.17?

1.19 Repeat Problem 1.17 with value of the resistor decreased by a factor of 100. How do the waveforms compare with that in Problem 1.17?

1.20 Compute the frequency response behavior of the RC circuit shown in Fig. P1.10 using Spice for a 1 V AC input signal. Plot both the magnitude and phase behavior of the voltage across the resistor and the voltage across the capacitor over a frequency range of 0.1 Hz to 10 MHz. Use 20 points per decade in your plot.

1.21 Compute the frequency response behavior of the RL circuit shown in Fig. P1.11 using Spice for a 1 A AC input signal. Plot both the magnitude and phase behavior of the current through the resistor and inductor over a frequency range of 1 mHz to 1 MHz. Use 10 points per decade in your plot.

1.22 Compute the frequency response behavior of the RLC circuit shown in Fig. P1.17 using Spice for a 1 V AC input signal. Plot both the magnitude and phase behavior of the voltage across the resistor, inductor, and capacitor over the frequency range of 1 Hz to 1 kHz. Use 10 points per octave in your plot.

1.23 Compute the frequency response behavior of the RLC circuit shown in Fig. P1.17 using Spice with R having values of 10, 100, and 1 kΩ. Plot the magnitude and phase response of each case and compare them. Select an appropriate frequency range and number of points that best illustrate your results.

2

Operational Amplifiers

This chapter will reinforce understanding of linear circuits constructed with operational amplifiers. A simple voltage-controlled voltage source (VCVS) representation of the op amp from which to study different types of op amp circuits will be developed; the complexity of the model will then be increased in order to reflect more realistic op amp behavior and its effect on closed-loop circuit operation. This chapter will also discuss new Spice concepts to describe the nonlinear circuit behavior of an op amp.

2.1

Modeling an Ideal Op Amp with Spice

An ideal op amp (Fig. 2.1) may be modeled as a voltage-controlled voltage source with an infinite voltage gain (i.e., $A \rightarrow \infty$). The input resistance is very high (infinite in fact), and the output resistance is considered to be zero because the output node is driven directly by a voltage source. The voltage gain is assumed to be independent of frequency. At first glance the model for the ideal op amp may seem to be trivial—apparently a one-line VCVS Spice statement. Unfortunately, Spice has no concept of infinity, so the infinite voltage gain cannot be specified to Spice. We must compromise

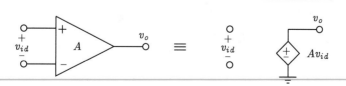

Figure 2.1 Equivalent circuit of the ideal op amp ($A \rightarrow \infty$).

our ideal model by specifying a large, but finite, voltage gain. Normally, 10^6 V/V is sufficient without any significant deviation from the ideal. Under this gain condition, we shall consider the op amp as pseudo-ideal.

2.2

Analyzing the Behavior of Ideal Op Amp Circuits

We can now use Spice to analyze the behavior of various types of op amp circuits and thus develop a better understanding of these circuits.

2.2.1 Inverting Amplifier

Consider the inverting op amp circuit in Fig. 2.2, consisting of one ideal op amp and two resistors, R_1 and R_2. We would like to determine the DC transfer function of this circuit when R_1 and R_2 assume values of 1 kΩ and 10 kΩ, respectively.

To perform this calculation using Spice, we make use of the transfer function (.TF) command mentioned in the last chapter. The transfer function analysis command computes the DC small-signal gain from the input of a circuit driven by some signal source to some prespecified network variable. It also calculates the input resistance of the circuit as seen by the input source and the output resistance as seen looking back into the circuit from the port formed by the output variable and ground. This command also can be viewed as calculating the Thevenin or Norton equivalent circuit of the network from the point of view of the input and output ports.

A general description of the syntax of the transfer function analysis command (.TF) is given in Table 2.1. The different fields of this command should be evident from the preceding discussion. The command line begins with the key word .TF followed by the output variable, either a voltage at a node or a current through a voltage source, and the name of the input signal source to which the output will be referenced. The results of the .TF command are sent directly to the Spice output file, much the same as the results of the .OP command. No .PRINT or .PLOT statement is required in the input file in order to view the results of the .TF command.

Looking at the circuit in Fig. 2.2, we can create the Spice input file in Fig. 2.3. Here the op amp is modeled as a VCVS with a voltage gain of 10^6 V/V. A 1 V DC

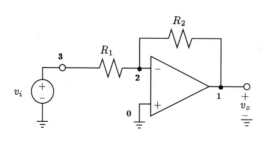

Figure 2.2   Inverting amplifier circuit.

Table 2.1   Small-signal transfer function analysis request.

Analysis Request	Spice Command
Small-signal transfer function	.TF *output_variable input_source_name*

```
Inverting Amplifier Configuration
** Circuit Description **
* signal source
Vi 3 0 DC 1v
* inverting amplifier circuit description
R1 3 2 1k
R2 2 1 10k
Eopamp 1 0 0 2 1e6
** Analysis Requests **
.TF V(1) Vi
** Output Requests **
* none required
.end
```

Figure 2.3   Spice input deck for calculating the small-signal characteristics of the circuit shown in Fig. 2.2.

signal is applied to the input of the circuit and a .TF analysis request is included to compute the DC small-signal voltage gain.

The results of the transfer function calculations are listed here with the small-signal bias solution:

```
****      SMALL SIGNAL BIAS SOLUTION        TEMPERATURE =   27.000 DEG C

 NODE    VOLTAGE      NODE    VOLTAGE      NODE    VOLTAGE      NODE    VOLTAGE

(    1)   -9.9999  (    2) 10.00E-06  (    3)    1.0000

****      SMALL-SIGNAL CHARACTERISTICS

      V(1)/Vi = -1.000E+01

      INPUT RESISTANCE AT Vi =  1.000E+03

      OUTPUT RESISTANCE AT V(1) =  0.000E+00
```

Spice calculates the voltage gain from the DC input source to the op amp output at -10, which agrees with the expected voltage gain determined by the ratio $-R_2/R_1$. In addition, the input and output resistances are listed. Combining this information with the gain calculation allows us to represent the op amp circuit of Fig. 2.2 with the equivalent circuit model shown in Fig. 2.4.

The small-signal bias solution shows that the negative terminal of the op amp (node 2) is not exactly at ground potential. This error is caused by the finite DC gain

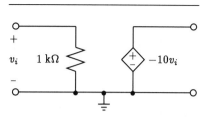

Figure 2.4 Equivalent circuit model of the inverting amplifier configuration of Fig. 2.2 as calculated by Spice.

used to model the terminal behavior of the ideal op amp and would be expected to decrease as the op amp DC gain increases. In most practical cases, an error of this magnitude is considered insignificant.

2.2.2 *The Miller Integrator*

Another important inverting op amp configuration is made by replacing R_2 in Fig. 2.2 with a capacitor C_2 to configure the inverting or Miller integrator circuit shown in Fig. 2.5. We wish to determine the transient response of a Miller integrator with $R_1 = 1$ kΩ and $C_2 = 10$ μF, subject to a 1 V step input. Then we would like to determine its AC frequency response.

The Spice input file used to calculate the transient response is given in Fig. 2.6. Most statements in this file should be self-explanatory, but we need to clarify the description provided for the step function. We approximate it with a series of piecewise linear segments. The pulse is held at 0 V for 1 ms, then made to rise to 1 V with a rise time of 1 μs, and then held at 1 V for 9 ms. If the rise time of this pulse were made to equal zero, then we would have realized a step function exactly, but Spice will not accept a waveform having a rise time of zero. We could decrease the rise time and more closely approximate the step function, but this only increases the time to complete a simulation. For this particular example a rise time of 1 μs was found to be sufficient.

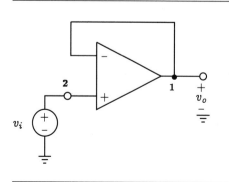

Figure 2.5 Miller integrator.

```
The Miller Integrator
** Circuit Description **
* signal source
Vi 3 0 PWL (0 0V 1ms 0V 1.001ms 1V 10ms 1V)
* components of the Miller integrator
R1 2 3 1k
C2 2 1 10uF
Eopamp 1 0 0 2 1e6
** Analysis Requests **
.TRAN 100us 5ms 0ms 100us
** Output Requests **
.PRINT TRAN V(3) V(1)
.probe
.end
```

Figure 2.6   Spice input deck for computing the step response of the circuit shown in Fig. 2.5.

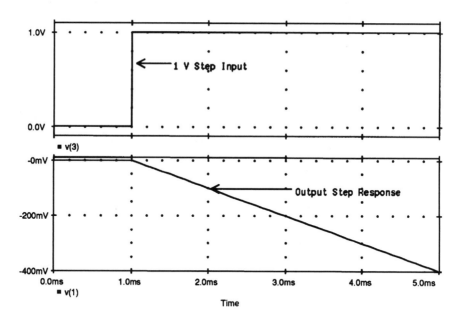

Figure 2.7   Step response of the Miller integrator circuit shown in Fig. 2.5 when $R_1 = 1 \text{ k}\Omega$ and $C_2 = 10 \text{ }\mu\text{F}$.

The results of the transient simulation are shown in Fig. 2.7. The top curve represents the input step signal and the bottom curve represents the output of the integrator. Clearly, the output is the time integral of the input (i.e., the integral of a step function is a ramp function). The ramp output decreases at -400 mV/4 ms or -100 V/s. The magnitude of this rate can easily be shown to be equal to V_I/C_2R_1, where V_I is the magnitude of the input step. It should be obvious why this circuit is called an integrator.

The AC frequency response can be computed by changing the input source in Fig. 2.6 from a time-varying PWL voltage source to an AC voltage source with the following syntax:

```
Vi 3 0 AC 1V
```

In addition, we must replace the .TRAN statement by an .AC statement specifying the range of frequencies that we are interested in. For this particular case, we are interested in a fairly broad frequency range of 1 Hz to 1 kHz, so we decided to use a log sweep of the input frequency using the .AC analysis statement

```
.AC DEC 5 1Hz 1kHz
```

The AC frequency response of the Miller integrator as calculated by Spice is displayed in Fig. 2.8. The magnitude of the output node voltage (V(1)) is large at low frequencies and rolls off at a rate of -20 dB for each decade increase in frequency. Using the Probe postprocessor available with PSpice, we found that the frequency at which the magnitude of the output voltage crosses the 0 dB level is 15.9 Hz. This corresponds exactly with the result of substituting the circuit parameters into the expression $1/(2\pi R_1 C_2)$.

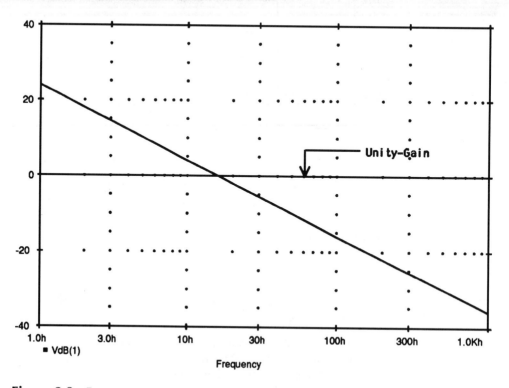

Figure 2.8 Frequency response of the Miller integrator circuit shown in Fig. 2.5 when $R_1 = 1$ kΩ and $C_2 = 10$ μF.

2.2.3   A Damped Miller Integrator

The DC instability of a Miller integrator is a result of its very high (ideally infinite) DC gain. The DC gain may be made finite by connecting a feedback resistor across integrating capacitor C_2. This nonideal integrator is known as a *damped integrator*. To obtain near-ideal response over a large frequency range, the feedback resistor should be as large as possible.

Next we compare the step response of an ideal integrator to that of a damped integrator with feedback resistors of 1 MΩ and 100 kΩ. The Spice damped integrator input file varies from that seen in Fig. 2.6 in the following few ways. First, the amplitude of the input step is reduced from 1 V to 1 mV to keep the output signal level within practical limits (i.e., between typical power supply levels). The element statement for this step input signal would be:

```
Vi 3 0 PWL ( 0 0V 1ms 0V 1.001ms 1mV 2s 1mV )
```

Second, the element statement for the feedback resistor is added to the Spice deck. For the 1 MΩ feedback resistor, we would add

```
R2 1 2 1Meg
```

and for the 100 kΩ feedback resistor

```
R2 1 2 100k
```

Third, a separate Spice deck is created for each case, and they are concatenated into one file, enabling us to compare the final results on a single graph using the Probe postprocessing facility from PSpice. As the final change, the transient analysis command statement is changed to read

```
.TRAN 100ms 2s 0s 100ms
```

We add the Spice deck for the Miller integrator shown in Fig. 2.6 (with appropriate changes made) to the file containing the two damped integrator decks so that we can make direct comparisons. The complete file containing the three Spice decks is in Fig. 2.9. Notice that no space separates the end of one Spice deck, denoted by .end, and the start of the next one. If this format is not adhered to, the file probably will be rejected by Spice.[†]

Spice processing of this input results in the three step responses shown in Fig. 2.10: one for the ideal integrator and two for the damped integrators. Up to about 0.1 s, the step responses are almost identical. Then the integrator damped with 100 kΩ begins to deviate and eventually settles toward -100 mV. The 1 MΩ damped integrator behaves similarly, but on a different time scale. It begins to significantly deviate from the ideal after about 1 s; if we were to have the transient response calculation run longer, we would see it settle to a level of -1 V.

[†]Spice versions 2G6 and later have a built-in command called .ALTER that allows the user to specify changes to the circuit without having to retype the entire file as we do here. Unfortunately, the student version of PSpice does not have this command or one that accomplishes the same thing, so we have opted to re-create a new Spice deck for each circuit change and concatenate them into one file for processing. This way, we can view the results together using Probe.

```
The Miller Integrator

** Circuit Description **
* signal sources
Vi 3 0 PWL (0 0V 1ms 0V 1.001ms 1mV 2s 1mV)
* components of the Miller integrator
R1 2 3 1k
C2 2 1 10uF
Eopamp 1 0 0 2 1e6
** Analysis Requests **
.TRAN 100ms 2s 0ms 100ms
** Output Requests
.PLOT TRAN V(1)
.probe
.end
The Damped Miller Integrator (R=1M)

** Circuit Description **
* signal sources
Vi 3 0 PWL (0 0V 1ms 0V 1.001ms 1mV 2s 1mV)
* components of the Miller integrator
R1 2 3 1k
R2 1 2 1Meg
C2 2 1 10uF
Eopamp 1 0 0 2 1e6
** Analysis Requests **
.TRAN 100ms 2s 0ms 100ms
** Output Requests
.PLOT TRAN V(1)
.probe
.end
The Damped Miller Integrator (R=100k)

** Circuit Description **
* signal sources
Vi 3 0 PWL (0 0V 1ms 0V 1.001ms 1mV 2s 1mV)
* components of the Miller integrator
R1 2 3 1k
R2 1 2 100k
C2 2 1 10uF
Eopamp 1 0 0 2 1e6
** Analysis Requests **
.TRAN 100ms 2s 0ms 100ms
** Output Requests
.PLOT TRAN V(1)
.probe
.end
```

Figure 2.9 Complete Spice deck for computing the step response of the two damped integrator circuits and one ideal integrator circuit, consisting of three separate Spice decks concatenated into one file.

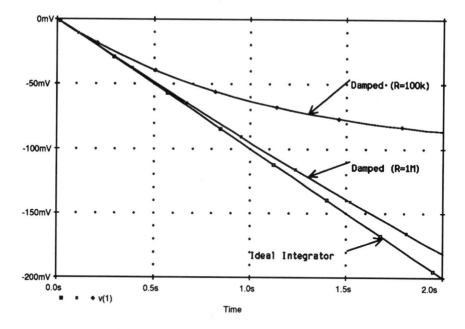

Figure 2.10   Comparing the 1 mV step response of two differently damped integrator circuits with that of an ideal Miller integrator circuit.

In either of the two damped integrator cases, we see that the step response deviates significantly from the ideal situation in about one-tenth the time constant formed by the integrating capacitor C_2 and the damping resistor R_2 (i.e., $C_2 R_2/10$). We can therefore conclude that the output of a damped integrator behaves much like an ideal integrator for times less than one tenth of such a time constant.

It is also interesting to observe the magnitude response behavior of the two damped integrators as a function of frequency and to compare them with that obtained from the Miller integrator. Add the following AC source statement and analysis command to each Spice input deck in Fig. 2.9:

```
Vi 3 0 AC 1V
.AC DEC 5 1mHz 1kHz
```

The AC analysis gives us the results shown in Fig. 2.11. In contrast to the results of the ideal integrator, the two damped integrators have a magnitude response that consists of two parts: a low-frequency component that is essentially independent of frequency and a second component that rolls off linearly with frequency at a rate of -20 dB/decade. The frequency point that divides the two regions is approximately the reciprocal of the time constant formed by the feedback resistor and capacitor [i.e., $1/(2\pi C_2 R_2)$]. For frequencies about 10 times larger than the corresponding break or 3 dB frequency of the damped integrator, both the ideal and the damped integrators have essentially identical frequency response behavior. Thus for input signal frequencies larger than 10 times the 3 dB frequency of the damped integrator, the response of the damped integrator closely approximates that of the ideal Miller integrator.

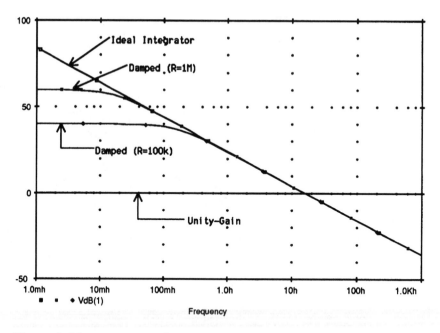

Figure 2.11 Comparing the magnitude response behavior of two damped integrator circuits with that of an ideal Miller integrator.

2.2.4 The Unity-Gain Buffer

This example will repeat the DC transfer function analysis previously performed for the inverting amplifier on a unity-gain buffer. This analysis will be used to highlight an apparent problem with Spice and the technique used to alleviate it.

Consider the unity-gain buffer shown in Fig. 2.12. The output of the op amp is fed directly back to its negative terminal, and the input signal generator v_i is connected to

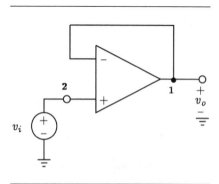

Figure 2.12 Unity-gain buffer.

```
Unity-Gain Buffer
** Circuit Description **
* signal source
Vi 2 0 DC 1V
* op amp in unity-gain configuration
Eopamp 1 0 2 1 1e6
** Analysis Requests **
.TF V(1) Vi
** Output Requests **
* none required
.end
```

Figure 2.13   Spice input deck for calculating the DC small-signal gain of the circuit shown in Fig. 2.12. Spice rejects this file because of the lack of two connections at nodes 1 and 2.

the positive terminal. The input file for this circuit is listed in Fig. 2.13, and the Spice results are found in the output file (excluding the input circuit description)

```
ERROR: Less than 2 connections at node      2
ERROR: Less than 2 connections at node      1
```

Obviously, something went wrong. Spice complains about having fewer than two connections at nodes 1 and 2. This is a topological restriction imposed by Spice to guarantee a unique solution. Prior to the start of any simulation, Spice performs a check during the equation formulation phase to ensure that this restriction is not violated. If it is, Spice reports the offending nodes.

On counting the number of connections made at nodes 1 and 2, there appear to be exactly two connections. So why does Spice complain? Spice does not consider the input port of a voltage-controlled dependent source as an element. To get on with the simulation, we must "fool" Spice into thinking that it has at least two connections at nodes 1 and 2. An elegant solution is to connect zero-valued current sources between node 1 and ground and between node 2 and ground. Each current source acts as an open circuit and has no effect on the operation of the circuit. Another solution is to connect high-valued resistors between each node in question and ground. If the resistors added are large enough relative to other resistors in the circuit, their presence will be insignificant.

Here we chose to add the two zero-valued current sources. The revised Spice input file is shown in Fig. 2.14, and the small-signal characteristics calculated by Spice are shown here:

```
****      SMALL SIGNAL BIAS SOLUTION        TEMPERATURE =   27.000 DEG C
```

NODE	VOLTAGE	NODE	VOLTAGE	NODE	VOLTAGE	NODE	VOLTAGE
(1)	1.0000	(2)	1.0000				

```
Unity-Gain Buffer
** Circuit Description **
* signal source
Vi 2 0 DC 1V
* op amp in unity-gain configuration
Eopamp 1 0 2 1 1e6
Iopen1 1 0 0A    ; redundant current sources to eliminate problem of
Iopen2 2 0 0A    ; fewer than two connections at nodes 1 and 2
** Analysis Requests **
.TF V(1) Vi
** Output Requests **
* none required
.end
```

Figure 2.14   Revised Spice input deck for the unity-gain buffer shown in Fig. 2.12, including two zero-valued current sources to rid the circuit of the problem of fewer than two connections at nodes 1 and 2.

```
****       SMALL-SIGNAL CHARACTERISTICS

       V(1)/Vi =  1.000E+00

       INPUT RESISTANCE AT Vi =  1.000E+12

       OUTPUT RESISTANCE AT V(1) =  0.000E+00
```

The input resistance of the unity-gain buffer is not infinite, as it should be, but rather 10^{12} Ω. This is an artifact of the algorithm used by Spice to calculate the DC bias solution of a network. For all practical purposes, in such a circuit a resistance of 10^{12} Ω can be considered equivalent to infinity.

2.2.5   Instrumentation Amplifier

Our next example is a two-stage instrumentation amplifier (Fig. 2.15) consisting of three op amps. Such an amplifier is usually employed as the front end of an instrument that measures a differential signal between the amplifier input terminals (20 mV in Fig. 2.15). We would like to investigate the effect of the 60 Hz common-mode signal on such a measurement. This example involves more op amps and usually poses some difficulty for the user because of the amount of effort that is required to accurately type the correct circuit description into a computer file.

To simplify matters, and reduce errors, provision has been made in Spice for defining *subcircuits.* A subcircuit is considered separate and isolated from the main circuitry except through connections through specific nodes. It can, of course, be used repeatedly in the same main circuit and can be reused for other circuits constructed from the same building blocks. Subcircuits provide a convenient way of creating a library of basic circuit components for future use. The concept is analogous to the subroutine concept found in most programming languages such as FORTRAN or C.

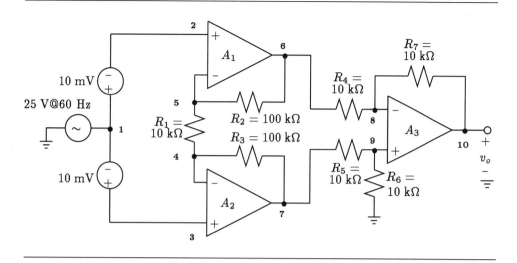

Figure 2.15   A two-stage, three–op amp instrumentation amplifier.

Spice input files are easier to read and simpler to debug when a large circuit is described using subcircuits.

The format and syntax of a subcircuit definition is displayed in Fig. 2.16. The .SUBCKT statement is followed by a unique alphanumeric name and a list of the internal nodes of the subcircuit that will be allowed to connect to the main circuit. The description of the subcircuit itself uses exactly the same language and syntax as the main circuit. The nodes of the subcircuit are local to that subcircuit and can have the same numbers as those used in the main circuit or any other subcircuit. One exception is the ground node (node 0), which is common to all circuits. Similarly, the names of the elements making up the subcircuit are also local to the subcircuit and can be the same as those used for other elements in the main circuit. To conclude the definition of the subcircuit, the final statment must be an .ENDS statement, which includes an appended S to distinguish it from the end of the Spice input file. The name of the subcircuit may also be included to clearly mark the end of the subcircuit.

Once a subcircuit has been defined, it can be incorporated into the main circuit in much the same way as a circuit element. The statement begins with a unique alphanumeric name prefixed with the letter X and is followed by a list of the main circuit nodes that connect to it. The final field specifies the name of the subcircuit that is being referenced. Obviously, there must be a one-to-one correspondence between the nodes given in this statement and the ones listed in the .SUBCKT statement. The syntax of this statement is displayed in Fig. 2.17.

.SUBCKT *subcircuit_name list_of_nodes*

    Circuit Description

        Power Supplies / Signal Sources
        Element Descriptions
        Model Statements

.ENDS [*subcircuit_name*]

Figure 2.16   Subcircuit format and syntax.

X*name*   *node_connections_to_subcircuit*   *subcircuit_name*

Figure 2.17   Accessing a subcircuit by the main circuit.

We see that the instrumentation amplifier displayed in Fig. 2.15 consists of three identical op amps: A_1, A_2, and A_3. Consider the op amp equivalent circuit (Fig. 2.1) and the following corresponding subcircuit description:

```
.subckt ideal_opamp 1 2 3
* connections:        | | |
*                 output | |
*               +ve input |
*                -ve input

Eopamp 1 0 2 3 1e6
Iopen1 2 0 0A        ; redundant connection made at +ve input terminal
Iopen2 3 0 0A        ; redundant connection made at -ve input terminal
.ends ideal_opamp
```

One element statement is used to specify the VCVS, and the other two are used to make redundant circuit connections at the floating nodes of the VCVS. A set of comment statements clarifies the list of nodes that connect the subcircuit and the main circuit. Every subcircuit needs such a description for easy reference.

Each op amp of the instrumentation amplifier can now be described to Spice using the following subcircuit calls:

```
Xop_A1   6 2 5 ideal_opamp
Xop_A2   7 3 4 ideal_opamp
Xop_A3  10 9 8 ideal_opamp
```

The complete input file is displayed in Fig. 2.18. In order to determine whether the 60 Hz common-mode signal is affecting the DC measurement performed by the instrumentation amplifier, a transient analysis is to be performed by Spice.

In our Spice results (Fig. 2.19) the voltage waveform appearing at the output of the instrumentation amplifier (node 10) is a constant DC signal of 420 mV, even though the voltages appearing at the inputs of the instrumentation amplifier (nodes 2 and 3)

```
Instrumentation Amplifier

* op amp subcircuit
.subckt ideal_opamp 1 2 3
* connections:        | | |
*               output | |
*              +ve input |
*              -ve input
Eopamp 1 0 2 3 1e6
Iopen1 2 0 0A        ; redundant connection made at +ve input terminal
Iopen2 3 0 0A        ; redundant connection made at -ve input terminal
.ends ideal_opamp

** Main Circuit **
* signal sources
Vcm 1 0 SIN (0 25V 60Hz)
Vdc1 1 2 DC 10mV
Vdc2 3 1 DC 10mV
* instrumentation amplifier
Xop_A1 6 2 5 ideal_opamp
Xop_A2 7 3 4 ideal_opamp
Xop_A3 10 9 8 ideal_opamp
R1 5 4 10k
R2 5 6 100k
R3 4 7 100k
R4 6 8 10k
R5 7 9 10k
R6 9 0 10k
R7 8 10 10k
** Analysis Requests **
.TRAN 0.1ms 66.68ms 0 0.1ms
** Output Requests **
.PRINT TRAN V(2) V(3) V(10)
.probe
.end
```

Figure 2.18 Spice input deck for calculating the transient response of the instrumentation amplifier shown in Fig. 2.15.

are time-varying voltage waveforms riding on the very small DC signal of 10 mV. We compare the DC level found at the output with that predicted[†] for the instrumentation amplifier with the constraints $R_2 = R_3$, $R_4 = R_5$, and $R_6 = R_7$:

$$v_o = -\frac{R_6}{R_4}\left(1 + \frac{2R_2}{R_1}\right)(v_1 - v_2)$$

With appropriate substitution, we are in perfect agreement. There is no evidence of the 60 Hz common-mode signal at the output. An important feature of the instrumentation

[†] Sedra and Smith in Example 2.5 of Section 2.6

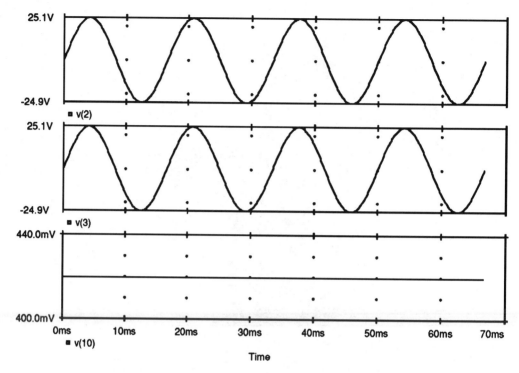

Figure 2.19   Input and output signals of the instrumentation amplifier. Notice that no AC signal is appearing at the output (V(10)) even though it is present at the two input nodes (V(2) and V(3)).

amplifier is its ability to *reject* common-mode signals while amplifying much smaller difference signals.

In the above simulation we assumed that the appropriate pairs of resistors (R_4 and R_5, R_6 and R_7) were perfectly matched. This is rarely the case, and mismatching affects the common-mode rejection capability of the instrumentation amplifier. To illustrate this, we alter one of the resistors of the difference amplifier, R_5, by 1% of its nominal value and repeat the preceding simulation with R_5 at 10.1 kΩ. The simulation results in Fig. 2.20 show a 250 mV peak-to-peak 60 Hz frequency component riding on a DC level of 420 mV at the output. Clearly, the common-mode signal is now present at the output. This illustrates the importance of having closely matched resistors in the difference amplifier portion of the instrumentation amplifier.

2.3

Nonideal Op Amp Performance

In the rest of this chapter we incorporate nonidealities of practical monolithic op amps into our simulations and obtain results that more nearly represent closed-loop op amp behavior. In Chapter 10 we shall simulate detailed op amp circuitry at the transistor level, but in this section we develop an equivalent circuit representation for the op amp that relies only on the information found on the manufacturer's data sheet.

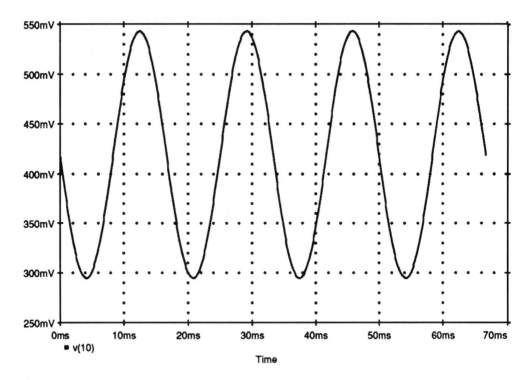

Figure 2.20 Output voltage signal of the instrumentation amplifier when there is a 1% mismatch between R_5 and R_6 in the difference amplifier of Fig. 2.15.

One method of incorporating nonidealities into the circuit model involves constructing a circuit whose terminal behavior closely resembles actual op amp terminal behavior. Obviously, we try to find a simple circuit that captures the op amp nonideal behavior. Alternatively, with newer versions of Spice, specifically PSpice, terminal behavior can be specified using a mathematical expression in either functional or piecewise linear form. This new approach to circuit modeling has come to be known as *analog behavior modeling* and forms a very elegant and powerful means of specifying nonlinear circuit behavior. We use this approach to investigate the effect of large-signal properties of an op amp on the closed-loop response. For the small-signal performance we use a lumped circuit model (an equivalent circuit).

2.3.1 Small-Signal Frequency Response of Op Amp Circuits

The differential small-signal open-loop gain of an internally compensated op amp can be mathematically described as

$$A(s) = \frac{A_0}{1 + s/\omega_b} \tag{2.1}$$

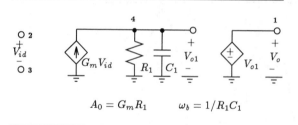

$$A_0 = G_m R_1 \qquad \omega_b = 1/R_1 C_1$$

Figure 2.21 A one-pole circuit representation of the small-signal open-loop frequency response of an internally compensated op amp.

where A_0 denotes the DC gain and ω_b is the 3 dB break frequency. Typically A_0 is very large, on the order of 10^6 V/V for modern bipolar op amps such as the 741 op amp, and ω_b typically ranges between 1 and 100 rad/s. The single-capacitor circuit in Fig. 2.21 has infinite input resistance and zero output resistance, much like the ideal op amp, and it can be shown that it has the following single-pole transfer function:

$$\frac{V_o}{V_{id}}(s) = \frac{G_m R_1}{1 + sR_1 C_1} \qquad (2.2)$$

Clearly, if we let $G_m R_1 = A_0$ and $R_1 C_1 = 1/\omega_b$, then the circuit in Fig. 2.21 can be used to model the small-signal frequency response of the op amp in Spice. As an example, the typical frequency response parameters for the 741 op amp are a DC gain of 2.52×10^5 V/V and a 3 dB frequency of 4 Hz. Using Eqs. (2.1) and (2.2) we can write two equations in terms of three unknowns. We can exercise the single degree of freedom to obtain the other two circuit parameters. That is, if we let $C = 30$ pF, then we can solve for $G_m = 0.190$ mA/V and $R_1 = 1.323 \times 10^9 \ \Omega$.

It is imperative to check that the op amp model behaves as expected in order to avoid later false conclusions. We perform this check in our next example about how the limited op amp gain and bandwidth affect the closed-loop gain.

Consider the calculation of the frequency response of the inverting amplifier shown in Fig. 2.2 using nominal gains of $-1, -10, -100,$ and -1000 and the foregoing one-pole op amp model. We use Probe to contrast the frequency response obtained in these four closed-loop cases with the open-loop response of the op amp. We concatenate the input decks into one file and submit this larger file to Spice as if it were a single job. The results of all the analyses are then found in one output file and are all accessible by Probe. We list only the Spice description for the inverting amplifier having a gain of -1 and for the subcircuit used to represent the op amp (Fig. 2.22). The description for the other amplifiers would be identical, except that R_2 would reflect the gain in each case. Also included in this concatenated file is a description for computing the open-loop frequency response of the op amp (i.e., from the circuit of Fig. 2.2 with R_1 shorted and R_2 removed).

The frequency response behavior of the inverting amplifier under different gain settings is in Fig. 2.23, with the op amp open-loop frequency response. We see the effect of increasing amplifier gain on its bandwidth. Moreover, the gain and bandwidth do not exceed those values of the open-loop frequency response.

```
Inverting Amplifier with Gain -1

* op amp subcircuit
.subckt small_signal_opamp 1 2 3
* connections:             | | |
*                   output | |
*                   +ve input |
*                       -ve input
Ginput 0 4 2 3 0.19m
Iopen1 2 0 0A       ; redundant connection made at +ve input terminal
Iopen2 3 0 0A       ; redundant connection made at -ve input terminal
R1 4 0 1.323G
C1 4 0 30p
Eoutput 1 0 4 0 1
.ends small_signal_opamp

** Main Circuit **
* signal source
Vi 3 0 AC 1V 0Degrees
Xopamp 1 0 2 small_signal_opamp
R1 3 2 1k
R2 2 1 1k
** Analysis Requests **
.AC DEC 5 0.1Hz 100MegHz
** Output Requests **
.PRINT AC V(3) V(1)
.probe
.end
```

Figure 2.22   Spice input deck for investigating the small-signal frequency response behavior of the inverting amplifier shown in Fig. 2.2 with a gain of −1. Other Spice decks can be appended to this one to enable comparison of results with Probe.

2.3.2   *Modeling the Large-Signal Behavior of Op Amps*[†]

The preceding subcircuit model is limited to circuits with small op amp output voltages. In this section we elaborate on the subcircuit so that it can model op amp behavior with both small and large input signals. The internal structure of an actual op amp consists of three parts (Fig. 2.24). The front-end stage consists of a differential-input transconductance amplifier, followed by a high-gain voltage amplifier with gain $-\mu$, together with a feedback compensation capacitor. The final stage is simply an output unity-gain buffer providing a low output resistance. By modeling the terminal

---

[†] Study of this section relies heavily on the analog behavior modeling feature of PSpice. Those readers who do not have access to a version of Spice equipped with this feature can still perform the op amp simulations presented here, provided the op amp macromodel presented in Section 12.1 is substituted for the op amp subcircuit developed in this section, albeit with the appropriate modifications to the subcircuit call.

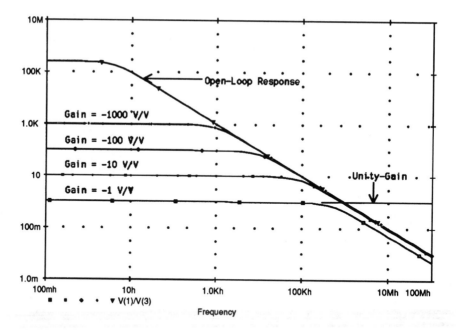

Figure 2.23 Frequency response of an inverting amplifier having nominal closed-loop gains of −1, −10, −100, and −1000. Also shown is the open-loop frequency response of the op amp used in the inverting amplifier ($A_0 = 2.52 \times 10^5$ V/V and $f_b = 4$ Hz).

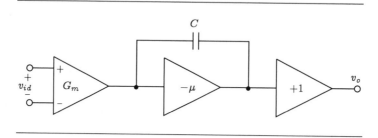

Figure 2.24 Block diagram of the internal structure of an internally compensated op amp.

behavior of each stage of the op amp, accommodating both linear and nonlinear behavior, realistic op amp characteristics can be captured by Spice without simulating detailed internal circuitry.

To describe the behavior of these three stages, we use the equivalent circuit shown in Fig. 2.25. The gain of each dependent source is expressed as an unspecified function of the controlling signal. The gains of these stages are written this way to convey to the reader that both the linear and nonlinear behavior of each of the internal stages is to be

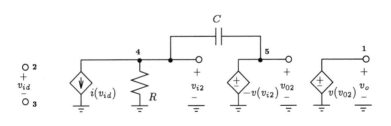

Figure 2.25 Equivalent circuit representation of the internal behavior of an internally compensated op amp. In the small-signal region of the op amp $i(v_{id}) = G_m v_{id}$, $v(v_{i2}) = \mu v_{i2}$, and $v(v_{o2}) = v_{o2}$.

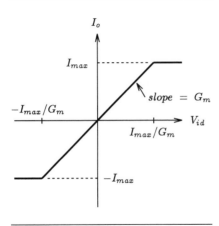

Figure 2.26 Transfer characteristic of the input transconductance amplifier of an op amp.

captured by these dependent sources. Spice has provisions for specifying nonlinear dependent sources, but describing arbitrary functions is difficult because the function must be expressible in polynomial form. Instead, newer versions of Spice, specifically PSpice, allow users to specify the control function either as a mathematical expression or in piecewise linear form.

Limited Current Output of the Front-End Transconductance Stage

The first stage of an op amp is a circuit that converts a differential-input voltage signal into a corresponding current. The operation of this stage is largely linear with transconductance G_m, but the maximum output current that it can source or sink is limited to a value I_{max}. Conversely, the transconductance stage will behave linearly, provided that the input voltage levels are restricted to the range $-I_{max}/G_m$ to I_{max}/G_m. Figure 2.26 will help the reader visualize these transfer characteristics. Capturing this behavior with a set of polynomials would be very difficult, if not impossible, but it can be easily incorporated into the circuit representation using the analog behavior modeling capability of PSpice.

The analog behavior modeling feature of PSpice is a set of extensions made to the VCVS and VCCS statements. These extensions allow the user to specify the controlled signal (voltage or current) in terms of the controlling variable (again, voltage or current) either as a mathematical expression or in piecewise linear form. For example, the level of a VCCS with its output terminals connected between node 2 and ground can be made dependent on the square of the input voltage V(1) by using the following PSpice description:

```
Gsquare 2 0 VALUE = { V(1)*V(1) }
```

The first three fields of the VCCS statement are the same as those described in Chapter 1: a unique name beginning with the letter G, followed by the nodes to which the current source output is connected. The subsequent fields are PSpice extensions. The key word VALUE indicates to PSpice that this VCCS has a functional description defined by the field enclosed by the braces {} found after the equal sign. In this particular case, the functional description is the product of the voltage at node 1 with itself. In general, the functional description consists of an arithmetic expression in terms of the network variables and arbitrary constants. The arithmetic operators allowable are +, −, *, and /. Parentheses also may be used to simplify the notation. Functions can also be included in each expression, but that discussion is beyond the scope of this text. Consult the *PSpice Users' Manual* for details.

Another extension to the PSpice voltage-controlled source statement allows the transfer function characteristics of the dependent source to be described in terms of a table of values. For example, the transfer characteristic of the input transconductance amplifier of the op amp shown in Fig. 2.26 is described using the PSpice statement

```
Ginput 2 0 TABLE { V(1) } = (-Imax/Gm,-Imax) (+Imax/Gm,+Imax)
```

The first three fields of this statement should be obvious. The key word TABLE indicates that the controlled source has a tabular description, in which the field between the braces specifies the control variable, V(1), and the table of ordered pairs on the right of the equal sign specifies the controlled source's value. The pairs are points on the graph of input versus output. Values that fall between the specified points are computed by linear interpolation. The table of ordered pairs can be viewed as points connected by a straight line. In this case, a straight line with a slope of G_m connects the points $(-I_{max}/G_m, -I_{max})$ and $(+I_{max}/G_m, +I_{max})$. If an input falls outside the limits of the table, its corresponding output is considered to be equal to the output of the smallest (or largest) specified input, thus forming the two saturating limits of the amplifier.

Output Saturation

Op amps behave linearly over a limited range of output voltages, usually bounded by the voltage levels of the power supplies. We can specify some piecewise linear function for the voltage gain of the middle-stage voltage amplifier or of the output buffer, or of both, in a manner similar to that used for the input transconductance stage just discussed. An example of a PSpice large-signal op amp subcircuit is presented later in this chapter.

Frequency Response

Within the linear region of the equivalent circuit devised for the op amp in Fig. 2.25 (i.e., $i(v_{id}) = G_m v_{id}$, $v(v_{i2}) = \mu v_{i2}$, and $v(v_{o2}) = v_{o2}$) the input-output transfer function is given by

$$\frac{V_o}{V_{id}}(s) = \frac{\mu G_m R}{1 + sC(1 + \mu)R} \qquad (2.3)$$

Comparison of this equation with the one-pole model of the frequency response for the op amp given in Eq. (2.1) shows that $A_0 = \mu G_m R$ and $\omega_b = 1/C(1 + \mu)R$. Multiplying these expressions yields the gain–bandwidth product ω_t given by

$$\omega_t = \frac{G_m}{C} \frac{\mu R}{(1 + \mu)R} \qquad (2.4)$$

Usually $\mu \gg 1$, so

$$\omega_t \approx \frac{G_m}{C} \qquad (2.5)$$

A PSpice Large-Signal Op Amp Subcircuit

If we combine the nonlinear effects described earlier with the op amp limited frequency response behavior, we can create a subcircuit description for the op amp that is valid under both large and small signal conditions. We use the op amp equivalent circuit in Fig. 2.25.

Consider an op amp characterized by a DC gain of 2.52×10^5 V/V, a unity-gain frequency of 1 MHz, and a slew rate of 0.633 V/μs. Further, we assume that the internal compensation capacitor C is 30 pF and that the op amp output stage saturates at ± 10 V. The slew rate of an op amp is related to the transconductance-stage current limit I_{max} and to the capacitor C according to

$$SR = \frac{I_{max}}{C} \qquad (2.6)$$

Thus, for this op amp example I_{max} is limited to 19 μA. Likewise, the transconductance of the first stage is deduced from Eq. (2.5) to be 0.19 mA/V. The two parameters μ and R remain, but unfortunately they cannot be uniquely determined from the information provided. We assume $R = 2.5 \times 10^6$ Ω and derive μ from $A_0 = \mu G_m R$ to get $\mu = 529$ V/V.

A subcircuit capable of capturing both the small- and large-signal behavior of the op amp described earlier is listed here (refer to Fig. 2.25):

```
.subckt large_signal_opamp 1 2 3
* connections:              | | |
*                    output | |
*                  +ve input |
*                    -ve input
R 4 0 2.5Meg
C 4 5 30p
Ginput 4 0 Table {V(2)-V(3)} = (-0.1V,-19uA) (+0.1V,19uA)
```

```
Emiddle 5 0 4 0 -529
Eoutput 1 0 Table {V(5)} = (-10V,-10V)  (10V,10V)
.ends large_signal_opamp
```

2.4

The Effects of Op Amp Large-Signal Nonidealities
on Closed-Loop Behavior

Now that we have created a subcircuit for the op amp that accounts for several of its nonidealities, let us explore some of the idiosyncracies of the op amp in various closed-loop configurations.

2.4.1 DC Transfer Characteristic of an Inverting Amplifier

Consider the inverting amplifier circuit shown in Fig. 2.2 with resistors $R_1 = 1$ kΩ and $R_2 = 10$ kΩ. Let us calculate the DC transfer characteristic of this circuit using PSpice, assuming that the op amp is nonideal and modeled as described in the last section for large-signal operation. The PSpice input file is given in Fig. 2.27.

The results of the DC sweep calculations are displayed in Fig. 2.28. Input signals of magnitude less than 1 V will experience a signal gain of -10 without distortion, as inferred from the slope of the line in this region. Signals exceeding this limit will not be amplified, but the output will be held at a constant voltage of ± 10 V depending on the sign of the input signal. Thus, linear circuit operation is limited to input signals less than a volt in magnitude.

To see how these transfer characteristics manifest themselves in the time domain, consider applying a sine wave of 400 mV peak amplitude at a 1 kHz frequency to the input of the amplifier, and then repeat the same analysis using an input signal that has an amplitude larger than the 1 V limit. Here we use a signal level of 1.1 V peak. We can use the same PSpice input deck as in Fig. 2.27 to calculate the amplifier DC transfer characteristic if we replace the DC source statement for the first case with the following:

```
Vi 3 0 SIN (0 400mV 1kHz)
```

and change the 400 mV parameter to 1.1 V for this new case. We replace the DC sweep command with a transient analysis command:

```
.TRAN 10u 5ms 0s 10u
```

The results of these two transient analyses are shown in Fig. 2.29. When the input level exceeds the 1 V limit, the output signal becomes clipped and the other signal is amplified without any distortion.

2.4.2 Slew-Rate Limiting

Internal amplifier saturation effects also cause slew-rate limiting, which plays an important role in determining the high-frequency operation of op amp circuits. To

```
DC Transfer Characteristics of an Inverting Amplifier with Gain -10

* op amp subcircuit
.subckt large_signal_opamp 1 2 3
* connections:              | | |
*                    output | |
*                  +ve input |
*                    -ve input
R 4 0 2.5Meg
C 4 5 30p
Ginput 4 0 Table V(2)-V(3) = (-0.1V,-19uA) (0.1V,19uA)
Emiddle 5 0 4 0 -529
Eoutput 1 0 Table V(5) = (-10V,-10V) (10V,10V)
.ends large_signal_opamp

** Main Circuit **
* signal source
Vi 3 0 DC 1V
Xopamp 1 0 2 large_signal_opamp
R1 3 2 1k
R2 2 1 10k
** Analysis Requests **
.DC Vi -15V +15V 100mV
** Output Requests **
.PLOT DC V(1)
.probe
.end
```

Figure 2.27 Spice input deck for calculating the DC transfer characteristic of an inverting amplifier containing a nonideal op amp.

better understand this effect, let us simulate a commonly used experimental setup for characterizing op amp slewing: an op amp in a unity-gain configuration and a generator supplying a voltage step as an input signal. This is the configuration in Fig. 2.12, albeit the signal source was a sine-wave generator instead of a PWL generator. We use the same large-signal op amp subcircuit developed in the last section, but we add additional terminals to the op amp subcircuit so that we can monitor the output current of the first stage. Op amp slew-rate behavior is fully explained by the behavior of this current.

For the first part of our simulation, apply a very small step input of 1 mV and observe the transient response at the output of the amplifier and the current that flows between the first and second stages. The PSpice input file is listed in Fig. 2.30, and the results are in Fig. 2.31(a). The top curve of Fig. 2.31(a) displays both the step input voltage signal and the corresponding output response. The curve below it represents the voltage appearing between the input terminals of the amplifier, and the bottom curve represents the output current of the first stage. As evident, the output voltage

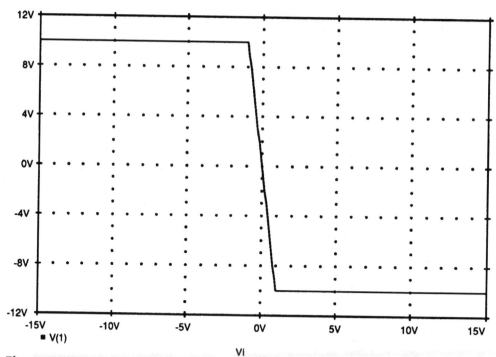

Figure 2.28 DC transfer characteristic of the inverting amplifier in Fig. 2.2 as calculated by Spice.

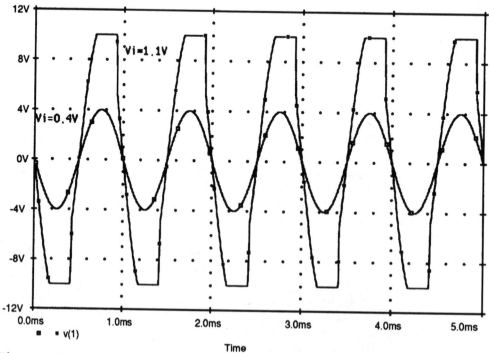

Figure 2.29 Evidence of output voltage clipping when the input signal level is too high.

```
Investigating Op Amp Slew-Rate Limiting

* op amp subcircuit
.subckt large_signal_opamp 1 2 3 6 4
* connections:              | | | | |
*                  output | | \ /
*               +ve input |   |
*                  -ve input   |
*        current monitor of 1st stage
Iopen1 2 0 0A  ; redundant connection made at +ve input terminal
Iopen2 3 0 0A  ; redundant connection made at -ve input terminal
R 4 0 2.5Meg
C 4 5 30p
Ginput 6 0 Table V(2)-V(3) = (-0.1V,-19uA) (0.1V,19uA)
Emiddle 5 0 4 0 -529
Eoutput 1 0 Table V(5) = (-10V,-10V) (10V,10V)
.ends large_signal_opamp

** Main Circuit **
* signal source
Vi 2 0 PWL (0,0V 1us,0V 1.01us,1mV 1s,1mV )
Xopamp 1 2 1 4 5 large_signal_opamp
Vmonitor 4 5 0

** Analysis Requests **
.TRAN 10ns 5us 0s 10ns
** Output Requests **
.PLOT TRAN V(2) V(1) I(Vmonitor)
.probe
.end
```

Figure 2.30 Spice input deck for investigating op amp slew-rate limiting.

signal increases in an exponential manner toward its final state. The voltage appearing between the input terminals of the amplifier and the output current of the first stage follow similar exponential patterns. Moreover, we see that these two signals are proportional to one another, albeit with a negative sign. These results are expected of an op amp whose dynamic behavior is modeled as a single–time constant network.

If a 1 V step input is applied to the input terminal, instead of an exponential increase in the output voltage, the output voltage ramps up at a constant rate as shown in the top graph of Fig. 2.31(b). This rate, of course, is the slew rate of the op amp at 0.633 V/μs. To help understand this effect, refer to the voltage and current waveforms shown in the two graphs below the top one. We see from the bottom graph that, for the most part, the current is saturated at a constant level of $-19\ \mu$A, the maximum current deliverable by the first-stage I_{max}. In the middle graph the transconductance stage enters its linear region, and the current generated begins to decrease exponentially only after the voltage between the input terminals goes between $\pm I_{max}/G_m$, or ± 100 mV.

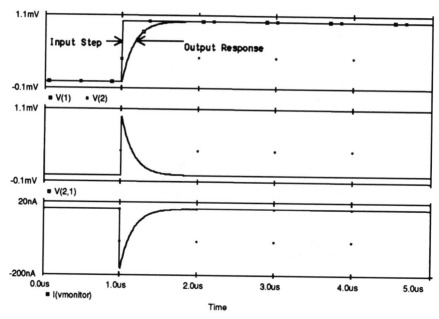

(a) a small input voltage step of 1 mV.

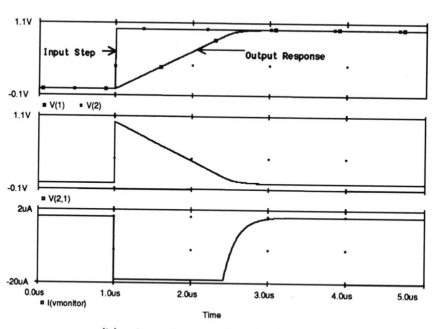

(b) a large input voltage step of 1 V.

Figure 2.31   Input and output waveforms of the unity-gain amplifier when both (a) a small and (b) a large voltage step input are applied. The middle curve of each graph is the voltage between the two input terminals of the op amp. Also shown in the lower curve of each graph is the current supplied by the front-end transconductance stage.

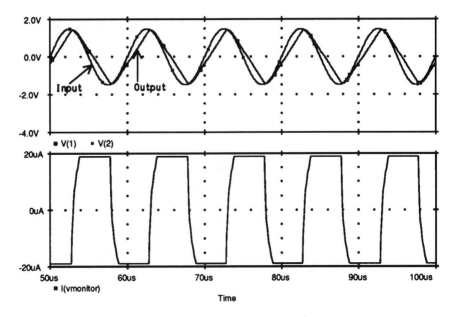

Figure 2.32 The top graph shows the input and output waveforms of the unity-gain amplifier subjected to a 100 kHz sinusoidal input signal of 1.5 V amplitude. The lower waveform is of the current supplied by the front-end transconductance stage.

In terms of sinusoidal inputs, slew-rate limiting manifests itself in the output as a distorted sine wave. Consider the application of a 100 kHz sine wave of 1.5 amplitude to the input of the unity-gain buffer described in Fig. 2.30. This requires that the step input statement be changed to a sine-wave input using the following source statement:

```
Vi 2 0 SIN (0 1.5V 100kHz)
```

The results of the PSpice simulation are illustrated in Fig. 2.32. The input and output voltage waveforms show evidence of distortion in the output signal. The cause of this distortion should be clear from the waveform of the current signal delivered by the front-end transconductance stage. Under small-signal conditions, this current waveform should be sinusoidal like the input, but obviously the front-end stage is being pushed beyond its linear capability.

2.4.3   Other Op Amp Nonidealities

A practical op amp deviates from its ideal behavior in many ways other than those just discussed. Some of these additional nonidealities include common-mode signal gain, finite input impedances, nonzero output impedance, and DC bias and offset signals. Problem 2.11 at the end of this chapter illustrates the effect of DC bias and offset signals on op amp circuit behavior. Readers are encouraged to try this problem.

2.5

Spice Tips

- An ideal op amp can be modeled as a voltage-controlled voltage source with a large DC gain of at least 10^6 V/V.

- Nodes that do not have a DC path to ground are considered floating nodes. Spice will not run with floating nodes. To circumvent this problem, connect a large resistor between each floating node and ground.

- At least two connections must be made at each node in a circuit for Spice to run. Situations often arise in which this condition is violated when using controlled sources. This problem can be avoided either by connecting a large resistor between the node in question and ground or by connecting a zero-valued current source between the node in question and ground. The latter method has the advantage that it does not disturb the operation of the network in any way.

- To simplify the writing of Spice input files, subcircuits can be used as basic building blocks of the main circuit to separate different portions of the circuit into smaller, more manageable circuit blocks.

- When concatenating several Spice decks together into one file, there cannot be any blank lines that separate the end of one Spice deck (denoted by an .end statement) from the start of another Spice deck.

- Newer versions of Spice have the capability of describing the terminal behavior of a dependent source in either functional or piecewise-linear form. This capability is known as analog behavior modeling and provides a very elegant means of describing circuit behavior to Spice.

- The initial conditions of a transient analysis can come from three sources: (1) DC operating point, (2) the assumption that all node voltages and branch currents are zero, or (3) predefined initial conditions. One should be aware of the initial conditions used by Spice during a transient analysis so that the results are meaningful.

- As a general rule, when performing a transient analysis of a circuit that does not contain any time-varying sources, a UIC (use initial conditions) command should be included on a .TRAN statement.

2.6

Bibliography

PSpice Users' Manual, MicroSim Corporation, Irvine, CA, Jan. 1991.

2.7

Problems

2.1 A Miller integrator incorporates an ideal op amp, a resistor R of 100 kΩ, and a capacitor C of 0.1 μF. Using the AC analysis capability of Spice together with the ideal op amp represented by a high-gain VCVS, determine the following:

(a) At what frequency are the input and output signals equal in amplitude?

(b) At this frequency how does the phase of the output sine wave relate to that of the input?

(c) If the frequency is lowered by a factor of 10 from that found in (a), by what factor does the output voltage change and in what direction (smaller or larger)?

(d) What is the phase relation between the input and output in situation (c)?

Confirm each of these situations by applying a 1 V peak sine wave at the appropriate frequency, and compare the voltage waveform appearing at the output using the transient analysis of Spice.

2.2 Consider a Miller integrator having a time constant of 1 ms and an initial output of zero when fed with a string of pulses of 10 μs duration and a 1 V amplitude rising from 0 V (see Fig. P2.2). Use Spice to obtain a plot of the output voltage waveform. How many pulses are required for an output voltage change of 1 V?

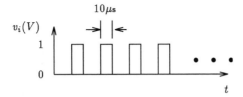

Figure P2.2

2.3 A differentiator utilizes a pseudo-ideal op amp, a 10 kΩ resistor, and a 0.01 μF capacitor. Using the AC analysis command of Spice, determine the frequency f_o at which its input and output sine-wave signals have equal magnitude. What is the output signal for a 1 V peak-to-peak sine-wave input with frequency equal to $10 f_o$?

2.4 An op amp differentiator with a 1 ms time constant is driven by the rate-controlled step shown in Fig. P2.4. Initializing the output voltage at 0 V, compute the voltage-time waveform that appears at the output using Spice over a time interval of at least 5 ms.

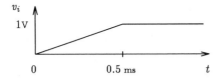

Figure P2.4

2.5 In an instrumentation system, there is a need to take the difference between two signals: $v_1 = 3 \sin(2\pi \times 60t) + 0.01 \sin(2\pi \times 1000t)$ volts and $v_2 = 3 \sin(2\pi \times 60t) - 0.01 \sin(2\pi \times 1000t)$ volts. Design a difference amplifier that meets these requirements using two op amps. In addition, amplify the resulting difference by a factor of 10. Verify your design using Spice by simulating the transient behavior of your circuit. Plot both the input and output signals. Model each op amp with a high-gain VCVS.

2.6 The circuit shown in Fig. P2.6 is intended to supply current to floating loads while making greatest possible use of the available power supplies. With a 1 V peak-to-peak, 1 Hz sine wave applied to its input, plot the voltage waveform appearing at nodes B and C. Also plot v_o. What is the voltage gain v_o/v_i?

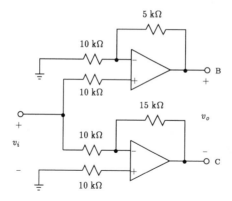

Figure P2.6

2.7 Measurements performed on an internally compensated op amp show that at low frequencies its gain is 4.2×10^4 V/V and has a 3 dB frequency located at 100 Hz. Create a small-signal equivalent circuit model of this op amp and verify, using the AC frequency analysis of Spice, that it satisfies the given measurements.

2.8 A noninverting amplifier with a nominal gain of $+20$ V/V employs an op amp having a DC gain of 10^4 and a unity-gain frequency of 10^6 Hz. Model this behavior using an equivalent circuit, and with the aid of Spice plot the magnitude response of the closed-loop amplifier and determine its 3 dB frequency f_{3dB}. What is the gain at $0.1 f_{3dB}$ and at $10 f_{3dB}$?

2.9 To demonstrate the many trade-offs that a designer faces when designing with op amps, investigate the limitations imposed on a noninverting amplifier having a nominal gain of 10 with an op amp that has a unity-gain bandwidth (f_t) of 2 MHz, a slew rate (SR) of 1 V/μs, and an output saturation voltage ($V_{o\,max}$) of 10 V. Model the op amp using the large-signal macromodel described in Section 2.3. Assume a sine-wave input with peak amplitude V_i.
   (a) If $V_i = 0.5$ V, what is the maximum frequency of the input signal that can be applied to this amplifier before the output signal shows visible distortions?
   (b) If the frequency of the input signal is 20 kHz, what is the maximum value of V_i before the output distorts?
   (c) If $V_i = 50$ mV, what is the useful frequency range of operation?
   (d) If $f = 5$ kHz, what is the useful input voltage range?

2.10 An op amp for which $R_{iCM} = 50$ MΩ, $R_{id} = 10$ kΩ, $A_0 = 10^4$, and $f_t = 10^6$ Hz is used to design a noninverting amplifier with a nominal closed-loop gain of 10. Model the op amp with a single-pole small-signal equivalent circuit model and create a Spice subcircuit for it before arranging the resistor feedback around it. With the aid of the AC analysis capability of Spice, apply a 1 V AC voltage signal to the amplifier input and plot the magnitude of the admittance seen by this source over a frequency interval of 0.1 Hz to 10^7 Hz.

2.11 An op amp is connected in a closed loop with gain of $+100$, utilizing a feedback resistor of 1 MΩ. With the aid of Spice, answer the following:
   (a) If the input bias current is 100 nA and everything else about the op amp is assumed ideal, what output voltage results with the input grounded?
   (b) If the input offset voltage is ± 1 V and the input bias current as in (a), what is the largest possible output that can be observed with the input grounded?
   (c) If bias current compensation is used, what is the value of the required resistor? Verify that this indeed reduces the output offset voltage.

3

Diodes

The semiconductor junction diode is the most fundamental nonlinear element of electronic circuits. It is important in both discrete and integrated circuits, so Spice has provided a built-in model for it. This chapter concerns accessing the diode model, altering the model parameters to suit particular applications, and investigating the behavior of some commonly used diode circuits.

3.1

Describing Diodes to Spice

A *pn* junction diode is described to Spice using two statements. One statement specifies the diode type and the manner in which it is connected to the rest of the network, and the other statement specifies the parameter values for the built-in model called by the first statement.

3.1.1 Diode Element Description

The presence of a *pn* junction diode in a circuit is described in the input file using an element statement beginning with the letter D. If more than one diode exists in a circuit, then a unique name must be appended to D to identify each diode. This name is followed by the two nodes to which the anode and cathode of the diode are connected. On the same line, the name of the model that will be used to characterize this particular diode is given. The model name must correspond to the name given in a model statement containing the model parameter values. Last, the number of diodes connected in parallel may be optionally specified; this is a convenient way of scaling the cross-sectional area of the device. Figure 3.1 shows the syntax of the Spice statements pertaining to the *pn* junction diode.

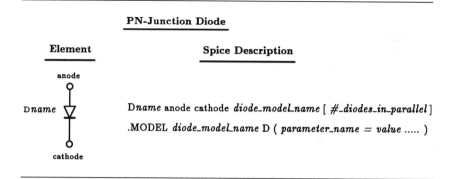

PN-Junction Diode

Element	Spice Description

anode

D*name*

cathode

D*name* anode cathode *diode_model_name* [*#_diodes_in_parallel*]

.MODEL *diode_model_name* D (*parameter_name* = *value* .....)

Figure 3.1 Spice element description for the *pn* junction diode and the general form of the associated diode model statement.

3.1.2 Diode Model Description

As shown in Fig. 3.1, the model statement for the *pn* junction begins with the key word .MODEL followed by the name of the model used by a diode element statement, the letter D to indicate that it is a diode model, and a list of the values of the model parameters enclosed between parentheses. Spice associates quite a few parameters with the *pn* junction diode model, and their individual meanings are rather involved, so we shall consider here only the parameters that are relevant to our introductory study of diodes in this chapter.

The large-signal model used for the semiconductor junction diode in Spice is shown in Fig. 3.2. The DC characteristic of the diode is modeled by the nonlinear current source i_D, which depends on v_D according to the following equation:

$$i_D = I_S(e^{v_D/nV_T} - 1) \tag{3.1}$$

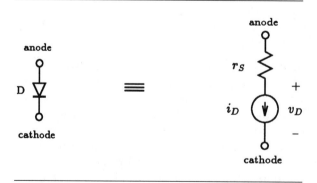

Figure 3.2 Spice large-signal *pn* junction diode model.

where I_S, n, and V_T are device parameters: I_S is referred to as the saturation current, n as the emission coefficient, and V_T as the thermal voltage. Both I_S and n are related to the physical makeup of the diode, in contrast to the thermal voltage, V_T, which depends on the temperature of the device and on two physical constants, namely,

$$V_T = \frac{kT}{q} \tag{3.2}$$

Here k is Boltzmann's constant ($k = 1.381 \times 10^{-23}$ V C/K), T is the absolute temperature in kelvins, and q is the charge of an electron ($q = 1.602 \times 10^{-19}$). In many approximate analyses, V_T is assumed to equal 25 mV at a room temperature of 290 K. Spice does not approximate this quantity but instead uses the exact values of k, T, and q to determine V_T.

Finally, the series resistance r_S in the diode model shown in Fig. 3.2 models the lump resistance of the silicon on both sides of the semiconductor junction. The value of r_S may range from 10 to 100 Ω in low-power diodes.

Under small-signal conditions, Spice adopts the diode equivalent circuit shown in Fig. 3.3. Here r_d is the incremental resistance of the diode around its quiescent operating point and is expressed in terms of the DC bias current I_D as follows:

$$r_d = \left(\left. \frac{\partial i_D}{\partial v_D} \right|_{OP} \right)^{-1} = \frac{nV_T}{I_D} \tag{3.3}$$

Spice does not use the name r_d for the small-signal resistance of the diode; instead it refers to this resistance in the output file as REQ.

Under large reverse-bias conditions (i.e., $v_D \ll 0$), the operation of the semiconductor diode is dominated by physical effects other than those which give rise to Eq. (3.1). The semiconductor diode enters its breakdown region and begins to conduct a

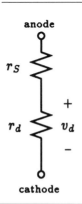

Figure 3.3 A small-signal equivalent circuit used by Spice to represent a semiconductor diode.

large reverse current. In Spice, the dependence of this reverse current on reverse voltage is modeled as an exponential function—much like that of the forward-bias region. In fact, for voltages below a commonly specified reverse-bias voltage $-V_{ZK}$ (breakdown voltage), the $i-v$ characteristic is a near-vertical straight line. Correspondingly, at this breakdown voltage the diode is said to conduct a reverse current of $-I_{ZK}$. Thus, V_{ZK} and I_{ZK} specify the start of the breakdown region of the semiconductor diode. Note that these two quantities are positive numbers.

A partial listing of the parameters associated with the Spice pn junction diode model under static conditions is given in Table 3.1 (a complete listing is in Appendix A). The first column lists the symbol used to denote each parameter as described in this chapter. Most of these symbols correspond to those used in Chapter 3 of Sedra and Smith. The next column lists the corresponding name of the parameter used by Spice; this is the only name that can be used for this parameter in the list of model parameters in a .MODEL statement. Also listed are the associated default values that the parameter assumes if a particular value is not specified in the .MODEL statement. Parameter values may be specified by writing, for example, Is = 1e-3, n = 2, BV = 50, IBV = 1e-12, and so on.

3.2

Spice as a Curve Tracer

Whenever a model of a device is created, whether it be a model of an op amp, a diode, or some other electronic device, one should make certain that the expected terminal characteristics are actually captured by the model. In the laboratory, a curve tracer is commonly used to measure the DC terminal characteristics of semiconductor devices. Using Spice we can emulate the behavior of the curve tracer and determine the suitability of the DC model chosen to represent the diode. A curve tracer is designed to measure the $i-v$ characteristics of semiconductor devices over a wide range of voltages and currents. A typical curve tracer contains a set of voltage and current sources that can vary their levels beginning at some low value and proceeding to some maximum in discrete steps. Concurrently, the current supplied to—or the voltage that develops across—the externally attached device is measured and recorded. The results are displayed on the screen of a cathode ray tube (CRT) as a graph. To emulate this behavior, we use the DC sweep command with some DC source.

Table 3.1   Partial listing of the Spice parameters for a static pn junction diode model.

Symbol	Spice Name	Model Parameter	Units	Default
I_S	Is	Saturation current	Amps	1×10^{-14}
r_S	Rs	Ohmic resistance	Ohms	0
n	n	Emission coefficient		1
V_{ZK}	BV	Reverse-bias breakdown voltage	Volts	∞
I_{ZK}	IBV	Reverse-bias breakdown current	Amps	1×10^{-10}

For example, in Fig. 3.4 we display a voltage source (V_D) with a single diode D_{test} as its load. We shall assume that the diode is a 1 mA diode (meaning that it conducts a current of 1 mA at a forward-bias voltage of 0.7 V) and that its voltage drop changes by 0.1 V for every decade change in current. Thus, using the diode current equation (3.1), we can show that this diode is characterized by $I_S = 100$ pA and $n = 1.679$. To describe this particular 1 mA diode to Spice, we use the following model statement:

```
.model 1mA_diode D ( Is=100pA n=1.679 )
```

Now, to compute and display the forward-bias i–v characteristics of this 1 mA diode we shall sweep V_D in Fig. 3.4 from 0 V to 800 mV in 10 mV steps using the following DC sweep command:

```
.DC VD 0V 800mV 10mV
```

and then plot the diode current as a function of this voltage. Although the diode current is not directly accessible by Spice,[†] it is equal to the current supplied by the voltage source V_D, which is readily accessible by Spice and can be used for the plot. However, according to Spice conventions, the current supplied by a voltage source is negative, so that the current plotted by Spice will be opposite to that flowing through the diode. Fortunately, Probe enables us to plot $-$I(VD).

The Spice input deck for this particular example is listed in Fig. 3.5, and the resulting i–v characteristic for this particular diode is shown in Fig. 3.6 in two different forms. The top curve displays the diode i–v characteristic on a linear scale, and the bottom curve is on a semilogarithmic scale.

The reverse-bias characteristics of a diode are computed in exactly the same way as the forward-bias diode characteristics. In fact, both the forward and reverse-bias characteristics can be combined on one i–v plot. Compare the forward- and reverse-bias characteristics of the 1 mA diode used in the preceding paragraphs (for which default values were used because its breakdown region was not specified) to a similar

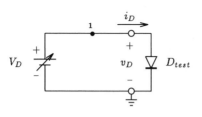

Figure 3.4 Simple circuit arrangement for determining the i-v characteristic of a 1 mA diode.

[†] PSpice does allow the user access to this current, as outlined in Chapter 1. In some examples the authors will make use of this extra feature of PSpice as a matter of convenience.

```
Spice as a Curve Tracer: Diode I-V Characteristics

** Circuit Description **
VD 1 0 DC 700mV
Dtest 1 0 1mA_diode
* diode model statement
.model 1mA_diode D (Is=100pA n=1.679)
** Analysis Requests **
* vary diode voltage and measure diode anode current
.DC VD 0V 800mV 10mV
** Output Requests **
.plot DC I(VD)
.probe
.end
```

Figure 3.5   Spice input deck for determining diode forward-bias characteristics.

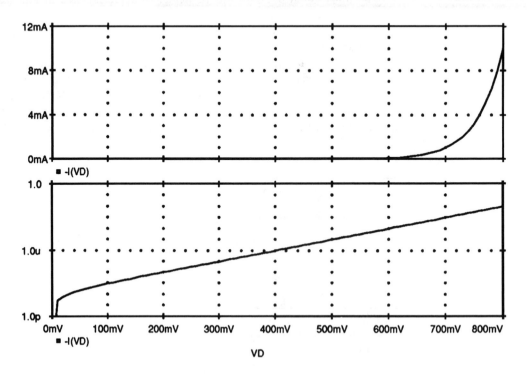

Figure 3.6   Forward-bias characteristics of a 1 mA diode with an emission coefficient of 1.679; upper curve: linear scale, lower curve: semilogarithmic.

one that has a breakdown region defined by $V_{ZK} = 10$ V and $I_{ZK} = 1$ nA. The Spice model statement for this new diode would be as follows:

```
.model bkdwn_diode   D (Is=100pA n=1.679 BV=10V IBV=1nA)
```

The resulting $i–v$ characteristics for these two diodes are displayed in Fig. 3.7. Clearly, both these diodes have identical forward-bias characteristics but very different reverse-bias behavior.

3.2.1   Extracting the Small-Signal Diode Parameters

If an operating-point (.OP) analysis command is included within the input file, then the small-signal parameters of each diode within the circuit described to Spice will be evaluated and listed in the output file. For example, the small-signal parameters in the preceding example (the 1 mA diode biased at 700 mV) would be found in the Spice output file as follows:

```
****          OPERATING POINT INFORMATION          TEMPERATURE =    27.000 DEG C
**** DIODES

NAME          Dtest
MODEL         1mA_diode
ID            1.00E-03
VD            7.00E-01
REQ           4.34E+01
CAP           0.00E+00
```

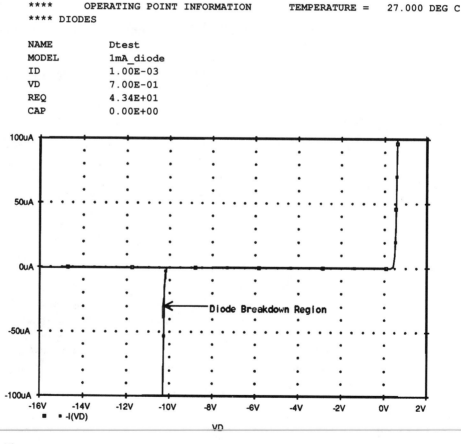

Figure 3.7   Comparing the $i–v$ characteristic of a 1 mA diode with a breakdown region specified by $V_{ZK} = 10$ V and $I_{ZK} = 1$ nA and one with no breakdown region specified.

Table 3.2 General syntax of the Spice command for setting the temperature of a circuit.

Analysis	Spice Command
Temperature analysis	.TEMP *temperature_list*

Included in this list of operating-point information is the name of the diode assigned by the user and the name of the model used to characterize the diode. This is followed by the DC bias point information (ID and VD) and the incremental resistance of the diode (REQ). At the bottom of this list is the small-signal capacitance (CAP) associated with this diode. (We shall defer discussion of (CAP) until Chapter 7.)

It would be highly instructive to check that the small-signal resistance of the diode in the operating-point information agrees with that computed by the simple formula $r_d = nV_T/I_D$. For this particular diode, $n = 1.679$, $V_T = 25.8$ mV, and $I_D = 1$ mA. Thus, on substituting these values into the expression for r_d, we get 43.47 Ω, almost in perfect agreement with the value computed by Spice (43.4 Ω).

In the preceding list of operating-point information, the temperature at which the analysis has been performed is also indicated—in this case, a room temperature of 27° C. The temperature at which the circuit is simulated by Spice can be changed by the user.

3.2.2 Temperature Effects

To investigate the effect of temperature variation on diode behavior we simply repeat the curve-tracer analysis at different temperatures by adding a .TEMP statement with the temperature (in degrees Celsius) at which the analysis should be performed. If more than one temperature is listed, the analysis will be repeated for each temperature. A general description of the syntax of this command is provided in Table 3.2. For the preceding example, let us compute the diode *i–v* characteristics for temperatures of 0°, 27°, and 125° C. The statement to be added to the deck in Fig. 3.5 is:

```
.TEMP 0 27 125
```

The job is then rerun; the results are shown in Fig. 3.8. To best illustrate the diode characteristics for all three temperatures on one graph, the results were restricted to a 0 to 1 mA current range by adjusting the scale of the *y*-axis. It should be evident from Fig. 3.8 that as the temperature increases the *i–v* curve for the diode shifts to the left. Close scrutiny with the PSpice Probe reveals that for a constant current of 0.4 mA the forward diode voltage decreases by about 1.7 mV for every degree Celsius increase in temperature.

3.3

Zener Diode Modeling

Although the start of the diode breakdown region defined by V_{ZK} and I_{ZK} can be specified on a diode model statement using the Spice parameters BV and IBV, no

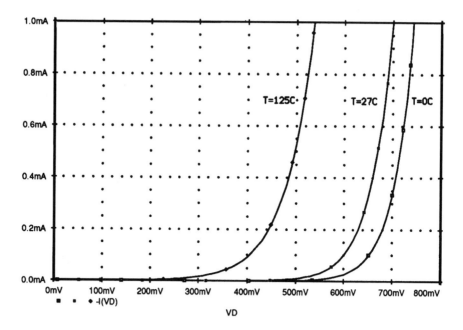

Figure 3.8   Temperature dependence of the forward-bias characteristics of a 1 mA diode.

control is provided to allow the user to specify the characteristics of the breakdown region (i.e., slope of the i–v curve in the breakdown region).[†] It is common for zener diode manufacturers to indicate the shape of the breakdown region by specifying the inverse of the slope of the almost-linear i–v curve at some operating point (V_Z, I_Z) inside the breakdown region. This inverse-slope parameter has dimensions of resistance and is known as the dynamic resistance of the zener, denoted by r_z.

To model a zener diode, the equivalent circuit shown in Fig. 3.9 is sometimes used. When $v_D > -V_{Z0}$, ideal diode D_2 is considered cut off and the terminal characteristics of the zener diode are determined solely by diode D_1. Conversely, when $v_D \leq -V_{Z0}$, ideal diode D_2 turns on, whereby a voltage of $v_D + V_{Z0}$ appears across resistor r_z. The resulting current that flows through this resistance will be much greater than the reverse-bias leakage current that flows through diode D_1, and therefore the current that dominates the breakdown region of the zener diode is given by the following equation:

$$i_D \approx \frac{v_D + V_{Z0}}{r_z} \tag{3.4}$$

[†]PSpice has since modified the built-in model for a diode so that this region of diode operation could be specified on the same model statement. However, this facility is unique to PSpice and will therefore not be used in this text.

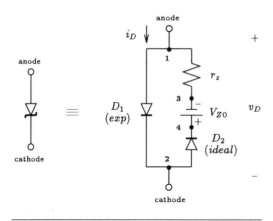

Figure 3.9   Circuit model of a zener diode.

The value of V_{Z0} is not specified directly by the zener diode manufacturer but can be derived from the operating-point information at which the dynamic resistance is given. It is found from the expression

$$V_{Z0} = V_Z - r_z I_Z \qquad (3.5)$$

The following example will illustrate a common application of a zener diode.

Voltage Regulation Using a Zener Diode

In Example 3.9 of Sedra and Smith a voltage regulator circuit was designed for an output voltage of approximately 7.5 V, assuming that the raw supply voltage fluctuates between 15 and 25 V and that the load current can vary between 0 and 15 mA (see Fig. 3.10). The zener diode available has a voltage drop of $V_Z = 7.5$ V at a current of 20 mA, and its r_z equals 10 Ω. The current-limiting resistor R in series with the zener diode was chosen at 383 Ω so that the minimum current through the diode never drops below 5 mA. Based on this design, the line and load regulation were found to be 25.4 mV/V and −9.7 mV/mA, respectively. We would like to use Spice with this zener diode model to confirm that the design requirements are indeed met. Further, we would like to check the line and load regulation directly from simulation results.

To carry out this investigation, we use the circuit setup shown in Fig. 3.11. Here the raw power supply level is modeled with two sources: a DC voltage source, V_S, to model the average value of the power supply level and a time-varying voltage source, v_{ripple}, to model the fluctuations of the power supply. The level of the DC source is set at 20 V, and the fluctuations are modeled as a sinusoidal signal having a peak amplitude of 5 V with the frequency arbitrarily set at 60 Hz. To mimic possible load current fluctuations, a single current source is connected across the output terminals of the voltage regulator. We shall begin our first simulation with

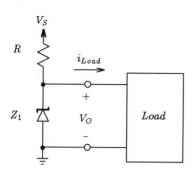

Figure 3.10 A simple voltage regulator circuit with load using a single zener diode.

this current source set to zero in order to determine the behavior of the regulator under no-load conditions. A zero-valued voltage source is connected in series with the zener diode and another with the current source i_{Load} to monitor the current through each of them.

The model for the zener diode is the equivalent circuit shown in Fig. 3.9. The value of r_z is still 10 Ω. From Eq. (3.5) and the data supplied for the zener diode (given earlier), we find $V_{Z0} = 7.3$ V. Diode D_1 will be modeled as a 1 mA diode with an emission coefficient of 1.679. The ideal diode, D_2, will be modeled with $I_S = 100$ pA and $n = 0.01$. The subcircuit describing this particular zener diode would then appear as follows:

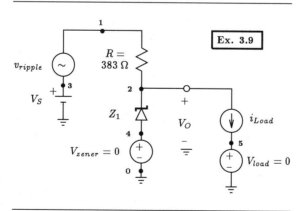

Figure 3.11 Circuit setup used to investigate the line and load regulation of the simple zener diode voltage regulator circuit of Fig. 3.10. Zero-valued voltage sources are placed in series with the zener diode and with the current source i_{Load} to monitor their respective currents. They play no part in the circuit operation.

```
.subckt zener_diode  1 2
* connections:          | |
*                  anode |
*                  cathode
Dforward 1 2 1mA_diode
Dreverse 2 4 ideal_diode
Vz0 4 3 DC 7.3V
Rz 1 3 10
* diode model statement
.model  1mA_diode  D (Is=100pA n=1.679 )
.model  ideal_diode D (Is=100pA n=0.01 )
.ends zener_diode
```

The complete Spice input file describing the circuit shown in Fig. 3.11 is listed in Fig. 3.12.

A transient analysis is requested so that six periods of the output voltage signal can be determined and shown in Fig. 3.13. Both the power supply voltage $V_S + v_{ripple}$ and the voltage across the zener diode are shown. The top graph displays the supply voltage, and the bottom graph displays the corresponding zener diode voltage waveform. As expected, the voltage fluctuation of the power supply is sinusoidal with a peak-to-peak amplitude of 10 V riding on a DC level of 20 V. The output voltage from the regulator circuit is also sinusoidal, having a peak-to-peak amplitude of 254.7 mV and riding on a DC level of 7.628 V. The precise values of these two levels were determined using the cursor facility of Probe. We notice that both the input and the output voltage waveforms are in phase, so we can conclude that a line voltage change of +10 V gives rise to an output voltage change of +254.7 mV. Thus, the line regulation given by

$$\text{Line regulation} = \frac{\Delta V_O}{\Delta V_S}$$

is calculated to be +25.47 mV/V, which agrees almost exactly with the value determined by the simple expression for line regulation given by $r_z/(r_z + R)$ derived in Chapter 3 of Sedra and Smith (i.e., 25.4 mV/V). This should not be too surprising, since the circuit has been operating entirely in its linear region.

We can perform another similar analysis but this time vary the load current between 0 and 15 mA in order to determine the load regulation by observing the output voltage waveform. The power supply voltage will be assumed constant at +20 V. We shall assume that the load current is triangular with a minimum value of 0 mA and a maximum value of 15 mA and the frequency arbitrarily set at 30 Hz, corresponding to a period of 33.33 ms. This signal will correctly model the minimum and maximum fluctuations of the load current. The statement in the deck of Fig. 3.12 for the output current source i_{Load} will need to be revised:

```
Iload 2 5 PULSE ( 0mA 15mA 0 16.66ms 16.66ms 1us 33.33ms )
```

Here a triangular waveform is emulated using the source PULSE statement. The rise time and fall time are set equal to one-half the period of the triangular waveform of 33.33 ms. The pulse width is assigned a very small value of 1 μs because Spice will not accept a zero value for the pulse width.

```
Zener Diode Voltage Regulator Circuit (No Load)

* zener diode subcircuit
.subckt zener_diode 1 2
* connections:     | |
*            anode |
*            cathode
Dforward 1 2 1mA_diode
Dreverse 2 4 ideal_diode
Vz0 4 3 DC 7.3V
Rz 1 3 10
* diode model statement
.model 1mA_diode D (Is=100pA n=1.679 )
.model ideal_diode D (Is=100pA n=0.01 )
.ends zener_diode

** Main Circuit **
* power supply
Vs 3 0 DC +20V
Vripple 1 3 sin ( 0V 5V 60Hz )
* zener diode voltage regulator circuit
R  1 2 383
XD1 4 2 zener_diode
Vzener 4 0 0
* simulated load condition
Iload 2 5 0A
Vload 5 0 0
** Analysis Requests **
.OP
.TRAN 0.5ms 100ms 0ms 0.5ms
** Output Requests **
.PLOT TRAN V(1) V(2)
.PROBE
.end
```

Figure 3.12   Spice input file for computing the time-varying no-load output voltage of the zener diode voltage regulator circuit shown in Fig. 3.11.

The amplitude of the ripple voltage superimposed on the DC supply voltage should be set to 0 V in order to eliminate its presence during this analysis. The revised statement is

```
Vripple 1 3 sin ( 0V 0V 60Hz )
```

The revised job results in the output voltage waveform V_O for the regulator circuit shown in the bottom graph of Fig. 3.14. The waveform shown in the top graph depicts the load current i_{Load}. Here the load current is triangular, as it should be, with a 15 mA peak-to-peak amplitude. The corresponding output voltage signal

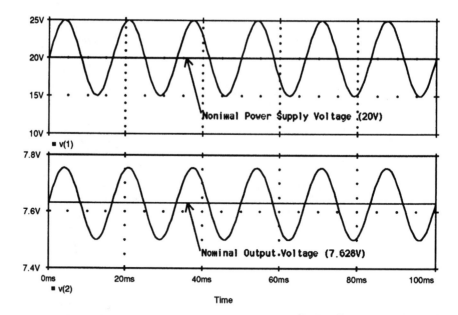

Figure 3.13 Several waveforms associated with the zener diode regulator circuit of Fig. 3.10 under no-load conditions. The top graph displays the voltage generated by the power supply, and the bottom graph displays the corresponding output voltage from the regulator circuit.

is also triangular with the same frequency and a peak-to-peak amplitude of 146.33 mV found using the cursor facility of Probe. For an average load current of 7.5 mA, the output voltage corresponds to 7.555 V. Notice that the phase of the output voltage waveform is opposite to that of the load current. This suggests that for a change in the load current of +15 mA, the output voltage changes by −146.33 mV, thus suggesting that the load regulation, expressed as

$$\text{Load regulation} = \frac{\Delta V_O}{\Delta I_{\text{Load}}}$$

would be −146.33 mV/15 mA, or −9.7 mV/mA. This agrees exactly with the value determined by the simple expression for load regulation given by $-r_z \parallel R$ derived in Chapter 3 of Sedra and Smith.

As a final check on this design, let us investigate the minimum current that flows through the zener diode. This diode current is minimum when the power supply voltage is at its minimum and the load current is at its maximum. In keeping with our earlier approach, we shall maintain the load current as a triangular wave varying linearly between 0 and 15 mA. The ripple voltage v_{ripple} will be set to a constant of level −5 V, which requires that the statement for this source be changed to

```
Vripple 1 3 DC -5V
```

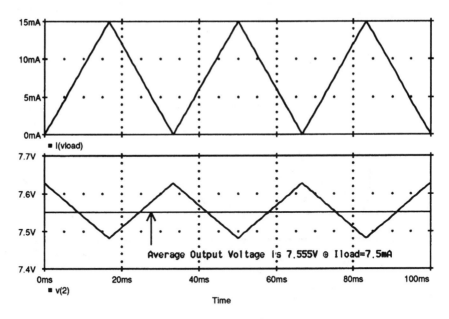

Figure 3.14 Output voltage of the zener diode regulator circuit of Fig. 3.10 when the load current varies between 0 and 15 mA. The top graph displays the current drawn by the load, and the bottom graph displays the corresponding output voltage from the regulator circuit.

Rerunning the job with the revised source statement results in the three waveforms shown in Fig. 3.15. The top graph displays the constant $+15$ V level associated with the power supply and the graph below it displays the load current waveform. The bottom graph displays the current waveform associated with the zener diode, which varies between 5 mA and 20 mA.

For all practical purposes, based on the foregoing Spice results, we can conclude that all aspects of the design did indeed meet the required conditions.

3.4

A Half-Wave Rectifier Circuit

One of the most important applications of semiconductor diodes is in the design of rectifier circuits. We now shall investigate the half-wave rectifier circuit.

A half-wave rectifier circuit is shown in Fig. 3.16. It consists of a transformer with a 14:1 turns ratio, a single diode D_1 of the commercial type 1N4148, and a load resistance R_{load} of 1 kΩ. The source resistance of 0.5 Ω of the AC line is also included in this circuit. The purpose of the transformer is to step down the main household AC power supply voltage of 120 V_{rms} to a 12 V_{peak} level. Spice does not make provision for an ideal transformer, probably for a good reason: One does not exist in practice. Instead, Spice allows coupled inductors to be described with a coefficient of coupling,

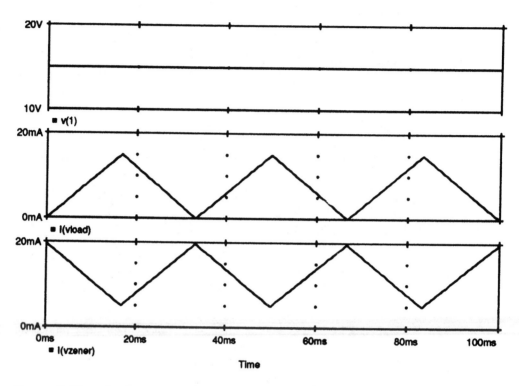

Figure 3.15 Observing a worst-case situation: The top graph displays the minimum power supply voltage, the middle graph displays the time-varying load current, and the bottom graph displays the current flowing through the zener diode.

k, less than 1. Two inductors, for example, L_P and L_S, which share a common magnetic path and have a coefficient of coupling k very close to unity, say 0.999, would be a reasonably good model of many practical transformers. The turns ratio N_P/N_S of such a transformer is given by the square root of the ratio of the primary to secondary inductance, $N_P/N_S = \sqrt{L_P/L_S}$.

To describe such a transformer to Spice, three element statements are required: one statement for each inductor and a statement that describes the coefficient of coupling between the two inductors. Inductor coupling is described using a new statement that begins with the letter K. If more than two inductors share a common magnetic path, a unique name is attached to K to identify each coefficient of coupling. This is then followed by the names of the two magnetically coupled inductors. These names must correspond to the names of two inductors described in the present deck. The final field of this statement describes the coefficient of coupling, k, which can take on a value between 0 and 1. Spice does not accept a k equal to unity, so we shall always use $k = 0.999$.

Spice adheres to the transformer dot convention. Observe that the dot of each transformer is located at the positive node ($n+$) of each inductor. Extension to three or more coupled inductors should be self-evident. N coupled inductors require N induc-

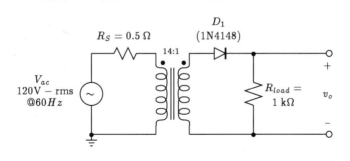

Figure 3.16 Half-wave rectifier circuit using a transformer with a 14:1 turns ratio to step down the line voltage of 120 V_{rms} to 12 V_{peak}.

tor element statements and $N(N+1)/2$ coefficients of coupling statements. Figure 3.17 illustrates a transformer and the three statements necessary to describe it to Spice.

We can create a Spice description of the half-wave rectifier circuit of Figure 3.16. We assume that the inductance of the primary side of the transformer is 10 mH and the inductance of secondary side is 51 μH, which provides an effective transformer turns ratio of 14:1. The alert student will quickly realize that the circuit on the secondary side of the transformer has no DC path to ground and therefore will be rejected by Spice.

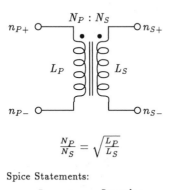

Spice Statements:

```
Lp np+ np– Lp_value
Ls ns+ ns– Ls_value
Kname Lp Ls 0.999
```

Figure 3.17 General syntax of the Spice statements used to describe a (nonideal) transformer. The transformer turns ratio $N_P:N_S$ is determined by the appropriate selection of primary and secondary inductor values, L_P and L_S, respectively.

Between ground and one point on the secondary side, we add a large resistor with a value chosen so that it does not interfere significantly with the operation of the circuit. Figure 3.18 illustrates the addition of a 100 MΩ resister between ground and node 4 of the rectifier. Also shown in this figure is a zero-valued voltage source placed in series with the rectifier diode D_1 to enable us to monitor the current flowing through the diode. The resulting deck for this modified circuit is in Fig. 3.19. A transient analysis is requested to compute the voltage appearing across the load resistance, the voltage appearing across the primary and secondary sides of the transformer, and finally the AC line voltage. The Spice model of the commercial diode, 1N4148, was obtained from a library of Spice models for various electronic components included in PSpice.[†]

The analysis results are shown in Fig. 3.20. The top graph displays the voltage waveform of the AC line voltage (V_{ac}) and the voltage appearing across the primary side of the transformer. Here we see that the voltage across the transformer experiences a short transient effect, quickly settling into its steady state with the transformer voltage slightly lagging behind the line voltage. The bottom graph displays the rectified voltage appearing across the load resistance and the voltage appearing across the secondary side of the transformer. Figure 3.21 is a blown-up view of a half period of the rectified output voltage and the transformer secondary-side voltage.

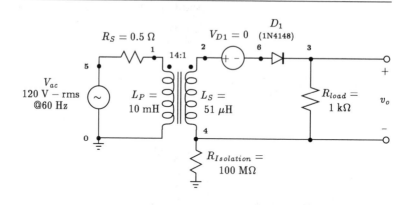

Figure 3.18 Preparing the half-wave rectifier circuit shown in Fig. 3.16 for Spice analysis. A large-valued isolation resistor (100 MΩ) is placed between the secondary side of the transformer and ground. This provides a DC path between the secondary side of the transformer and the common reference node (0). Also added is a zero-valued voltage source in series with the rectifier diode. This will allow indirect access to the diode current.

[†] These devices are in a file named NOM.LIB in the same directory as the PSpice program.

```
Half-Wave Rectifier Circuit

** Circuit Description **

* ac line voltage
Vac 5 0 sin (0 169V 60Hz)
Rs 5 1 0.5
* transformer section
Lp 1 0 10mH
Ls 2 4 51uH
Kxfrmr Lp Ls 0.999
* isolation resistor (allows secondary side to pseudo-float)
Risolation 4 0 100Meg
* diode current monitor
VD1 2 6 0
* rectifier circuit
D1 6 3 D1N4148
Rload 3 4 1kOhm
* diode model statement
.model D1N4148 D (Is=0.1pA Rs=16 CJO=2p Tt=12n Bv=100 Ibv=0.1p)

** Analysis Requests **
.TRAN 0.5ms 100ms 0ms 0.5ms
** Output Requests **
.plot TRAN V(3,4) V(2,4) V(1)
.plot TRAN V(6,3)
.plot TRAN I(VD1)
.probe
.end
```

Figure 3.19 Spice input file for calculating the transient behavior of the half-wave rectifier circuit shown in Fig. 3.18.

Diode current-handling capability, which is an important consideration in the design of rectifier circuits, is determined by the largest current that the diode must conduct and by the peak inverse voltage (PIV) that the diode must be able to withstand without breakdown. In Fig. 3.22, we display both the current conducted by the diode and the voltage across it. The maximum current that the diode must conduct is seen to be about 11.1 mA according to the cursor facility of Probe. The data sheets of the 1N4148 indicate that this diode can handle a peak current of no more than 100 mA. We see that the PIV of this particular rectifier circuit is 12 V. Because the diode has not broken down, we can assume that the breakdown voltage of the 1N4148 commercial diode is larger than 12 V. In fact, data sheets of the 1N4148 indicate that its breakdown voltage is about 100 V. Thus our rectifier design is well within the limits of the 1N4148.

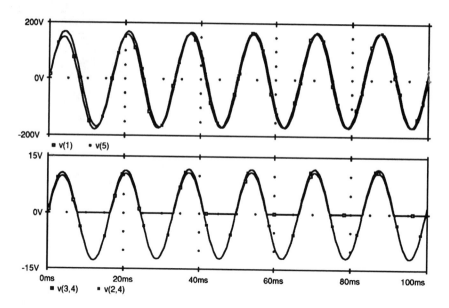

Figure 3.20 Various voltage waveforms associated with the half-wave rectifier circuit shown in Fig. 3.18. The top graph displays both the AC line voltage and the voltage appearing across the primary side of the transformer. The bottom graph displays the voltage appearing across the load resistor and the voltage appearing across the secondary side of the transformer.

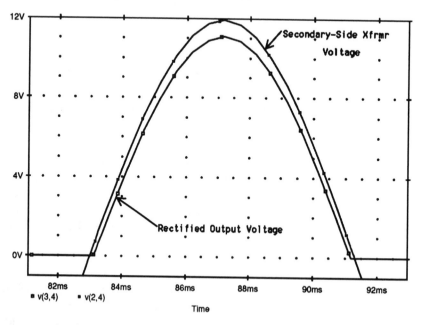

Figure 3.21 Zooming in on a half cycle of the voltage waveform appearing across the load resistor and comparing it to the voltage developed across the secondary side of the transformer.

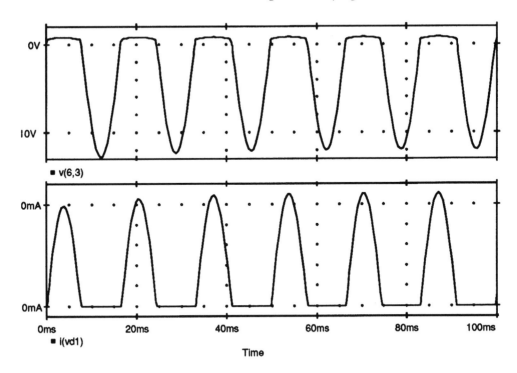

Figure 3.22 The voltage and current waveform associated with diode D_1. The peak inverse voltage (PIV) is seen to be 12 V and the maximum diode current is 11.1 mA.

3.5

Limiting and Clamping Circuits

Using Spice, we shall simulate and analyze the circuit operations of a back-to-back diode limiter, a DC restorer, and a voltage doubler—all commonly used diode circuits.

3.5.1   A Diode Limiter Circuit

In Fig. 3.23 we present a simple back-to-back diode limiter circuit constructed with two diodes of the 1N4148 type, and we would like to observe its transfer characteristic. Figure 3.24 shows the deck for this circuit with a sweep of the input DC source v_I between -1 and $+1$ V.

In the results of this analysis (Fig. 3.25) we see that the transfer characteristic exhibits rather soft limiting. The linear region, which has a slope of unity, ranges between -0.5 V and $+0.5$ V.

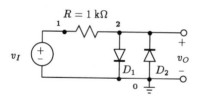

Figure 3.23   A back-to-back diode limiter circuit.

```
A Diode Limiter Circuit

** Circuit Description **

Vi 1 0 DC 0V
R 1 2 1k
D1 2 0 D1N4148
D2 0 2 D1N4148
* diode model statement
.model D1N4148 D (Is=0.1pA Rs=16 CJO=2p Tt=12n Bv=100 Ibv=0.1p)

** Analysis Requests **
* sweep the input voltage level from -5 V to +5 V in 100 mV increments
.DC Vi -5V 5V 100mV
** Output Requests **
.PLOT DC V(2)
.probe
.end
```

Figure 3.24   Spice input file for calculating the DC transfer characteristic of the back-to-back diode limiter circuit of Fig. 3.23.

3.5.2   A DC Restorer Circuit

In Fig. 3.26 we present a DC restorer, or a clamped capacitor circuit, for which we will analyze the transient behavior with a square-wave input having a 10 V peak-to-peak amplitude, a +2 V DC offset, and a 1 kHz frequency. The diode is assumed to be of the 1N4148 type, and the capacitor has a 1 μF value. The deck is shown in Fig. 3.27, where the square-wave input is described by the following source PULSE statement:

```
Vi 1 0 PULSE ( -3 7 0s 10us 10us 0.490ms 1ms )
```

which specifies input between −3 and 7 V with 0 s delay, a rise and a fall time of 10 μs, a pulse width of 0.490 ms, and a period of 1 ms.

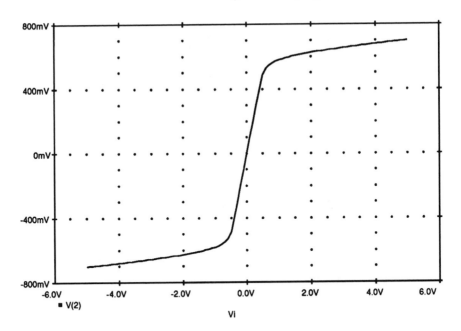

Figure 3.25 DC transfer characteristic of the back-to-back diode limiter shown in Fig. 3.23.

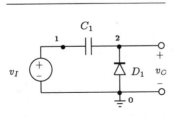

Figure 3.26 DC restorer circuit.

The results of the Spice transient analysis are shown in Fig. 3.28. The top graph displays the input 10 V square-wave signal and the bottom graph shows the corresponding output signal. We see that it is also a 10 V square wave but its DC level has now changed to a 5 V level (one-half the peak-to-peak value).

3.5.3   Voltage Doubler Circuit

We wish to use Spice to observe the transient behavior of the voltage doubler circuit in Fig. 3.29 with an input sine-wave signal having a 10 V amplitude and a 1 kHz frequency. The diodes are of the 1N4148 type and the two capacitors are each 1 μF. The input deck is shown in Fig. 3.30.

```
A DC Restorer Circuit

** Circuit Description **
Vi 1 0 PULSE ( -3 7 0s 10us 10us 0.490ms 1ms )
C1 1 2 1u
D1 0 2 D1N4148
* diode model statement
.model D1N4148 D (Is=0.1pA Rs=16 CJO=2p Tt=12n Bv=100 Ibv=0.1p)

** Analysis Requests **
.TRAN 100u 10m 0m 100u
** Output Requests **
.PLOT TRAN V(1) V(2)
.probe
.end
```

Figure 3.27   Spice input file for calculating the time-varying output voltage from the DC restorer circuit shown in Fig. 3.26.

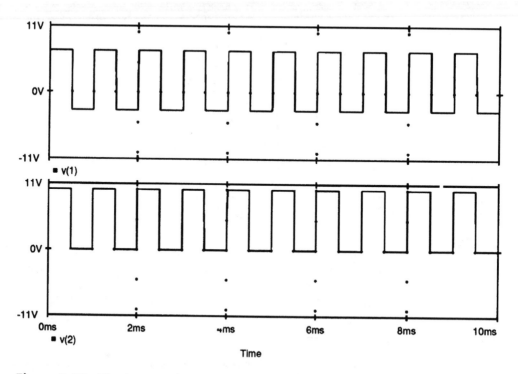

Figure 3.28   The input and output voltage waveforms associated with the DC restorer circuit of Fig. 3.26.

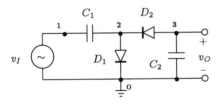

Figure 3.29   Voltage doubler circuit.

```
A Voltage Doubler Circuit

** Circuit Description **
Vi 1 0 sin ( 0 10V 1kHz )
C1 1 2 1u
C2 3 0 1u
D1 2 0 D1N4148
D2 3 2 D1N4148
* diode model statement
.model D1N4148 D (Is=0.1pA Rs=16 CJO=2p Tt=12n Bv=100 Ibv=0.1p)

** Analysis Requests **
.TRAN 100u 10m 0m 100u
** Output Requests **
.PLOT TRAN V(1) V(2) V(3)
.probe
.end
```

Figure 3.30   Spice input file for calculating the transient response of the voltage doubler circuit shown in Fig. 3.29.

In the results shown in Fig. 3.31, the top graph displays the input 10 V_{peak} sine wave, the middle graph displays the voltage that appears across diode D_1, and the bottom graph shows the voltage that appears at the output of the doubler circuit. Looking at the output voltage waveform in the bottom graph, we see that it experiences a transient stage that lasts for about 7 ms and then settles to a constant level of -18.8 V. The magnitude of this signal is approximately twice the peak value of the input sine-wave signal. The voltage across diode D_1 settles into a 10 V_{peak} sine-wave signal with a -10 V DC offset.

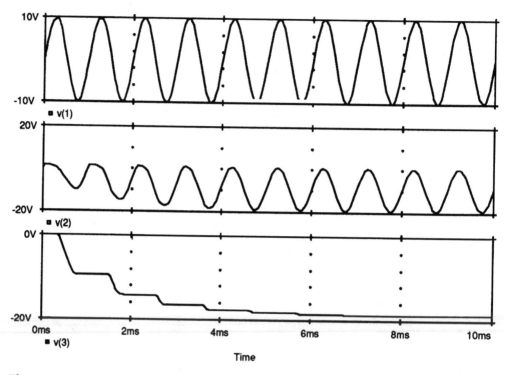

Figure 3.31   Various voltage waveforms of the voltage doubler circuit shown in Fig. 3.29. The top graph displays the input sine-wave voltage signal, the middle graph displays the voltage across diode D_1, and the bottom graph displays voltage that appears at the output.

3.6

Spice Tips

- In some circuit situations, a DC path does not exist between a node in the circuit and the ground reference node 0. This causes a problem for Spice that can be circumvented by placing a large resistor between this node and ground. The value of this resistor should be large enough that it does not interfere with the circuit operation.
- Spice can emulate the behavior of a laboratory curve tracer by use of the DC sweep command.
- Spice can investigate the effect of temperature on a circuit.
- An ideal diode can be very well approximated by setting the emission coefficient n of the diode model between 0.01 and 0.001.
- A zener diode cannot be properly represented by the diode model statement found in Spice. Instead a zener diode is represented by a subcircuit consisting of several diodes, a voltage source, and a resistor (see Fig. 3.9).

- An ideal transformer is not represented in Spice; instead it is closely approximated by two inductors with a coefficient of mutual coupling very close to one (i.e., $k = 0.999$).
- Commercial vendors of electronic components are making available Spice models of their devices and circuits. This should greatly improve the range of circuits that can be simulated by Spice.
- The student version of PSpice provides a series of device models for several commercial parts. These are found in a file named NOM.LIB located in the same directory as the PSpice program.
- When performing long transient runs, to ensure that Spice does not stop after 5000 iterations of the transient analysis, the option parameter ITL5 can reset this limit to some other value. Because in most circuit simulations one does not know *a priori* the number of iterations the transient analysis will require, it is best to set this limit to infinity by indicating ITL5 = 0.

3.7

Problems

3.1 Assume that the diodes in the circuits of Fig. P3.1 are modeled with parameters $I_S = 10^{-14}$ and $n = 2$. Determine, with the aid of Spice, values of the labeled voltages and currents. Repeat with diode parameters $I_S = 10^{-12}$ and $n = 1$.

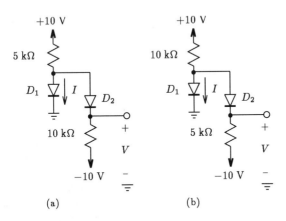

Figure P3.1

3.2 Consider the battery charger circuit shown in Fig. P3.2. If v_S is a 60 Hz sinusoid with 24 V peak amplitude, find the fraction of each cycle during which the diode conducts. Also find the peak value of the diode current and the maximum reverse-bias voltage that appears across the diode. Assume that the diode has Spice model parameters $I_S = 10^{-12}$ and $n = 1.6$.

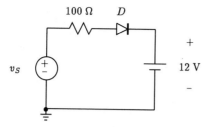

Figure P3.2

3.3 For the circuit shown in Fig. P3.3 with $R_S = 1\ k\Omega$, $C = 1.0\ \mu F$, and I having values of 1 mA, 0.1 mA, and 1 μA, verify that the signal component in the output voltage for each case is given by

$$v_O = v_S \frac{nV_T}{nV_T + IR_S}$$

Using Spice, simulate the transient behavior of this circuit with an input 100 Hz sine-wave signal of 1 mV peak amplitude. Assume that the diode has model parameters $I_S = 10\ fA$ and $n = 2$.

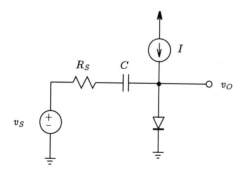

Figure P3.3

3.4 A voltage regulator consisting of two diodes in series fed with a constant current source is used as a replacement for a single carbon–zinc (battery) cell of nominal voltage 1.5 V. The regulator load current varies from 2 to 7 mA. Compare this regulator circuit for three different current source levels of 5, 10, and 15 mA as the load current varies over its full range. What is the change in the output voltage for each case? Assume that the diodes have a 0.7 V drop at a 1 mA current and $n = 2$.

3.5 A zener shunt regulator of the type shown in Fig. 3.10 has been designed to provide a regulated voltage of about 10 V. The zener diode is of the type 1N4740, which is specified to have a 10 V drop at a test current of 25 mA. At this current its $r_z = 7\ \Omega$. The raw supply available has a nominal value of 20 V but can vary by as much as $\pm25\%$. The regulator is required to supply a load current of 0 to 20 mA. Assuming that the minimum zener current is to be 5 mA, the resistance R was determine to be 200 Ω. Using Spice, simulate the behavior of this circuit and determine the following:

(a) Find the load regulation. By what percentage does V_O change from the no-load to full-load condition?

(b) Find the line regulation. What is the change in V_O, expressed as a percentage, corresponding to the $\pm 25\%$ change in V_S?

(c) What is the maximum current that the zener diode has to conduct? What is the maximum zener power dissipation?

3.6 The circuit in Fig. P3.6 implements a complementary-output rectifier. Simulate the behavior of this circuit using Spice and plot the voltage that appears across the two output terminals. Assume that the diodes have model parameters $I_S = 100$ fA and $n = 2$. What is the PIV of each diode?

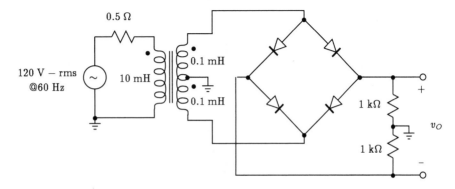

Figure P3.6

3.7 Using Spice, plot the transfer characteristic v_O versus v_I for the four limiter circuits shown in Fig. P3.7. Assume that the diodes have model parameters $I_S = 100$ fA and $n = 1.6$.

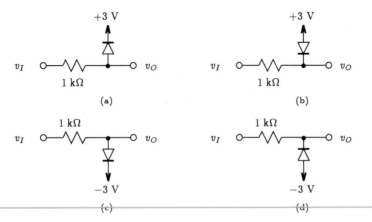

Figure P3.7

3.8 Using Spice, plot the transfer characteristic v_O versus v_I of the circuit in Fig. P3.8 for -20 V $\leq v_I \leq +20$ V. Assume that the diodes have model parameters $I_S = 100$ fA and $n = 1.6$, that the zener diode has a reverse-bias voltage drop of 8.2 V at a current of 10 mA, and that $r_z = 20$ Ω.

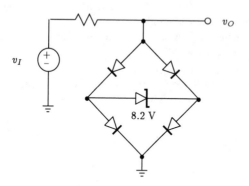

Figure P3.8

3.9 For the circuits in Fig. P3.9, each utilizing diodes of the 1N4148 type, plot the output waveform of the circuit for a 10 V peak amplitude input square wave having a frequency of 1 kHz.

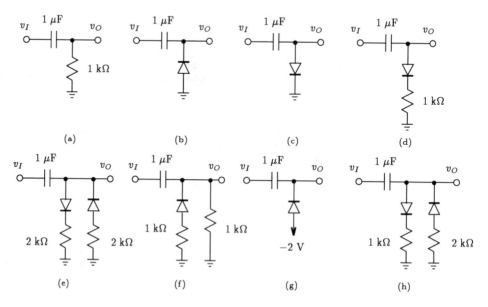

Figure P3.9

4

Bipolar Junction Transistors (BJTs)

This chapter describes the two statements that are required to specify a bipolar junction transistor (BJT) to Spice and outlines the model used to represent the BJT. We shall use Spice to simulate the low-frequency behavior of various types of electronic circuits containing BJTs ranging from single *npn* and *pnp* transistor amplifiers to multiple-transistor amplifier circuits.

4.1

Describing BJTs to Spice

The first of the two statements required to describe any particular semiconductor device to Spice is a description of the nature of the device and the manner in which it is connected to the network. The second statement describes the parameters of the built-in model of the semiconductor device specified by the first statement. We shall look at these two statements as they apply to the BJT.

4.1.1   BJT Element Description

The presence of a BJT in a circuit is described through the input file using an element statement beginning with the letter Q. If more than one transistor exists in a circuit, a unique name must be appended to Q to identify each one. This name is followed by a list of the three nodes connected to the BJT's collector, base, and emitter, and the node connected to the BJT's substrate, if it is an integrated transistor. The name of the model that will characterize this particular BJT is given on the same line. The model name must correspond to the name given on the model statement containing the parameter values that characterize this transistor. An option is available to specify the number of BJTs that are connected in parallel.

For quick reference, in Fig. 4.1 we depict the syntax for the Spice statement describing the BJT and for the model statement (.MODEL) necessary for any input file that makes reference to the built-in BJT models. The model statement defines the terminal characteristics of the BJT by specifying the values of particular parameters of the model.

4.1.2 BJT Model Description

Figure 4.1 shows the model statement for either the *npn* or *pnp* transistor, which begins with the key word .MODEL and is followed by the name of the model used by a BJT element statement, the nature of the BJT (i.e., *npn* or *pnp*), and, in brackets, a list of the parameters characterizing the BJT's terminal behavior. Some 40 parameters are associated with the Spice model of the BJT, and their individual meanings are rather involved. Here we shall simply outline the ones that are relevant to this chapter. Appendix A provides a list and a short description of their meanings.

The BJT model is illustrated schematically in Fig. 4.2. The ohmic resistances of the base, collector, and emitter regions are lumped into three linear resistances: r_B, r_C, and r_E, respectively. The DC characteristics of the intrinsic BJT are determined by the nonlinear dependent current sources i_B and i_C.[†] For a transistor operated in its

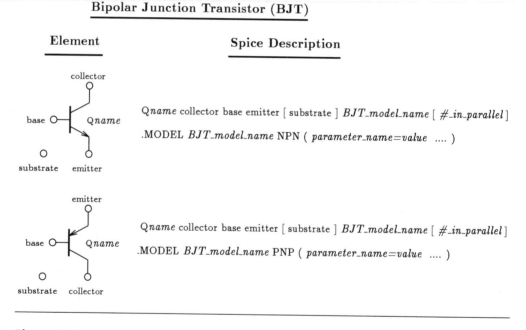

Figure 4.1 Spice element description for the *npn* and *pnp* BJT and for the general form of the associated BJT model statement.

[†]The functional descriptions of these two currents—as adopted by Spice—are rather complex and will not be given here. Interested readers can consult [Nagel, 1975].

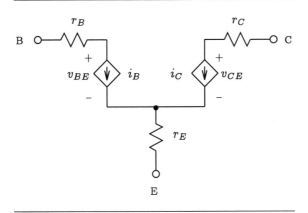

Figure 4.2   Spice large-signal BJT model for DC analysis.

active mode, a first-order representation of these two currents can be described by

$$i_B = \frac{I_S}{\beta_F} e^{v_{BE}/V_T} \qquad (4.1)$$

and

$$i_C = I_S e^{v_{BE}/V_T} \left(1 + \frac{v_{CE}}{V_{AF}}\right) \qquad (4.2)$$

Here I_S is the saturation current (similar to the diode saturation current) and V_T is the thermal voltage. The constant β_F is the forward common-emitter current gain. In Sedra and Smith, this constant is designated as β. Spice attaches the subscript F to distinguish β_F from β_R, the common-emitter current gain of the same transistor in reverse mode (i.e., with the emitter and collector interchanged). Finally, V_{AF} is the forward Early voltage (denoted V_A in Sedra and Smith).

Table 4.1 is a partial listing of the parameters for the BJT model under static conditions and their associated default values; if its value is not specified in the .MODEL statement, a parameter assumes its default value. Parameter values are specified by writing, for example, Is=1e-14, Bf=100, and so on.

Table 4.1   Partial listing of the Spice parameters for a static BJT model.

Symbol	Spice Name	Model Parameter	Units	Default
I_S	Is	Saturation current	Amps	1×10^{-16}
β_F	Bf	Forward current gain		100
V_{AF}	VAf	Forward Early voltage	Volts	∞
r_B	Rb	Base ohmic resistance	Ohms	0
r_C	Rc	Collector ohmic resistance	Ohms	0
r_E	Re	Emitter ohmic resistance	Ohms	0

4.1.3   *Verifying npn Transistor Circuit Operation*

We wish to verify the *npn* transistor circuit designed in Example 4.1 of Sedra and Smith and duplicated here in Fig. 4.3. It was designed to have a collector current of 2 mA and a collector voltage of +5 V.

For design purposes, the *npn* transistor was assumed to have a $\beta_F = 100$ and to exhibit a v_{BE} of 0.7 V at $i_C = 1$ mA. The first condition can be directly specified on the BJT model statement using $\beta_F = 100$; however, the latter condition needs to be translated into a BJT parameter. From Eq. (4.2) we can write

$$1 \times 10^{-3} = I_s e^{0.7/25.89 \times 10^{-3}} \left(1 + \frac{v_{CE}}{V_{AF}} \right) \tag{4.3}$$

To make matters simpler, we shall assume that $V_{AF} = \infty$, thus reducing the preceding equation to

$$1 \times 10^{-3} = I_s e^{0.7/25.89 \times 10^{-3}} \tag{4.4}$$

We can then solve to obtain $I_S = 1.8104 \times 10^{-15}$ A. Our model statement for this is then described to Spice using the following statement:

```
.model npn_transistor npn (Is=1.8104e-15 Bf=100)
```

Notice that we do not specify the value of V_{AF} in the list of parameters, but rely on its default value. Of course, we could do the same for the β_F.

The input file in Fig. 4.4 corresponds to the circuit shown in Fig. 4.3. This file includes an operating-point analysis command (.OP) to tell Spice to calculate the circuit's DC operating point. Spice produces the following DC operating-point information:

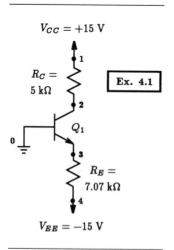

Figure 4.3   Transistor circuit design created by Sedra and Smith in Example 4.1 of their text. Spice is used to calculate the DC operating point of this circuit.

```
Verifying Transistor Circuit Design
** Circuit Description **
Vcc 1 0 DC +15V
Vee 4 0 DC -15V
Q1 2 0 3 npn_transistor
Rc 1 2 5k
Re 3 4 7.07k
* transistor model statement
.model npn_transistor npn (Is=1.8104e-15 Bf=100)
** Analysis Requests **
* calculate DC bias-point information
.OP
** Output Requests **
* none required
.end
```

Figure 4.4   Spice input file for calculating the collector current and voltage of the transistor circuit shown in Fig. 4.3.

```
****       SMALL SIGNAL BIAS SOLUTION          TEMPERATURE =    27.000 DEG C
*****************************************************************************

NODE    VOLTAGE       NODE    VOLTAGE       NODE    VOLTAGE       NODE    VOLTAGE

(    1)    15.0000  (    2)     4.9990  (    3)     -.7172  (    4)   -15.0000

     VOLTAGE SOURCE CURRENTS
     NAME            CURRENT

     Vcc            -2.000E-03
     Vee             2.020E-03

     TOTAL POWER DISSIPATION   6.03E-02  WATTS
```

The collector voltage of the transistor (node 2) is 4.9990 V and the collector current I_C is 2.000 mA, as inferred from the current supplied by the voltage source V_{CC}. (The negative sign results from a Spice convention.) The collector voltage deviates very little (1 mV) from +5 V. If we backtrack to find why this error occurred, we find that in the design phase we assumed $V_T = 25$ mV. Spice assumed a circuit temperature of 27° C, which produced $V_T = 25.89$ mV. If we repeat the design procedure in this example with $V_T = 25.89$ mV, we get an emitter resistance of $R_E = 7.0703$ kΩ and a collector voltage much closer to +5 V.

4.2

Using Spice as a Curve Tracer

A typical curve tracer arrangement for measuring the i_C–v_{CE} characteristics of a transistor is illustrated in Fig. 4.5. Here two independent sources, v_{CE} and i_B, are varied, and the collector current of the transistor is calculated. The collector current would

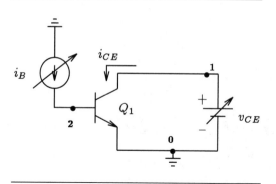

Figure 4.5   Spice curve tracer arrangement for calculating the i_C–v_{CE} characteristics of a BJT.

then be plotted as a function of v_{CE} and i_B. For example, consider plotting the forward i_C–v_{CE} characteristics of an *npn* transistor characterized by $I_S = 1.8104 \times 10^{-15}$ A, $\beta_F = 100$, and a forward Early voltage $V_{AF} = 35$ V, for a base current of 10 μA. Using the circuit displayed in Fig. 4.5 as a guide, we can create the input file shown in Fig. 4.6. We use the DC sweep command to vary the collector–emitter voltage of transistor Q_1 from 0 V to +10 V in 100 mV steps. The resulting i_C–v_{CE} characteristic as calculated by Spice is displayed in Fig. 4.7.

Typically, it is useful to vary the base current i_B while also varying v_{CE} of the transistor. We augment the DC sweep command with another source name and the range of values it should be stepped through. For example, the DC sweep command required to sweep v_{CE} from 0 V to +10 V in 100 mV steps while at the same time sweeping i_B from 1 μA to 10 μA in 1 μA steps is

```
.DC Vce 0V +10V 100mV Ib 1u 10u 1u
```

In essence, this command tells Spice to perform the voltage sweep for each value specified by the current sweep. (This command should remind those familiar with computer programming of a set of programming loops: the inner sweep being nested within the outer sweep.)

We revise the Spice input deck, replacing the old DC sweep command with this augmented one. The i_C–v_{CE} characteristics displayed in Fig. 4.8 are obtained.

Clearly, other arrangements of the two independent sources are possible, thus allowing investigation of other characteristics of the transistor. The reader is encouraged to try some, using the approach just outlined.

4.3

Spice Analysis of Transistor Circuits at DC

We are ready to investigate the DC operating point of several simple transistor circuits and to use Spice to determine the working mode of operation of the transistor. In this

```
Spice as a Curve Tracer:   BJT I-V Characteristics

** Circuit Description **
Vce 1 0 DC 0V
Ib 0 2 DC 10uA
* device under test
Q1 1 2 0 npn_transistor
* transistor model statement
.model npn_transistor NPN (Is=1.8104e-15A Bf=100 VAf=35V)
** Analysis Requests **
* vary Vce from 0 V to 10 V in steps of 100 mV
.DC Vce 0V +10V 100mV
** Output Requests **
.plot DC I(Vce)
.probe
.end
```

Figure 4.6 Spice input file for calculating the collector current of the transistor circuit shown in Fig. 4.5 for a given base current and collector–emitter voltage.

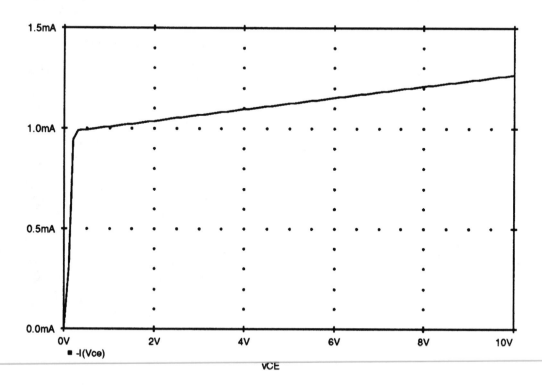

Figure 4.7 The i_C–v_{CE} curve at a base current of 10 μA for a transistor characterized by $I_S = 1.8104 \times 10^{-15}$ A, $\beta_F = 100$, and $V_{AF} = 35$ V.

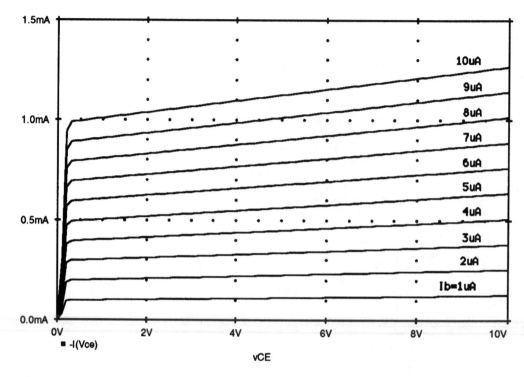

Figure 4.8 A family of i_C–v_{CE} curves for a base current varied between 1 μA and 10 μA in steps of 1 μA for a transistor characterized by $I_S = 1.8104 \times 10^{-15}$ A, $\beta_F = 100$, and $V_{AF} = 35$ V.

section we shall assume that the transistor is characterized by a $\beta_F = 100$, exhibits a v_{BE} of 0.7 V at $i_C = 1$ mA, and that its Early voltage is infinite.

4.3.1 Transistor Modes of Operation

Depending on the bias condition imposed across the emitter–base junction (EBJ) and the collector–base junction (CBJ), different modes of operation of the BJT are obtained, as shown in Table 4.2.

Active Region

Consider the circuit shown in Fig. 4.9, which is the same one hand-analyzed by Sedra and Smith in Example 4.2 of their text. Transistor Q_1 was shown to be operating in the

Table 4.2 BJT modes of operation.

Mode	EBJ	CBJ
Cutoff	Reverse	Reverse
Active	Forward	Reverse
Saturation	Forward	Forward

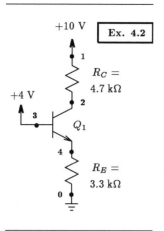

Figure 4.9   A simple *npn* transistor circuit (Example 4.2 of Sedra and Smith).

active region. We shall use Spice to repeat the example, to calculate the DC operating
point, and to show that, indeed, transistor Q_1 is operating in the active mode.

A Spice description of this circuit is listed in Fig. 4.10, and a DC operating-point
analysis request is included. The results of the analysis are partially listed below.

```
****     SMALL SIGNAL BIAS SOLUTION        TEMPERATURE =   27.000 DEG C
***************************************************************************

NODE   VOLTAGE       NODE   VOLTAGE      NODE   VOLTAGE     NODE   VOLTAGE

(   1)   10.0000   (    2)    5.3452   (   3)    4.0000   (    4)    3.3009

****     OPERATING POINT INFORMATION       TEMPERATURE =   27.000 DEG C
***************************************************************************
**** BIPOLAR JUNCTION TRANSISTORS

NAME           Q1
MODEL          npn_transistor
IB             9.90E-06
IC             9.90E-04
VBE            6.99E-01
VBC            -1.35E+00
VCE            2.04E+00
```

One should notice among this operating point information that the base–emitter junction is forward biased with V_{BE} = 0.699 V and that the collector–base junction is
reverse biased with V_{BC} = −1.35 V, thus confirming that transistor Q_1 is indeed operating in its active region.[†] One could have deduced this same information from the
DC node voltages, but this is tedious for multiple-transistor circuits. Also note that the
collector and base current values agree reasonably well with those of the hand analysis.

[†] Other data are included in the output of operating-point information but are not shown here. Also, we do
not show the small-signal parameters associated with the hybrid-pi model of the transistor, discussed later
in this chapter.

```
NPN Transistor Operated in Active Mode

** Circuit Description **
Vcc 1 0 DC +10V
Vbb 3 0 DC +4V
Q1 2 3 4 npn_transistor
Rc 1 2 4.7k
Re 4 0 3.3k
* transistor model statement
.model npn_transistor npn (Is=1.8104e-15 Bf=100)
** Analysis Requests **
* calculate DC bias-point information
.OP
** Output Requests **
* none required
.end
```

Figure 4.10   Spice input file for calculating the DC operating point of the circuit shown in Fig. 4.9.

Saturation Region

If we increase the base voltage of the circuit of Fig. 4.9 from +4 V to +6 V, transistor Q_1 leaves the active region and moves into the saturation region. To see this, change the value of Vbb in the input file shown in Fig. 4.10 to +6 V and rerun the job. The results are as follows:

```
****     SMALL SIGNAL BIAS SOLUTION        TEMPERATURE =   27.000 DEG C
****************************************************************************

NODE    VOLTAGE        NODE    VOLTAGE      NODE    VOLTAGE      NODE    VOLTAGE

(    1)    10.0000  (    2)    5.3147  (    3)    6.0000  (    4)    5.2808

****     OPERATING POINT INFORMATION        TEMPERATURE =   27.000 DEG C
****************************************************************************
**** BIPOLAR JUNCTION TRANSISTORS

NAME            Q1
MODEL           npn_transistor
IB              6.03E-04
IC              9.97E-04
VBE             7.19E-01
VBC             6.85E-01
VCE             3.39E-02
```

Both V_{BE} and V_{BC} are forward-biased, suggesting that Q_1 is operating in its saturation region.

Cutoff Region

Finally, if we reduce the base voltage to zero volts, the transistor becomes cut off. Altering the Spice input deck to reflect this (i.e., setting Vbb = 0) and rerunning the job has the following results:

```
****        SMALL SIGNAL BIAS SOLUTION          TEMPERATURE =   27.000 DEG C
**************************************************************************

    NODE   VOLTAGE     NODE   VOLTAGE     NODE   VOLTAGE     NODE
 VOLTAGE
 (    1)   10.0000 (     2)   10.0000 (     3)    0.0000 (      4) 33.01E-09

    ****      OPERATING POINT INFORMATION          TEMPERATURE =   27.000 DEG C
**************************************************************************
 BIPOLAR JUNCTION TRANSISTORS

 NAME          Q1
 MODEL         npn_transistor
 IB           -1.00E-11
 IC            2.00E-11
 VBE          -3.30E-08
 VBC          -1.00E+01
 VCE           1.00E+01
```

Here, both V_{BE} and V_{BC} are reverse biased, indicating that Q_1 is now cut off. Notice that even under cutoff conditions, the transistor is still conducting a base and a collector current (and, of course, an emitter current). These currents are mainly leakage currents associated with the two reverse-biased junctions of the transistor.

4.3.2 Computing DC Bias of a pnp Transistor Circuit

The preceding examples were of *npn* transistors only. Now we shall calculate the DC operating point of a circuit with a *pnp* transistor. As far as Spice is concerned, *pnp* transistor circuits are no more complicated than *npn* transistor circuits.

Consider the circuit shown in Fig. 4.11, which corresponds to Example 4.5 of Sedra and Smith. The input file for this circuit is listed in Fig. 4.12, and the results of the DC analysis are shown here:

```
****        SMALL SIGNAL BIAS SOLUTION          TEMPERATURE =   27.000 DEG C
**************************************************************************

    NODE   VOLTAGE     NODE   VOLTAGE     NODE   VOLTAGE     NODE   VOLTAGE

 (    1)   10.0000 (     2)    .7387 (     3)   -5.4152 (      4)  -10.0000

    VOLTAGE SOURCE CURRENTS
    NAME           CURRENT

    Vcc          -4.631E-03
    Vee           4.585E-03

    TOTAL POWER DISSIPATION    9.22E-02  WATTS

    ****      OPERATING POINT INFORMATION          TEMPERATURE =   27.000 DEG C
**************************************************************************

 **** BIPOLAR JUNCTION TRANSISTORS

 NAME          Q1
 MODEL         pnp_transistor
 IB           -4.58E-05
 IC           -4.58E-03
```

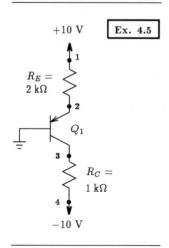

Figure 4.11   A simple *pnp* transistor circuit (Example 4.5 of Sedra and Smith).

```
PNP Transistor DC Operating-Point Calculation

** Circuit Description **
Vcc 1 0 DC +10V
Vee 4 0 DC -10V
Q1 3 0 2 pnp_transistor
Re 1 2 2k
Rc 3 4 1k
* transistor model statement
.model pnp_transistor pnp (Is=1.8104e-15 Bf=100)
** Analysis Requests **
* calculate DC bias-point information
.OP
** Output Requests **
* none required
.end
```

Figure 4.12   Spice input file for calculating the DC operating point of the circuit shown in Fig. 4.11.

```
    VBE             -7.39E-01
    VBC              5.42E+00
    VCE             -6.15E+00
```

From these results, it is obvious that the transistor is operating in the active mode. Notice that the polarity of the transistor currents are opposite to those of an *npn* transistor operating in its active region (see the previous section). Further, be aware that Spice defines a positive collector current of a *pnp* transistor as the current flowing *into* the collector, the opposite of Sedra and Smith's convention.

The foregoing Spice results compare well with those generated by Sedra and Smith's simple hand calculation in Example 4.5; there is only about a 0.4% difference.

In practice the Early effect can never be ignored as it was in the preceding analysis. Let us assume an Early voltage V_{AF} of 100 V. Using the deck in Fig. 4.12, replace the given *pnp* transistor model statement with the following one:

```
.model pnp_transistor pnp (Is=1.8104e-15 Bf=100 Vaf=100V).
```

The Spice output file shows the following results:

```
****      SMALL SIGNAL BIAS SOLUTION          TEMPERATURE =   27.000 DEG C
*******************************************************************************

   NODE    VOLTAGE     NODE    VOLTAGE     NODE    VOLTAGE     NODE    VOLTAGE

(    1)    10.0000  (    2)     .7373  (    3)   -5.4122  (    4)  -10.0000

        VOLTAGE SOURCE CURRENTS
        NAME            CURRENT

        Vcc            -4.631E-03
        Vee             4.588E-03


****      OPERATING POINT INFORMATION         TEMPERATURE =   27.000 DEG C
*******************************************************************************

**** BIPOLAR JUNCTION TRANSISTORS

NAME           Q1
MODEL          pnp_transistor
IB             -4.35E-05
IC             -4.59E-03
VBE            -7.37E-01
VBC             5.41E+00
VCE            -6.15E+00
```

We see from these results that Spice analysis with the Early effect results in a collector current of 4.59 mA, in contrast to 4.58 mA when the Early effect is ignored. In fact, Sedra and Smith's hand calculation without the Early effect only varied from Spice's complete analysis by 0.2%. For most practical engineering situations, a 5% to 10% accuracy obtained from a simple hand analysis is usually acceptable. Furthermore, accounting for the Early effect during hand analysis would complicate the analysis so significantly that it probably would mask any insight gained by doing it. If great precision is necessary, we can do a detailed Spice analysis with the transistor Early effect included.

4.4

BJT Transistor Amplifiers

Transistors operated in their active regions are important as linear amplifiers, and their design and analysis are facilitated by the use of small-signal BJT models.

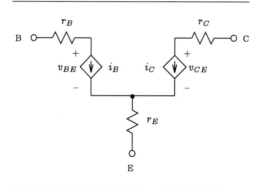

Figure 4.13 Small-signal BJT Spice model under static conditions.

4.4.1 BJT Small-Signal Model

Under static and small-signal conditions, the linearized BJT Spice model is the familiar hybrid-pi equivalent circuit shown in Fig. 4.13. This model applies to both the *npn* and the *pnp* transistor, so we make no distinction between the two. The transconductance g_m is related to the DC bias current I_C according to the following:

$$g_m = \frac{\partial i_C}{\partial v_{BE}}\bigg|_{OP} + \frac{\partial i_C}{\partial v_{BC}}\bigg|_{OP} + \frac{\partial i_B}{\partial v_{BC}}\bigg|_{OP} \approx \frac{I_C}{V_T} \tag{4.5}$$

Similarly, the input and output resistances are also related to the transistor's DC operating point, according to the following:

$$r_\pi = \left(\frac{\partial i_B}{\partial v_{BE}}\bigg|_{OP}\right)^{-1} \approx \frac{V_T}{I_B} \tag{4.6}$$

$$r_o = \left(\frac{\partial i_C}{\partial v_{BC}}\bigg|_{OP} + \frac{\partial i_B}{\partial v_{BC}}\bigg|_{OP}\right)^{-1} \approx \frac{V_A}{I_C} \tag{4.7}$$

and r_μ is approximated as

$$r_\mu = \left(\frac{\partial i_B}{\partial v_{BC}}\bigg|_{OP}\right)^{-1} \approx 10\frac{V_A}{I_C}\beta_F \tag{4.8}$$

Also included in the hybrid-pi model are the ohmic resistances of the three junctions: r_C, r_B (r_x), and r_E. Note that r_E is *not* the r_e used in the text by Sedra and Smith.

The small-signal parameters of a transistor are usually computed by Spice prior to most analyses, because modeling the small-signal behavior of the BJT is paramount to many analyses. Spice will list the small-signal model parameters of all transistors in a given circuit when an .OP command is included in the input file. Because of the importance of the base resistance r_B on small-signal operation, Spice also will list this value among the small-signal parameters in the output file, denoted by Rx.

The following operating-point information for transistor Q_1 from our previous example in section 4.3.2 shows the small-signal parameters that an .OP analysis command produces in the output file:

```
**** BIPOLAR JUNCTION TRANSISTORS

NAME              Q1
MODEL              pnp_transistor
IB                -4.58E-05
IC                -4.58E-03
VBE               -7.39E-01
VBC                5.42E+00
VCE               -6.15E+00
BETADC             1.00E+02
GM                 1.77E-01
RPI                5.64E+02
RX                 0.00E+00
RO                 1.00E+12
CBE                0.00E+00
CBC                0.00E+00
CBX                0.00E+00
CJS                0.00E+00
BETAAC             1.00E+02
FT                 2.82E+18
```

The list contains the DC operating-point and small-signal model parameters for the *pnp* transistor in the circuit of Fig. 4.11. Besides the parameters already explained, this list also contains other parameters, such as capacitances CBE, CBC, CBX, and CJS, that model dynamic effects of the transistor, and BETAAC and FT, which are derived from the dynamic small-signal model of the transistor. These are discussed in Chapter 7.

4.4.2 Single-Stage Voltage-Amplifier Circuits

Consider the voltage amplifier circuit shown in Fig. 4.14, which was analyzed for its voltage gain by Sedra and Smith in Example 4.9. We shall use Spice to calculate

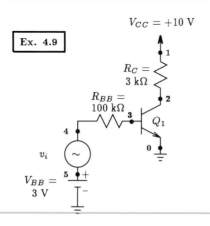

Figure 4.14 A simple common-emitter voltage amplifier (Example 4.9 of Sedra and Smith).

the voltage gain of this circuit, assuming that the transistor has $\beta_F = 100$ and $I_S = 1.8104 \times 10^{-15}$ A. The other parameters of the BJT model will take their default values.

In Fig. 4.15 we list the input file for this circuit. A transfer function command (.TF) is included to calculate the voltage gain v_o/v_i, where v_o is taken at node 2. In addition, we are also requesting an operating-point calculation (.OP) so that we can cross-check the results found using the .TF command with those computed by hand in Sedra and Smith's Example 4.9.

The results of the .TF command are as follows:

```
****      SMALL-SIGNAL CHARACTERISTICS

        V(2)/Vi = -2.966E+00

        INPUT RESISTANCE AT Vi =  1.011E+05

        OUTPUT RESISTANCE AT V(2) =  3.000E+03
```

The voltage gain of this circuit is -2.966 V/V, and the input and output resistances are 101.1 kΩ and 3 kΩ, respectively. These resistance levels seem to be in line with what one would expect from the circuit in Fig. 4.14.

We can corroborate our voltage-gain calculation by performing the same gain calculation using the small-signal equivalent circuit representation of the amplifier

```
Calculating Voltage Gain of a Transistor Amplifier

** Circuit Description **
* dc supplies
Vcc 1 0 DC +10V
Vbb 5 0 DC +3V
* small-signal input
Vi 4 5 DC 1mV
* amplifier circuit
Q1 2 3 0 npn_transistor
Rc 1 2 3k
Rbb 4 3 100k
* transistor model statement
.model npn_transistor npn (Is=1.8104e-15 Bf=100)
** Analysis Requests **
* calculate small-signal transfer function: Vo/Vi
.TF V(2) Vi
* we shall also calculate the small-signal parameters of the transistor
.OP
** Output Requests **
* none required
.end
```

Figure 4.15 Spice input file for calculating the small-signal voltage gain of the circuit shown in Fig. 4.14.

circuit shown in Fig. 4.14. In this manner Sedra and Smith found that the voltage gain, in terms of the small-signal parameters of the transistor, is given by

$$\frac{v_o}{v_i} = -g_m R_C \frac{r_\pi}{r_\pi + R_{BB}} \tag{4.9}$$

In this derivation the output resistance of the transistor was assumed infinite. Spice's transistor operating-point information in the output file includes the small-signal parameters of the hybrid-pi model for this transistor. A partial listing of such operating-point information for transistor Q_1 is given here:

```
**** BIPOLAR JUNCTION TRANSISTORS

NAME            Q1
MODEL           npn_transistor
IB              2.28E-05
IC              2.28E-03
VBE             7.21E-01
VBC             -2.44E+00
VCE             3.16E+00
BETADC          1.00E+02
GM              8.82E-02
RPI             1.13E+03
RX              0.00E+00
RO              1.00E+12
```

From this list, we see that $g_m = 88.2$ mA/V and $r_\pi = 1.13$ kΩ. Substituting these values, together with $R_C = 3$ kΩ and $R_{BB} = 100$ kΩ, into Eq. (4.9), we find the voltage gain to be -2.96 V/V, which almost exactly agrees with the result computed using the Spice .TF command. In principle, these two calculations should be in perfect agreement because both calculate the voltage gain in exactly the same way, but Spice uses more significant digits here.

4.5

DC Bias Sensitivity Analysis

4.5.1 Sensitivity to Component Variations

Bias networks are used in transistor amplifier design to establish the proper DC operating point of each transistor. In order to maintain consistent operation, the DC operating point of each transistor must be held constant. This is normally achieved by a bias network that maintains the emitter current of each transistor relatively constant under potential circuit variations, such as variations in transistor β and so on.

Provisions have been made in Spice, through a DC sensitivity analysis, for investigating the DC stability of circuit behavior in the face of component variations. When a .SENS analysis command in the input file is invoked, Spice will compute the sensitivities (which are actually just derivatives) of selected circuit variables to most of the circuit components, including the sensitivities to many of the model parameters of the BJT, such as β and I_S. If diodes are present in the circuit, the sensitivities to the diode model parameters will also be computed and listed. Unfortunately, Spice does not compute the sensitivities to the FET model parameters.

Table 4.3 provides the general description of the syntax of the sensitivity analysis command (.SENS). The command line begins with the key word .SENS followed by a list of the variables whose derivatives will be computed with respect to the different components of the circuit. The .SENS command sends results directly to the output file in much the same way as does the operating point or transfer function analysis command. No .PRINT or .PLOT statement is required. It should be noted that Spice first computes the DC operating point of the circuit and then computes the appropriate derivatives. The next example demonstrates a Spice sensitivity analysis.

Consider the BJT amplifier shown in Fig. 4.16(a), which is biased using a single–power supply resistive network arrangement. Let us compute the sensitivities of the emitter current of Q_1, assuming that the BJT is modeled after the widely available commercial transistor 2N2222A. This requires that we place a zero-valued voltage source, $V_{emitter}$, in the emitter branch of the amplifier circuit to monitor the emitter current, as shown in Fig. 4.16(b).[†]

The Spice input file describing this circuit is provided in Fig. 4.17. The sensitivity analysis command

```
.SENS I(Vemitter)
```

will request that Spice compute the sensitivity of the emitter current to various components in the circuit. The results of this analysis are then found in the output file, together with some of the DC bias solution, as follows:

```
****      SMALL SIGNAL BIAS SOLUTION        TEMPERATURE =   27.000 DEG C
*****************************************************************************

NODE   VOLTAGE       NODE    VOLTAGE      NODE    VOLTAGE      NODE   VOLTAGE

(   1)   12.0000   (    2)    8.1567   (    3)    3.8346   (    4)    3.1912
(   5)    0.0000

VOLTAGE SOURCE CURRENTS
NAME            CURRENT

Vcc            -1.063E-03
Vemitter        9.670E-04
```

Table 4.3 General form of the DC sensitivity analysis request.

Analysis Request	Spice Command
DC sensitivity	.SENS *output_variables*

[†] The model parameters for the 2N2222A were derived from information on the manufacturer's data sheets, but the details of converting manufacturer's data on supplied transistor parts to a Spice model are beyond the scope of this text. We note, however, that software to greatly simplify this task is now available from various development houses, such as MicroSim Corporation, the supplier of PSpice. An interesting paper [Malik, 1990] describes a step-by-step procedure for determining numerical parameter values for the Spice model of a BJT from manufacturer's measured data.

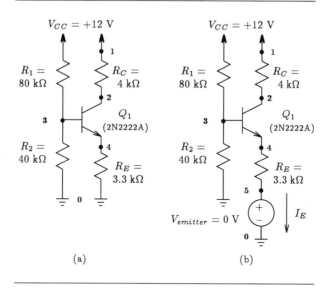

Figure 4.16 (a) An amplifier biased using a single–power supply resistive network arrangement. (b) Monitoring the emitter current of the amplifier circuit by including a zero-valued voltage source ($V_{emitter}$) in series with the emitter of the transistor.

```
Investigating the Sensitivity of the Emitter Current to Amplifier Components

** Circuit Description **
* power supply
Vcc 1 0 DC +12V
* amplifier circuit
Q1 2 3 4 Q2N2222A
Rc 1 2 4k
R1 1 3 80k
R2 3 0 40k
Re 4 5 3.3k
Vemitter 5 0 0
* transistor model statement for the 2N2222A
.model Q2N2222A NPN (Is=14.34f Xti=3 Eg=1.11 Vaf=74.03 Bf=255.9 Ne=1.307
+                    Ise=14.34f Ikf=.2847 Xtb=1.5 Br=6.092 Nc=2 Isc=0 Ikr=0 Rc=1
+                    Cjc=7.306p Mjc=.3416 Vjc=.75 Fc=.5 Cje=22.01p Mje=.377 Vje=.75
+                    Tr=46.91n Tf=411.1p Itf=.6 Vtf=1.7 Xtf=3 Rb=10)
** Analysis Requests **
.SENS I(Vemitter)
** Output Requests **
* none required
.end
```

Figure 4.17 Spice input file for calculating the DC sensitivities of the emitter current of the BJT amplifier shown in Fig. 4.16.

```
            TOTAL POWER DISSIPATION    1.28E-02  WATTS

    ****      DC SENSITIVITY ANALYSIS              TEMPERATURE =    27.000 DEG C
    *****************************************************************************

    DC SENSITIVITIES OF OUTPUT I(Vemitter)
```

ELEMENT NAME	ELEMENT VALUE	ELEMENT SENSITIVITY (AMPS/UNIT) (AMPS/PERCENT)	NORMALIZED SENSITIVITY
Rc	4.000E+03	-6.110E-10	-2.444E-08
R1	8.000E+04	-9.727E-09	-7.782E-06
R2	4.000E+04	1.827E-08	7.309E-06
Re	3.300E+03	-2.771E-07	-9.144E-06
Vcc	1.200E+01	9.594E-05	1.151E-05
Vemitter	0.000E+00	-2.865E-04	0.000E+00

Q1

RB	1.000E+01	-1.773E-09	-1.773E-10
RC	1.000E+00	-6.110E-10	-6.110E-12
RE	0.000E+00	0.000E+00	0.000E+00
BF	2.559E+02	1.053E-07	2.695E-07
ISE	1.434E-14	-1.395E+09	-2.001E-07
BR	6.092E+00	-2.946E-15	-1.795E-16
ISC	0.000E+00	0.000E+00	0.000E+00
IS	1.434E-14	1.584E+09	2.272E-07
NE	1.307E+00	2.913E-04	3.808E-06
NC	2.000E+00	0.000E+00	0.000E+00
IKF	2.847E-01	5.546E-07	1.579E-09
IKR	0.000E+00	0.000E+00	0.000E+00
VAF	7.403E+01	-3.712E-08	-2.748E-08
VAR	0.000E+00	0.000E+00	0.000E+00

This list of sensitivities consists of two parts: The top portion of the table lists the sensitivities to the DC sources and passive components of the circuit. The bottom part of the table lists the sensitivities to the model parameters of the active components in the circuit. In this particular case, there is only one active component in the circuit, Q_1.

The first column indicates the element with respect to which the sensitivity of the emitter current I_E is taken. The second column indicates the nominal value of that element as it appears in the input file. The third column indicates the sensitivity quantity $\partial I_E/\partial x$, where x is the element named in the leftmost column. In general, the units of this sensitivity quantity are the units of the output variable specified on the .SENS statement divided by the units of the element x. For instance, in the case of the 3.3 kΩ emitter resistor R_E, denoted by Re, the fourth element from the top, the sensitivity quantity $\partial I_E /\partial R_E$ is expressed in A/Ω and has a value of -2.771×10^{-7} A/Ω. The final column on the right is a normalized sensitivity measure. It simply expresses the sensitivity in more convenient units of A/%. Mathematically, it is written as $\partial I_E/(\partial R_E/R_E)_\%$, where $(\partial R_E/R_E)_\%$ is the relative change of R_E expressed as a percentage. For R_E the normalized sensitivity measure is -9.144×10^{-6} A/%.

From the sensitivity analysis we can obtain estimates of how stable the emitter current I_E is in the face of component variations. For example, if the emitter resistor

R_E changes by 10%, we can expect that the emitter current will change approximately according to the relationship

$$\Delta I_E \approx \frac{\partial I_E}{(\partial R_E / R_E)_\%} \times (\Delta R_E / R_E)_\% \qquad (4.10)$$

Substituting the appropriate numerical values, we find that a 10% change in R_E results in a 91.4 μA change in the emitter current of Q_1. This is about 9.5% of its quiescent value of 967.0 μA.

Similarly, we can estimate the effect that a 20% variation in transistor β_F has on the emitter current of Q_1. From the preceding sensitivities, we see that $\partial I_E / (\partial \beta_F / \beta_F)_\%$ = 2.695×10^{-7} A/%. Thus, a 20% change in β_F gives rise to a 5.39 μA change in the emitter current of Q_1. This is about 0.56% of the quiescent value.

This same approach can be used for any of the components of the circuit in Fig. 4.16, whose sensitivities are given in the preceding table. One can also investigate the effect of different component changes on the emitter current by using the mathematical concept of a total derivative. An example in Section 5.5 applies to a MOSFET amplifier circuit.

```
Investigating Emitter Current Dependence on Temperature

** Circuit Description **
* power supply
Vcc 1 0 DC +12V
* amplifier circuit
Q1 2 3 4 Q2N2222A
* resistive biasing network accounting for temperature dependence
Rc 1 2 4k TC=1200u
R1 1 3 80k TC=1200u
R2 3 0 40k TC=1200u
Re 4 5 3.3k TC=1200u
Vemitter 5 0 0
* transistor model statement for the 2N2222A
.model Q2N2222A NPN (Is=14.34f Xti=3 Eg=1.11 Vaf=74.03 Bf=255.9 Ne=1.307
+                     Ise=14.34f Ikf=.2847 Xtb=1.5 Br=6.092 Nc=2 Isc=0 Ikr=0 Rc=1
+                     Cjc=7.306p Mjc=.3416 Vjc=.75 Fc=.5 Cje=22.01p Mje=.377 Vje=.75
+                     Tr=46.91n Tf=411.1p Itf=.6 Vtf=1.7 Xtf=3 Rb=10)
** Analysis Requests **
* vary temperature of circuit beginning at 0 C to 125 C in temperature steps of 25 C
.DC TEMP 0 125 25
** Output Requests **
.PLOT DC I(Vemitter)
.probe
.end
```

Figure 4.18 Spice input file for investigating the dependence of the emitter current on temperature in the amplifier circuit shown in Fig. 4.16.

4.5.2 Sensitivity to Temperature Variations

A different approach is required to investigate the bias stability of an amplifier subject to temperature changes. We make use of an extension to the DC sweep command that allows the user to sweep the temperature of the circuit over a specified range while repeatedly performing a DC analysis. The syntax of this command is identical to any other DC sweep command but with the key word TEMP replacing the field marked by *source_name*. The range of temperature values is specified in exactly the same way as for a voltage or current source. For the example shown in Fig. 4.18, we will monitor the emitter current over a temperature range beginning at 0° C and ending at 125° C in steps of 25° C.

The built-in model for the BJT accounts for variations in temperature. The same cannot be said for the resistors. To account for changes in resistance due to temperature variations, a resistor's temperature dependence must be indicated in each resistor statement in the input file using the two temperature coefficients TC_1 and TC_2. Spice calculates the value of a resistor at a temperature (*Temp*) other than 27° C using the equation

$$R(Temp) = R(27°)[1 + TC_1(Temp - 27°) + TC_2(Temp - 27°)^2] \qquad (4.11)$$

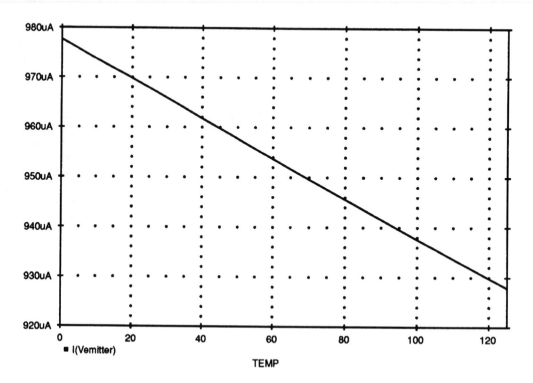

Figure 4.19 Temperature variation of the emitter current in the amplifier circuit shown in Fig. 4.16(a).

Here we shall consider only the linear dependence of resistance on temperature, so we shall consider $TC_2 = 0$. For this example $TC_1 = 1200$ ppm/°C, where *ppm* denotes parts per million, or a multiplication factor of 10^{-6}. After the field indicating the nominal value of the resistor, we specify the temperature dependence of the resistor by appending TC $= TC_1$. In this particular case, we write TC $= 1200$u, as indicated in Fig. 4.18.

The emitter current's dependence on temperature is summarized in Fig. 4.19. For a 125° C change in circuit temperature, the emitter current changed by about 50 μA. This change is sometimes expressed as a temperature coefficient. It consists of the ratio of (1) the change in emitter current divided by the 27° C emitter current to (2) the circuit temperature change that created the current change. Units are ppm/°C. In this case, the emitter bias current has a temperature coefficient of -413 ppm/°C.

4.6

The Common-Emitter Amplifier

This section considers the analysis of the common-emitter (CE) amplifier shown in Fig. 4.20. We will determine the small-signal parameters of this amplifier, such as input resistance R_i, output resistance R_o, voltage gain A_v, and current gain A_i, using the AC analysis facility of Spice. These results will then be compared to Sedra and Smith's small-signal hand analysis. The transistor will be assumed to be modeled after the commercial 2N2222A *npn* transistor.

To begin, let us compute the small-signal parameters of the CE amplifier. To obtain all four parameters, A_v, A_i, R_i, and R_o, we will have to run two separate Spice analyses: one for computing the input current and the output voltage for a known

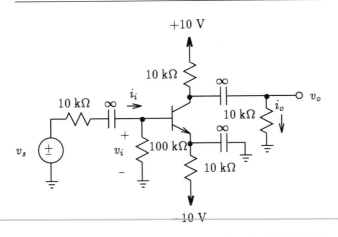

Figure 4.20   Common-emitter BJT amplifier.

voltage applied to the input of the amplifier and the other for computing the current supplied by a voltage source connected to the output terminal of the amplifier when the input voltage source is set to zero. These two situations are depicted in Fig. 4.21. To monitor the current flowing through the load resistor, we placed a zero-valued voltage source in series with it, as shown in Fig. 4.21(a). The input current to the amplifier can be determined by monitoring the current supplied by the input voltage source v_s. The

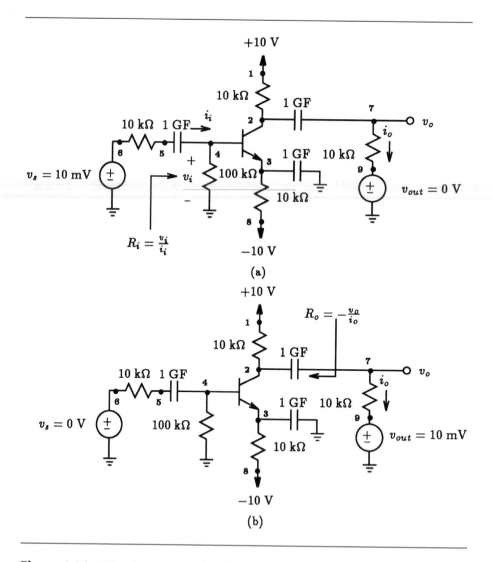

Figure 4.21 Circuit setups used to determine the amplifier's small-signal parameters: (a) circuit arrangement for computing i_i, v_o, and i_o; (b) circuit arrangement for computing i_o due to the voltage source connected to output terminal of amplifier.

four parameters of the amplifier can then be computed from these results according to the following:

$$A_v = \frac{v_o}{v_s} \tag{4.12}$$

$$A_i = \frac{i_o}{i_i} \tag{4.13}$$

$$R_i = \frac{v_i}{i_i} \tag{4.14}$$

$$R_o = -\frac{v_o}{i_o} \tag{4.15}$$

For the first case depicted in Fig. 4.21(a), consider the application of a 10 mV AC signal to the input of the amplifier. A DC input voltage signal would not be useful here because the input source to the amplifier is AC coupled. The frequency of the input signal should be chosen from the midband frequency range of the amplifier. With the choice of decoupling and bypass capacitors selected here (each selected very large at 1 GF),[†] an input frequency of 1 kHz is sufficiently midband. The input file describing this circuit setup is provided in Fig. 4.22. Both DC and AC analysis requests are specified. The results of the AC analysis will be used to indirectly calculate the small-signal parameters of the amplifier, as previously mentioned. The results of the DC analysis will provide us with the small-signal parameters of the transistor, which we can use in the formulae from Sedra and Smith to predict the small-signal parameters of the amplifier. We then can compare the results.

We find that Spice produces the magnitude and phase of the input and output voltages of the amplifier as follows:

```
****       AC ANALYSIS                        TEMPERATURE =   27.000 DEG C

   FREQ        VM(6)        VP(6)        VM(7)        VP(7)
  1.000E+03   1.000E-02    0.000E+00    5.068E-01    1.799E+02
```

We find that this amplifier has a midband voltage gain A_v of -50.68 V/V. Likewise, the input and output currents associated with this amplifier are found to be

```
****       AC ANALYSIS                        TEMPERATURE =   27.000 DEG C

   FREQ        IM(Vs)       IP(Vs)       IM(Vout)     IP(Vout)
  1.000E+03   6.803E-07    -1.796E+02    5.068E-05    1.799E+02
```

These two current results indicate that the midband current gain A_i of this amplifier is -74.49 A/A. Finally, the input resistance of the CE amplifier can be computed from the following data:

```
****       AC ANALYSIS                        TEMPERATURE =   27.000 DEG C

   FREQ        VM(4)        VP(4)        IM(Vs)       IP(Vs)
  1.000E+03   3.198E-03    -9.300E-01    6.803E-07    -1.796E+02
```

making the midband input resistance R_i of this amplifier 4.7 kΩ.

[†]A 1 GF capacitor is impossible to buy! We use such a large capacitance in the simulation in order to eliminate the effect of the coupling capacitor on the performance of the circuit. Circuit simulation has definite advantages compared with experimentation with actual electronic components.

```
Common-Emitter Amplifier Stage

** Circuit Description **
* power supplies
Vcc 1 0 DC +10V
Vee 8 0 DC -10V
* input signal
Vs 6 0 AC 10mV
Rs 5 6 10k
* amplifier
C1 4 5 1GF
Rb 4 0 100k
Q1 2 4 3 Q2N2222A
Rc 1 2 10k
Re 3 8 10k
C2 2 7 1GF
C3 3 0 1GF
* load + ammeter
Rl 7 9 10k
Vout 9 0 0
* transistor model statement for the 2N2222A
.model Q2N2222A NPN (Is=14.34f Xti=3 Eg=1.11 Vaf=74.03 Bf=255.9 Ne=1.307
+                    Ise=14.34f Ikf=.2847 Xtb=1.5 Br=6.092 Nc=2 Isc=0 Ikr=0 Rc=1
+                    Cjc=7.306p Mjc=.3416 Vjc=.75 Fc=.5 Cje=22.01p Mje=.377 Vje=.75
+                    Tr=46.91n Tf=411.1p Itf=.6 Vtf=1.7 Xtf=3 Rb=10)
** Analysis Requests **
* calculate DC bias point information
.OP
.AC LIN 1 1kHz 1kHz
** Output Requests **
*  voltage gain Av=Vo/Vs
.PRINT AC Vm(6) Vp(6) Vm(7) Vp(7)
*  current gain Ai=Io/Ii
.PRINT AC Im(Vs) Ip(Vs) Im(Vout) Ip(Vout)
* input resistance Ri=Vi/Ii
.PRINT AC Vm(4) Vp(4) Im(Vs) Ip(Vs)
.end
```

Figure 4.22 Spice input file for calculating the input current and the output voltage of the common-emitter amplifier shown in Fig. 4.21(a), subject to a 10 mV AC input signal.

To determine the output resistance of this amplifier we repeat this same process but set the level of the input voltage source to 0 V and increase the level of the voltage source in series with the load resistance to 10 mV AC. We modify the statements for these two AC sources in Fig. 4.22 to read as follows:

```
Vs   6 0 AC 0V
Vout 9 0 AC 10mV
```

Further, to access the resulting current supplied by the output voltage source V_{out} and the voltage developed across the output terminal of the CE amplifier, we include the following PRINT statement:

```
.PRINT AC Vm(7) Vp(7) Im(Vout) Ip(Vout)
```

Rerunning the Spice job, we find the following results in the output file:

```
****      AC ANALYSIS                           TEMPERATURE =   27.000 DEG C

   FREQ        VM(7)        VP(7)       IM(Vout)     IP(Vout)
  1.000E+03    4.728E-03    0.000E+00   5.272E-07    1.800E+02
```

From these results we compute the output resistance R_o to be 8.97 kΩ.

To compare the preceding results with those predicted by hand analysis, we give the expressions for A_v, A_i, R_i, and R_o in terms of the small-signal model parameters of the transistor, as derived in Section 4.11 of Sedra and Smith:

$$R_i = R_B \| r_\pi \tag{4.16}$$

$$R_o = R_C \| r_o \tag{4.17}$$

$$A_v \simeq -\frac{\beta(R_C \| R_L \| r_o)}{r_\pi + R_S} \tag{4.18}$$

$$A_i \simeq -\frac{\beta(R_C \| R_L \| r_o)}{R_L} \tag{4.19}$$

Here β is the small-signal transistor current gain[†] defined in terms of g_m and r_π:

$$\beta = g_m r_\pi \tag{4.20}$$

Table 4.4 Comparing the small-signal amplifier parameters of the common-emitter amplifier shown in Fig. 4.20(a) calculated by hand analysis with those indirectly computed using Spice.

Parameter	Formula	Hand Analysis	Spice	% Error
R_i	$R_B \| r_\pi$	4.69 kΩ	4.70 kΩ	-0.21
R_o	$R_C \| r_o$	8.97 kΩ	8.97 kΩ	0
A_v	$-\dfrac{\beta(R_C \| R_L \| r_o)}{r_\pi + R_S}$	-52.3 V/V	-50.68 V/V	3.2
A_i	$-\dfrac{\beta(R_C \| R_L \| r_o)}{R_L}$	-78.0 A/A	-74.49 A/A	4.7

[†]Spice computes two different βs and denotes them β_{ac} and β_{dc}. In general they are not equal to one another. β_{ac} refers to the small-signal current gain equal to $g_m r_\pi$ and β_{dc} is the DC current gain equal to I_C/I_B. Moreover, β_{dc} depends on the collector current level I_C, and its peak value corresponds with that specified in the BJT model statement using parameter Bf.

With the inclusion of the .DC operating point command in the spice deck, we find that the parameters of the hybrid-pi model for the 2N2222A transistor of the common-emitter amplifier are as follows:

```
**** BIPOLAR JUNCTION TRANSISTORS

NAME           Q1
MODEL          Q2N2222A
IB             5.84E-06
IC             8.72E-04
```

```
Common-Emitter Amplifier Stage with Sine-Wave Input

** Circuit Description **
* power supplies
Vcc 1 0 DC +10V
Vee 8 0 DC -10V
* input signal
Vs 6 0 SIN ( 0V 10mV 1kHz )
Rs 5 6 10k
* amplifier
C1 4 5 10uF
Rb 4 0 100k
Q1 2 4 3 Q2N2222A
Rc 1 2 10k
Re 3 8 10k
C2 2 7 10uF
C3 3 0 10uF
* load + ammeter
Rl 7 9 10k
Vout 9 0 0
* transistor model statement for the 2N2222A
.model Q2N2222A NPN (Is=14.34f Xti=3 Eg=1.11 Vaf=74.03 Bf=255.9 Ne=1.307
+                    Ise=14.34f Ikf=.2847 Xtb=1.5 Br=6.092 Nc=2 Isc=0 Ikr=0 Rc=1
+                    Cjc=7.306p Mjc=.3416 Vjc=.75 Fc=.5 Cje=22.01p Mje=.377 Vje=.75
+                    Tr=46.91n Tf=411.1p Itf=.6 Vtf=1.7 Xtf=3 Rb=10)
** Analysis Requests **
.OP
.TRAN 50us 8ms 5ms 50us
** Output Requests **
.Plot TRAN V(7) V(6)
.probe
.end
```

Figure 4.23 Spice input file for calculating the time-varying output signal of the common-emitter amplifier subject to a 10 mV, 1 kHz sinusoidal input signal. Unlike the AC analysis in Fig. 4.22, which is a small-signal analysis, a transient analysis performs a large-signal analysis even when the input level is small.

VBE	6.42E-01
VBC	-1.87E+00
VCE	2.51E+00
BETADC	1.49E+02
GM	3.36E-02
RPI	4.92E+03
RX	1.00E+01
RO	8.71E+04

Substituting the appropriate parameter values, together with values for the different circuit components, we find that $A_v = -52.3$ V/V, $A_i = -78.0$ A/A, $R_i = 4.69$ kΩ, and $R_o = 8.97$ kΩ. Table 4.4 compares the results of this hand analysis and the results obtained indirectly from Spice. As is evident from the table, the results agree quite well.

We can confirm the usefulness of small-signal analysis by applying a small-amplitude sinusoidal signal to the input of the amplifier and observing the time-varying signal that appears at the output. We can apply a 10 mV sinusoidal signal of 1 kHz frequency to the input, as shown in the Spice deck in Fig. 4.23. The transient analysis command is specified to compute the output waveform over three cycles of the input signal beginning at a time of 5 ms, ensuring that the circuit has had sufficient time to reach steady state. The DC-coupling and bypass capacitors have been

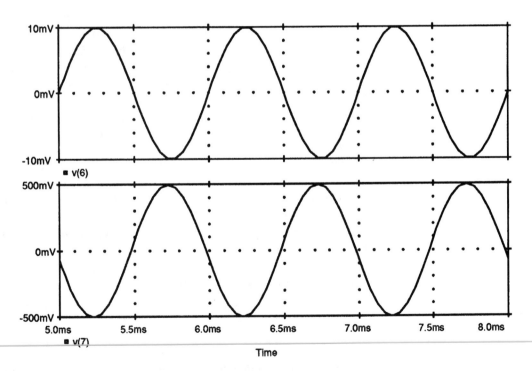

Figure 4.24 The input and output time-varying voltage waveforms associated with the common-emitter amplifier shown in Fig. 4.21(a).

assigned more practical values of 10 μF each, which drastically reduces the time to reach steady state. One should note that, by its very nature, the analysis performed here is a large-signal analysis and not a small-signal analysis, even though the input signal level is small.

The results of the transient analysis are shown in Fig. 4.24. The top graph displays the 10 mV peak, 1 kHz sinusoidal input signal and the bottom graph displays the corresponding output signal. With the aid of Probe, we found that the amplitude of the output signal is 500 mV. Thus, the voltage gain of this amplifier is seen to be about -50 V/V, which agrees with that predicted by the two preceding analyses (A_v).

4.7

Spice Tips

- A BJT is described to Spice using an element statement and a model statement.
- The element statement describes how the base, collector, and emitter of a transistor are connected to the rest of the network.
- The model statement contains a list of parameters describing the terminal characteristics of a BJT using the built-in model of Spice.
- There are 40 parameters associated with the built-in Spice model of the BJT. (We discussed only six of them; the remainder are described in Appendix A.)
- A specific transistor mode of operation is deduced from the DC voltages that appear across its emitter–base junction and its collector–base junction. This information can be found in the output file as a result of an operating-point (.OP) analysis.
- When an operating-point (.OP) calculation is included in the input file, a list of operating-point information for each transistor is obtained, which includes DC bias conditions for the transistor and the parameters associated with its small-signal model.
- Models of many commercial bipolar transistors are available from various semiconductor manufacturers. A library of transistor models for various commercial transistors is available with the PSpice program.
- Spice can be used to generate families of i–v curves for transistors, just as the laboratory curve tracer instrument can.
- A DC sweep command can be extended to include a sweep of temperature, allowing one to investigate the circuit's dependence on temperature.
- The DC sensitivities of a circuit can be computed using the .SENS analysis command. This command will compute the sensitivities of a particular circuit variable to most of the components in the circuit, including the sensitivities of many of the model parameters of the BJT.
- If the effect of decoupling and bypass capacitors is to be ignored, their values should be selected to be very large during the simulation (generally, a good value is 1 GF).

4.8

Bibliography

N. R. Malik, "Determining Spice Parameter Values for BJT's," *IEEE Transaction on Education,* Vol. 33, No. 4, pp. 366–368, Nov. 1990.

L. W. Nagel, *SPICE2—A Computer Program to Simulate Semiconductor Circuits,* Memorandum No. ERL-M520, Electronic Research Laboratory, University of California, Berkeley, May 1975.

4.9

Problems

4.1 Consider a *pnp* transistor having $I_S = 10^{-13}$ A and $\beta_F = 40$. If the emitter is connected to ground, the base is connected to a current source that pulls out of the base terminal a current of 10 μA, and the collector is connected to a negative supply of -10 V via a 10 kΩ resistor, find the base voltage, the collector voltage, and the emitter current using the operating point (.OP) command of Spice.

4.2 Using Spice as a curve tracer, plot the i_C–v_{CE} characteristics of an *npn* transistor having $I_S = 10^{-15}$ A and $V_A = 100$ V. Provide curves for $v_{BE} = 0.65, 0.70, 0.72, 0.73,$ and 0.74 V. Show the characteristics for v_{CE} up to 15 V.

4.3 For a BJT having an Early voltage of 200 V, what is its output resistance at 1 mA as calculated by Spice? At 100 μA?

4.4 The transistor in the circuit of Fig. P4.4 has a very high β_F (assume at least 10^6 in the Spice file). The other parameters of the transistor Spice model can assume default values. Find V_E and V_C for V_B equal to (a) $+3$ V, (b) $+1$ V, and (c) 0 V. What is the transistor mode of operation in each case?

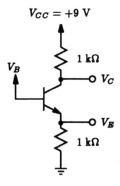

Figure P4.4

4.5 For the circuit in Fig. P4.4, with the aid of Spice and with V_B set equal to $+2$ V, find all node voltages for (a) β_F very high ($> 10^6$) and (b) $\beta_F = 99$.

4.6 In the circuit of Fig. 4.14, the input signal v_i is described by $0.004 \sin(\omega t)$ volts and the transistor is assumed modeled after the 2N2222A type. In addition, V_{BB} is reduced from 3 V to 1 V. Using Spice, plot the base and collector current of Q_1 as a function of time for at least one period of the input signal. Likewise, plot the collector voltage of Q_1. What is the voltage gain of this amplifier?

4.7 For the common-emitter amplifier shown in Fig. P4.7, let $V_{CC} = 9$ V, $R_1 = 27$ kΩ, $R_2 = 15$ kΩ, $R_E = 1.2$ kΩ, and $R_C = 2.2$ kΩ. The transistor has $\beta_F = 100$ and $V_A = 100$ V. Using Spice, calculate the DC bias current I_E. If the amplifier operates between a source for which $R_s = 10$ kΩ and a load of 2 kΩ, determine the values of R_i, g_m, R_o, A_v, and A_i.

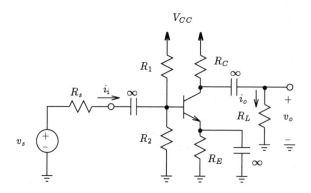

Figure P4.7

4.8 The amplifier of Fig. P4.8 consists of two identical common-emitter amplifiers connected in cascade.
   (a) For $V_{CC} = 15$ V, $R_1 = 100$ kΩ, $R_2 = 47$ kΩ, $R_E = 3.9$ kΩ, $R_C = 6.8$ kΩ, and $\beta_F = 100$, determine the DC collector current and collector voltage of each transistor using Spice. The amount of data that must be typed into the computer is reduced if a subcircuit is created for one of the stages and used twice to form the overall amplifier.
   (b) Find R_{in1} and v_{b1}/v_s for $R_s = 5$ kΩ.
   (c) Find R_{in2} and v_{b2}/v_{b1} for $R_s = 5$ kΩ.
   (d) For $R_L = 2$ kΩ, find v_o/v_s.

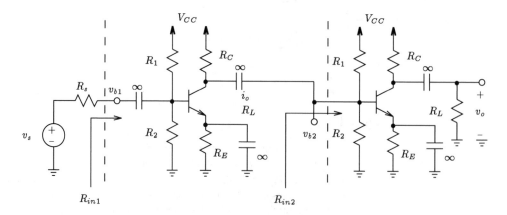

Figure P4.8

4.9 For the bootstrapped follower circuit shown in Fig. P4.9, where the transistor is assumed to be of the 2N2222A type, compare the input and output resistance and the voltage gain v_o/v_s, with and without the bootstrapping capacitor C_B in place.

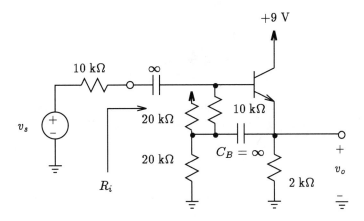

Figure P4.9

5

Field-Effect Transistors (FETs)

This chapter shows how Spice is used to simulate circuits containing field-effect transistors (FETs). Spice has built-in models for two of the three FET types considered here, metal-oxide-semiconductor FETs (MOSFETs) and junction FETs (JFETs). For metal-semiconductor FETs (MESFETs), our circuit simulations will use the built-in model of PSpice.

5.1

Describing MOSFETs to Spice

MOSFETs are described using two statements: One statement describes the nature of the FET and its connections to the rest of the circuit, and the other specifies the values of the parameters of the built-in FET model. Figure 5.1 shows the syntax for both statements. We also will describe some details of the built-in level 1 MOSFET model of Spice.

5.1.1 MOSFET Element Description

The presence of a MOSFET in a circuit is described in the input file using an element statement beginning with the letter M. If there is more than one MOSFET in a circuit, then a unique name must be appended to the M to identify each transistor. This name is followed by a list of the nodes to which the drain, gate, source, and substrate (body) of the MOSFET are connected. Next, the name of the model that will be used to characterize a particular MOSFET is given on the same line. The name of the model must correspond to the name given in the model statement with the parameter values for this MOSFET. Last, the length and width of the MOSFET are given.

5.1.2 MOSFET Model Description

Figure 5.1 shows the model statement for either the NMOS or the PMOS transistor, which begins with the key word .MODEL followed by the model name used in the

Metal-Oxide-Semiconductor FET (MOSFET)

| Element | Spice Description |

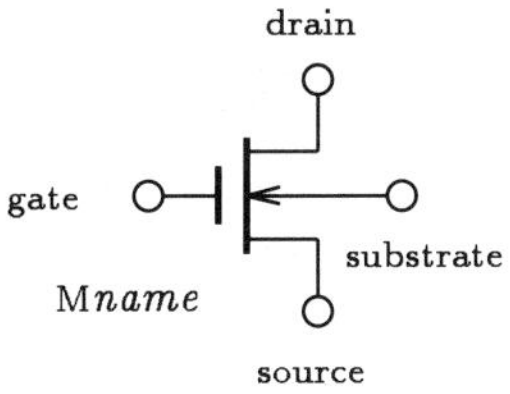

M*name* drain gate source substrate *MOS_model_name* L=*value* W=*value*

.MODEL *MOS_model_name* NMOS (*parameter_name=value* ....)

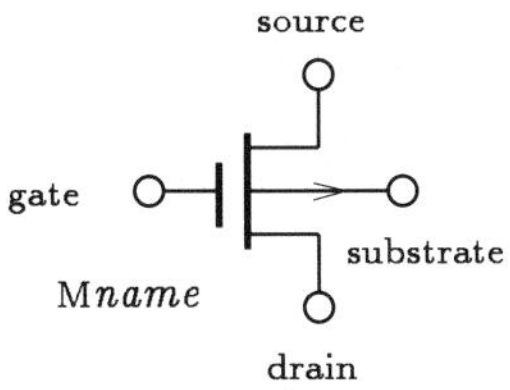

M*name* drain gate source substrate *MOS_model_name* L=*value* W=*value*

.MODEL *MOS_model_name* PMOS (*parameter_name=value* ....)

Figure 5.1   Spice element description for NMOS and PMOS MOSFETs and the general form of the associated MOSFET model statement. A partial listing of the parameter values applicable to either the NMOS or PMOS MOSFET is given in Table 5.1. Enhancement or depletion mode of operation is determined by the values assigned to these parameters.

element statement, the nature of the MOSFET (i.e., NMOS or PMOS), and a list of the model parameter values enclosed in parentheses. Parameters of the model that are not specified are assigned default values by Spice. The parameters associated with the Spice model of the MOSFET are numerous, and their meanings complicated; in addition, Spice has more than one large-signal model for the MOSFET. Parameters are classified as levels 1, 2, or 3, from simplest to most complicated. The MOSFET behavior described in Chapter 5 of Sedra and Smith is based on the level 1 MOSFET model. The other two models are more complicated, and their mathematical description will not be discussed here.[†]

The general form of the DC Spice model for an n-channel MOSFET is illustrated schematically in Fig. 5.2. The bulk resistance of both the drain and source regions

[†]The level 2 MOSFET model is a more complex version of the level 1 model and includes extensive second-order effects largely dependent on the geometry of the MOSFET. The level 3 MOSFET model is a semi-empirical model (having some model parameters that are not necessarily physically based) especially suited to short-channel MOSFETs (i.e., $L \leq 5$ μm). For more details on these models, interested readers can consult Vladimirescu [1980]. Also, an advanced VLSI textbook by Geiger, Allen, and Strader [1990] provides a good review of the three MOSFET models of Spice.

of the MOSFET are lumped into two linear resistances r_D and r_S, respectively. The DC characteristic of the intrinsic MOSFET is determined by the nonlinear dependent current source i_D, and the two diodes represent the two substrate junctions that define the channel region. A similar model exists for the p-channel device; the direction of the diodes, the direction of the current source, and the polarities of the terminal voltages are all reversed. The mathematical relationship that describes the DC behavior of the dependent current source varies, depending on which level of model is used. For the level 1 MOSFET model, the expression for drain current i_D, assuming that the drain is at a higher potential than the source, is described by[†]

$$i_D = \begin{cases} 0 & \text{for } v_{GS} < V_t \\ K\left[2(v_{GS} - V_t)v_{DS} - v_{DS}^2\right](1 + \lambda v_{DS}) & \text{for } v_{GS} > V_t \text{ and } v_{DS} \leq v_{GS} - V_t \\ K(v_{GS} - V_t)^2(1 + \lambda v_{DS}) & \text{for } v_{GS} > V_t \text{ and } v_{DS} \geq v_{GS} - V_t \end{cases} \quad (5.1)$$

where the device constant K is related to process parameters and device geometry according to

$$K = \frac{1}{2}\mu C_{OX}\left(\frac{W}{L}\right) \quad (5.2)$$

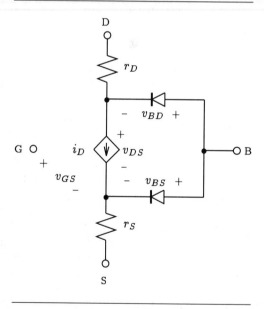

Figure 5.2 General form of the Spice large-signal model for an n-channel MOS-FET under static conditions.

[†]The expression for the drain current of the MOSFET in the triode region differs from that in Sedra and Smith in that it includes the factor $(1 + \lambda v_{DS})$. This ensures mathematical continuity between the triode and saturation regions.

and the threshold voltage V_t is given by

$$V_t = V_{t0} + \gamma \left[\sqrt{2\phi_f + V_{SB}} - \sqrt{2\phi_f} \right] \tag{5.3}$$

We see that the drain current equations are determined by the eight parameters W, L, μ, C_{OX}, V_{t0}, λ, γ, and $2\phi_f$. Both W and L define the dimensions of the device. These two parameters usually are specified on the element statement of the MOSFET, although if neither is specified, Spice will assume that both W and L are 100 μm. Parameters μ and C_{OX} are process-related parameters that are multiplied together to form the process transconductance coefficient kp, which is usually specified in the parameter list of the model statement. V_{t0}, the zero-bias threshold voltage, is positive for enhancement-mode n-channel MOSFETs and depletion-mode p-channel MOSFETs but is negative for depletion-mode n-channel MOSFETs and enhancement-mode p-channel MOSFETs. The parameter λ is the channel-length modulation parameter and represents the influence that drain–source voltage has on the drain current i_D when the device is in saturation. In Spice, unlike in Sedra and Smith, the sign of λ is always positive, regardless of the device type. The last two parameters, γ and $2\phi_f$, are the body-effect parameter and the surface potential, respectively.

A partial listing of the parameters for the Spice MOSFET model under static conditions is given in Table 5.1. (A complete listing is provided in Appendix A.) Also listed is the default value that a parameter assumes if a value is not specified for it. A parameter value is specified by writing, for example, level = 1, kp = 20u, Vto = 1V, and so on.

5.1.3 An Enhancement-Mode N-Channel MOSFET Circuit

Now we shall calculate the DC conditions of a circuit containing a FET and determine both the DC node voltages of the circuit and the DC operating-point information of the FET.

Consider the circuit shown in Fig. 5.3, which was presented in Example 5.1 of Sedra and Smith. Here we would like to confirm that the FET is indeed biased at a

Table 5.1 Partial listing of the Spice parameters for the level 1 MOSFET model.

Symbol	Spice Name	Model Parameter	Units	Default
	Level	Model type		1
μC_{OX}	kp	Transconductance coefficient	A/V^2	20 μ
V_{t0}	Vto	Zero-bias threshold voltage	V	0
λ	lambda	Channel-length modulation	V^{-1}	0
γ	gamma	Body-effect parameter	V$^{1/2}$	0
$2\phi_f$	phi	Surface potential	V	0.6
r_D	Rd	Drain ohmic resistance	Ω	0
r_S	Rs	Source ohmic resistance	Ω	0

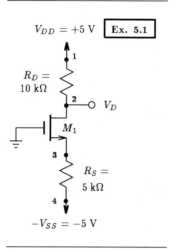

$V_{DD} = +5$ V   Ex. 5.1

$R_D = 10$ kΩ

V_D

M_1

$R_S = 5$ kΩ

$-V_{SS} = -5$ V

Figure 5.3   MOSFET circuit example.

current level of 0.4 mA and that the voltage appearing at the drain is $+1$ V. The NMOS transistor is assumed to have $V_t = 2$ V, $\mu_n C_{OX} = 20\ \mu\text{A/V}^2$, $L = 10\ \mu\text{m}$, and $W = 400\ \mu\text{m}$. Furthermore, the channel-length modulation effect is assumed zero (i.e., $\lambda = 0$). Assuming a level 1 MOSFET model, we can create the following Spice model statement:

```
.model nmos_enhancement_mosfet nmos (kp=20u Vto=+2V lambda=0)
```

We named this model *nmos_enhancement_mosfet*; the meaning should be obvious. Transistor M_1 and its geometry is then described using the following element statement:

```
M1 2 0 3 3 nmos_enhancement_mosfet L=10u W=400u
```

where we have assumed that the substrate (body) is connected to the source. We add these statements to the ones for the other devices of the circuit shown in Fig. 5.3 to create the input file shown in Fig. 5.4. Submitting this file with an operating-point (.OP) analysis request to Spice results in the following DC circuit information:

```
****      SMALL SIGNAL BIAS SOLUTION          TEMPERATURE =   27.000 DEG C
*****************************************************************************

NODE    VOLTAGE        NODE    VOLTAGE       NODE    VOLTAGE      NODE    VOLTAGE

(    1)     5.0000  (    2)     1.0000  (    3)    -3.0000  (    4)    -5.0000

        VOLTAGE SOURCE CURRENTS
        NAME            CURRENT

        Vdd           -4.000E-04
        Vss            4.000E-04
```

```
Simple Enhancement-Mode MOSFET Circuit

** Circuit Description **
* DC supplies
Vdd 1 0 DC +5V
Vss 4 0 DC -5V
* MOSFET circuit
M1 2 0 3 3 nmos_enhancement_mosfet L=10u W=400u
Rd 1 2 10k
Rs 3 4 5k
* MOSFET model statement (level 1 by default)
.model nmos_enhancement_mosfet nmos (kp=20u Vto=+2V lambda=0)
** Analysis Requests **
* calculate DC bias point
.OP
** Output Requests **
* none required
.end
```

Figure 5.4   Spice input file for computing the DC operating point of the circuit shown in Fig. 5.3.

```
TOTAL POWER DISSIPATION    4.00E-03  WATTS

****      OPERATING POINT INFORMATION      TEMPERATURE =    27.000 DEG C
*************************************************************************

**** MOSFETS

NAME            M1
MODEL           nmos_enhancement_mosfet
ID              4.00E-04
VGS             3.00E+00
VDS             4.00E+00
VBS             0.00E+00
VTH             2.00E+00
VDSAT           1.00E+00
```

As expected, these results confirm that the FET is biased at the intended current level of 0.4 mA and also that the drain of the FET is at the correct voltage level of $+1$ V. We see from the above results that M_1 is biased in its saturation region because $V_{DS} > V_{DS_{SAT}}$ where Spice uses the notation $V_{DS_{SAT}} = V_{GS} - V_t$.

5.1.4   *Observing the MOSFET Current–Voltage Characteristics*

The i_D–v_{DS} characteristics of a MOSFET are obtained by sweeping the drain-to-source voltage through a range of DC voltages while holding the gate-to-source voltage constant. The drain current of the MOSFET is then monitored and plotted against

the drain–source voltage. Consider the circuit shown in Fig. 5.5. Here two independent voltage sources, V_{GS} and V_{DS}, will be used to establish the different bias conditions on the enhancement-mode n-channel MOSFET whose source and body are
connected. It is assumed that the MOSFET has the same geometry and device parameters as in the last section, namely, the NMOS transistor is characterized by $V_t = +2$
V, $\mu_n C_{OX} = 20\ \mu\text{A/V}^2$, $L = 10\ \mu\text{m}$, and $W = 400\ \mu\text{m}$. The modulation factor will
be assumed to be zero (i.e., $\lambda = 0\ \text{V}^{-1}$). Let us set $V_{GS} = +3$ V and sweep V_{DS}
from 0 V to +10 V in steps of 100 mV so that the MOSFET will be taken through
both its triode and saturation regions. The resulting drain current is then monitored by
observing the current supplied by voltage source V_{DS}. The Spice input file is listed in
Fig. 5.6.

The results of this analysis are plotted in Fig. 5.7 as the curve marked *lambda*
= *0*. The behavior of the MOSFET is as expected from Eq. (5.1): For drain–source
voltages less than $V_{DS_{SAT}}$ ($V_{DS_{SAT}} = V_{GS} - V_t = +1$V), it is in its triode region. For
drain–source voltages above +1 V, the MOSFET saturates at a current level of 400
μA. The slope of the line drawn tangent to the output current in the saturation region
is zero, and the incremental drain–source resistance is infinite.

It is interesting to compare these results with the i_{DS}–v_{DS} characteristic of a MOS-
FET that is identical except for having a nonzero channel-length modulation coeffi-
cient of, say, $\lambda = 0.05\ \text{V}^{-1}$. This i_{DS}–v_{DS} curve also is shown superimposed on Fig.
5.7. The input deck for producing this curve is the same as Fig. 5.6, except the MOS-
FET model statement is modified:

```
.model nmos_enhancement_mosfet nmos (kp=20u Vto=+2V lambda=0.05)
```

We can see that the effect of a nonzero channel-length modulation coefficient results
in an output current that increases with increasing drain–source voltage at a rate of

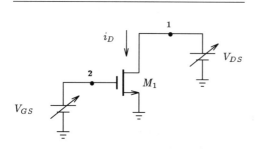

Figure 5.5   Spice curve-tracer arrangement for calculating the i–v characteristics
of a MOSFET. The i_D–v_{DS} characteristic of the MOSFET is obtained by sweeping
V_{DS} through a range of voltages while keeping V_{GS} constant at some value. The
i_D–v_{GS} characteristic of the MOSFET is obtained by sweeping V_{GS} through a range
of values while V_{DS} is held constant.

```
Enhancement-Mode N-Channel MOSFET Id-Vds Characteristics

** Circuit Description **
* bias conditions
Vds 1 0 DC +10V        ; this value is arbitrary, we are going to sweep it
Vgs 2 0 DC +3V
* MOSFET under test
M1 1 2 0 0 nmos_enhancement_mosfet L=10u W=400u
* MOSFET model statement (level 1 by default)
.model nmos_enhancement_mosfet nmos (kp=20u Vto=+2 lambda=0)
** Analysis Requests **
.DC Vds 0V 10V 100mV
** Output Requests **
.Plot DC I(Vds) V(1)
.Probe
.end
```

Figure 5.6   Spice input file for computing the i_D–v_{DS} characteristic of a MOSFET with device parameters $V_t = +2$ V, $\mu_n C_{OX} = 20\ \mu A/V^2$, $\lambda = 0$ V^{-1}, $L = 10\ \mu m$, and $W = 400\ \mu m$. The gate–source voltage V_{GS} is set equal to $+3$ V, and the drain–source voltage is swept between 0 and $+10$ V in 100-mV increments.

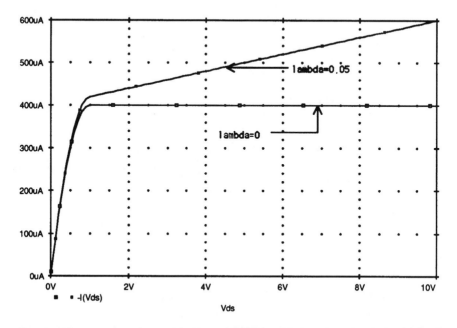

Figure 5.7   Comparing the i_D–v_{DS} characteristic of a MOSFET with a channel-width modulation factor $\lambda = 0$ and one with $\lambda = 0.05$ V^{-1}. The gate–source voltage is held constant at $+3$ V.

20.314 μA/V. The corresponding incremental drain–source resistance is 49.23 kΩ. It is reassuring that this increment resistance agrees closely with the value estimated by hand analysis; for example, for a drain current of about 400 μA, r_o, which is given by $1/(\lambda I_D)$, is estimated to be 50 kΩ.

Another current–voltage characteristic used to describe the behavior of a MOS-FET is i_D versus v_{GS}. The circuit arrangement illustrated in Fig. 5.5 can be used to obtain this current–voltage behavior also. This time the drain–source voltage is held constant and the gate–source voltage of the MOSFET is swept over a desired range. Let us set V_{DS} to +5 V and sweep V_{GS} from 0 V to +5 V in increments of 100 mV. This input deck is similar to that used previously and is shown in Fig. 5.8. Again, we have modeled the MOSFET with the channel-modulation coefficient at zero. For comparison, we have created another file identical to Fig. 5.8 except that the channel-modulation coefficient $\lambda = 0.05$ V^{-1}. These two Spice decks are then concatenated and submitted to Spice for analysis.

The results are shown in Fig. 5.9, where we have plotted the i_D–v_{GS} curves for $\lambda = 0$ and $\lambda = 0.05$ V^{-1}. The presence of a nonzero λ gives rise to a vertical shift in drain current. One interpretation of this is that for a given gate–source voltage, a MOSFET with a nonzero channel-length modulation factor will draw more current from a circuit than one with a zero channel-length modulation factor.

```
Enhancement-Mode N-Channel MOSFET Id-Vgs Characteristics

** Circuit Description **
* bias conditions
Vds 1 0 DC +5V
Vgs 2 0 DC +3V        ; this value is arbitrary, we are going to sweep it
* MOSFET under test
M1 1 2 0 0 nmos_enhancement_mosfet L=10u W=400u
* MOSFET model statement (level 1 by default)
.model nmos_enhancement_mosfet nmos (kp=20u Vto=+2 lambda=0)
** Analysis Requests **
.DC Vgs 0V 10V 100mV
** Output Requests **
.Plot DC I(Vgs) V(2)
.Probe
.end
```

Figure 5.8 Spice input file for computing the i_D–v_{GS} characteristic of a MOSFET with device parameters $V_t = +2$ V, $\mu_n C_{OX} = 20$ μA/V^2, $\lambda = 0$ V^{-1}, $L = 10$ μm, and $W = 400$ μm. The drain–source voltage V_{DS} is set equal to +5 V, and the gate–source voltage is swept between 0 and +5 V in 100 mV increments. The drain–source voltage is held constant at +5 V.

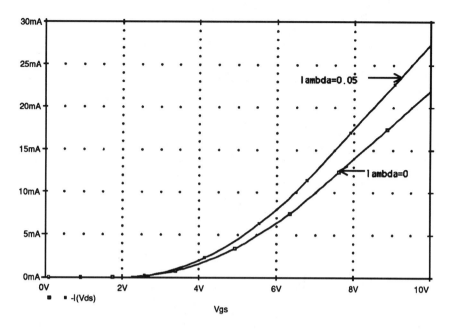

Figure 5.9 Comparing the i_D–v_{GS} characteristic of a MOSFET with a channel-length modulation factor $\lambda = 0$ and one with $\lambda = 0.05$ V^{-1}. The drain–source voltage is held constant at +5 V.

5.2

Spice Analysis of MOSFET Circuits at DC

MOSFETs are classified as n-channel or p-channel devices depending on the material used to form the channel. In addition, these devices are classified according to their mode of operation as enhancement- or depletion-type devices. As a result, the model statements characterizing these different FETs have subtle differences. In the following discussion we shall highlight these differences as we analyze the DC operating point of several simple MOSFET circuits from Chapter 5 of Sedra and Smith. Our results can then be compared with those derived by hand analysis in Sedra and Smith.

5.2.1   An Enhancement-Mode P-Channel MOSFET Circuit

Consider the circuit shown in Fig. 5.10. Here we have a single transistor circuit containing a p-channel enhancement-mode MOSFET. The resistors have been chosen so that the FET is biased in its saturation region, has a bias current of 0.5 mA, and has a drain voltage of +3 V. It is also assumed that the MOSFET has a threshold voltage of -1 V, has a process transconductance parameter $\mu_p C_{OX}$ equal to 1 mA/V^2, and does not exhibit any channel-length modulation effect (i.e., $\lambda = 0$). The details of this design can be found in Example 5.5 of Sedra and Smith. This circuit would be

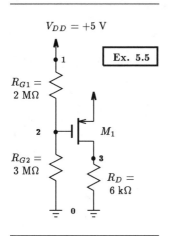

$V_{DD} = +5$ V

Ex. 5.5

$R_{G1} =$ 2 MΩ

M_1

$R_{G2} =$ 3 MΩ

$R_D =$ 6 kΩ

Figure 5.10 A transistor circuit designed in Example 5.5 of Sedra and Smith. Spice is used to calculate the DC operating point of this circuit.

described to Spice in the usual way; however, because the dimensions of the MOSFET are not given we shall assume arbitrarily that the length and width of it are both equal to 10 μm. The model statement describing the characteristics of the p-channel MOSFET appears as follows:

```
.model pmos_enhancement_mosfet pmos (kp=1m Vto=-1V lambda=0)
```

Here we have declared the MOSFET to be a PMOS device, and by assigning a negative threshold voltage we are indicating to Spice that this particular PMOS device is of the enhancement-mode type.

The Spice input deck for this example is listed in Fig. 5.11. We are requesting that Spice compute the DC operating point of this circuit. The results of this analysis are then found in the output file; some of its contents are displayed here:

```
****     SMALL SIGNAL BIAS SOLUTION          TEMPERATURE =   27.000 DEG C
*********************************************************************************

NODE   VOLTAGE     NODE   VOLTAGE     NODE   VOLTAGE     NODE   VOLTAGE

(    1)    5.0000  (    2)    3.0000  (    3)    3.0000

    VOLTAGE SOURCE CURRENTS
    NAME           CURRENT

    Vdd          -5.010E-04

    TOTAL POWER DISSIPATION    2.51E-03   WATTS

****     OPERATING POINT INFORMATION          TEMPERATURE =   27.000 DEG C
*********************************************************************************

**** MOSFETS
```

```
A Simple Enhancement-Mode PMOS Circuit  (Rd=6k)

** Circuit Description **
* DC supplies
Vdd 1 0 DC +5V
* MOSFET circuit
M1 3 2 1 1 pmos_enhancement_mosfet L=10u W=10u
Rd 3 0 6k
Rg1 1 2 2Meg
Rg2 2 0 3Meg
* MOSFET model statement (level 1 by default)
.model pmos_enhancement_mosfet pmos (kp=1m Vto=-1V lambda=0)
** Analysis Requests **
* calculate DC bias point
.OP
** Output Requests **
* none required
.end
```

Figure 5.11   Spice input file for calculating the DC operating point of the p-channel MOSFET circuit shown in Fig. 5.10.

NAME	M1
MODEL	pmos_enhancement_mosfet
ID	-5.00E-04
VGS	-2.00E+00
VDS	-2.00E+00
VBS	0.00E+00
VTH	-1.00E+00
VDSAT	-1.00E+00

As is evident from this output, the p-channel MOSFET is biased at a current level of 0.5 mA, and its drain is at +3 V. The sign of I_D is negative because of the convention adopted by Spice; positive drain current flows into the drain terminal of a FET regardless of the device type. This convention is different from that used in Sedra and Smith. We also know that the device is biased in its saturation region because $V_{DS} < V_{DS_{SAT}}$. It is interesting to note that if we alter the value of R_D from 6 kΩ to 8 kΩ and rerun the Spice input file, we find that M_1 is biased on the edge of saturation (i.e., $V_{DS} = V_{DS_{SAT}}$). This is evident from the results found in the Spice output file:

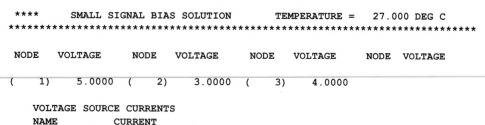

```
****        SMALL SIGNAL BIAS SOLUTION          TEMPERATURE =    27.000 DEG C
****************************************************************************

NODE    VOLTAGE      NODE    VOLTAGE      NODE    VOLTAGE      NODE   VOLTAGE

(    1)     5.0000  (     2)     3.0000  (    3)     4.0000

        VOLTAGE SOURCE CURRENTS
        NAME            CURRENT
        Vdd            -5.010E-04
```

```
            TOTAL POWER DISSIPATION    2.51E-03  WATTS

     ****       OPERATING POINT INFORMATION        TEMPERATURE =    27.000 DEG C
     ***********************************************************************************

     **** MOSFETS

     NAME          M1
     MODEL           pmos_enhancement_mosfet
     ID            -5.00E-04
     VGS           -2.00E+00
     VDS           -1.00E+00
     VBS            0.00E+00
     VTH           -1.00E+00
     VDSAT         -1.00E+00
```

Any increase in R_D above 8 kΩ will certainly cause M_1 to move out of the saturation region and into the triode region.

5.2.2   A Depletion-Mode P-Channel MOSFET Circuit

An example of a circuit incorporating a depletion-mode p-channel MOSFET is illustrated in Fig. 5.12. The depletion mode PMOS transistor is assumed to have $V_t = +1$ V, $\mu_p C_{OX} = 1$ mA/V^2, and $\lambda = 0$. We would like to compute the drain current and the corresponding drain voltage with Spice. The Spice input file for this circuit is listed in Fig. 5.13. Take note of the model statement describing the depletion-mode PMOS transistor in this file:

```
.model pmos_depletion_mosfet pmos (kp=1m Vto=+1V lambda=0)
```

In contrast to the enhancement-mode PMOS transistor of the last example, the threshold voltage for a depletion-mode PMOS transistor is made positive, but everything else remains the same.

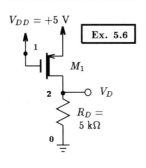

Figure 5.12   A depletion-mode p-channel MOSFET circuit.

```
A Depletion-Mode PMOS Transistor Circuit

** Circuit Description **
* DC supplies
Vdd 1 0 DC +5V
* MOSFET circuit
M1 2 1 1 1 pmos_depletion_mosfet L=10u W=10u
Rd 2 0 5k
* MOSFET model statement (level 1 by default)
.model pmos_depletion_mosfet pmos (kp=1m Vto=+1V lambda=0)
** Analysis Requests **
* calculate DC bias point
.OP
** Output Requests **
* none required
.end
```

Figure 5.13 Spice input file for calculating the DC operating point of the depletion-mode p-channel MOSFET circuit shown in Fig. 5.12.

Submitting the input file to Spice results in the following DC bias information:

```
****       SMALL SIGNAL BIAS SOLUTION        TEMPERATURE =   27.000 DEG C
****************************************************************************

   NODE    VOLTAGE      NODE    VOLTAGE      NODE    VOLTAGE      NODE   VOLTAGE

 (    1)    5.0000  (    2)     2.5000


     VOLTAGE SOURCE CURRENTS
     NAME            CURRENT

     Vdd           -5.000E-04


     TOTAL POWER DISSIPATION    2.50E-03   WATTS

 ****       OPERATING POINT INFORMATION       TEMPERATURE =   27.000 DEG C
****************************************************************************

**** MOSFETS

NAME          M1
MODEL         pmos_depletion_mosfet
ID            -5.00E-04
VGS            0.00E+00
VDS           -2.50E+00
VBS            0.00E+00
VTH            1.00E+00
VDSAT         -1.00E+00
```

The drain current and voltage of transistor M_1 are 0.5 mA and $+2.5$ V, respectively. We also see that the transistor is operating in the saturation region (i.e., $V_{DS} < V_{DS_{SAT}}$). These results are identical to those computed by hand in Example 5.6 of Sedra and Smith. In the preceding analysis the effect of channel-length modulation was assumed to be zero. In practice this is not the case. In the following example we repeat the preceding analysis assuming the transistor has a more realistic channel-length modulation coefficient of $\lambda = 0.02$ V^{-1}. We shall then compare the resulting transistor drain current with that obtained previously. This will give us some sense of how practical the assumption of neglecting the effect of channel-length modulation is when computing the DC bias point of a MOSFET circuit.

To carry out this task, we simply change the transistor model statement in Fig. 5.13 to the following:

```
.model pmos_depletion_mosfet pmos (kp=1m Vto=+1V lambda=0.02)
```

and resubmit the Spice deck for analysis. The results of the analysis are then found in the Spice output file. The pertinent details are shown here:

```
****     SMALL SIGNAL BIAS SOLUTION        TEMPERATURE =   27.000 DEG C
***********************************************************************


  NODE   VOLTAGE      NODE   VOLTAGE      NODE   VOLTAGE      NODE   VOLTAGE

(    1)    5.0000  (    2)    2.6190

   VOLTAGE SOURCE CURRENTS
   NAME            CURRENT

   Vdd           -5.238E-04

   TOTAL POWER DISSIPATION   2.62E-03  WATTS

  ****     OPERATING POINT INFORMATION      TEMPERATURE =   27.000 DEG C
***********************************************************************

**** MOSFETS

NAME            M1
MODEL           pmos_depletion_mosfet
ID              -5.24E-04
VGS             0.00E+00
VDS             -2.38E+00
VBS             0.00E+00
VTH             1.00E+00
VDSAT           -1.00E+00
```

We see that the transistor drain current is now 524 μA as a result of the channel-length modulation effect. This is about a 5% increase in the transistor drain current of 500 μA when no channel-length modulation effect was present. When performing rough "back-of-the-envelope" calculations, neglecting the channel-length modulation effect seems to be quite reasonable. When more accuracy is required, one resorts to the use of Spice.

5.2.3 A Depletion-Mode N-Channel MOSFET Circuit

As the final example of this section, we shall look at a circuit containing a depletion-mode NMOS transistor. Consider the circuit shown in Fig. 5.14, where M_1 is assumed to have the following parameters: $V_t = -1$ V, $\mu_n C_{OX} = 1$ mA/V^2, and $\lambda = 0$. A Spice listing of this circuit is shown in Fig. 5.15, and the results of an operating-point analysis (.OP) are presented here in part:

```
****        SMALL SIGNAL BIAS SOLUTION        TEMPERATURE =   27.000 DEG C
*******************************************************************************

   NODE   VOLTAGE        NODE   VOLTAGE      NODE   VOLTAGE      NODE   VOLTAGE

 (    1)   10.0000  (     2)    9.8956

       VOLTAGE SOURCE CURRENTS
       NAME              CURRENT

       Vdd            -9.896E-05

       TOTAL POWER DISSIPATION    9.90E-04   WATTS

   ****        OPERATING POINT INFORMATION        TEMPERATURE =   27.000 DEG C
*******************************************************************************

**** MOSFETS

NAME           M1
MODEL          nmos_depletion_mosfet
ID             9.90E-05
VGS            0.00E+00
```

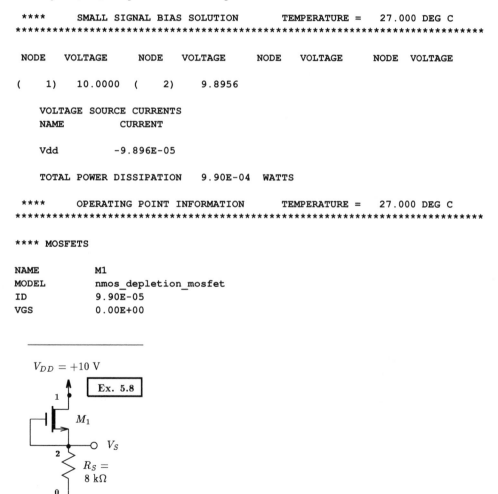

$V_{DD} = +10$ V

Ex. 5.8

M_1

V_S

$R_S = 8$ kΩ

Figure 5.14 A depletion-mode n-channel MOSFET circuit.

```
A Depletion-Mode NMOS Transistor Circuit

** Circuit Description **
* DC supplies
Vdd 1 0 DC +10V
* MOSFET circuit
M1 1 2 2 2 nmos_depletion_mosfet L=10u W=10u
Rd 2 0 100k
* MOSFET model statement (level 1 by default)
.model nmos_depletion_mosfet nmos (kp=1m Vto=-1V lambda=0)
** Analysis Requests **
* calculate DC bias point
.OP
** Output Requests **
* none required
.end
```

Figure 5.15 Spice input file for calculating the DC operating point of the depletion-mode n-channel MOSFET circuit shown in Fig. 5.14.

```
VDS             1.04E-01
VBS             0.00E+00
VTH            -1.00E+00
VDSAT           1.00E+00
```

Notice that in this case $V_{DS} < V_{DS_{SAT}}$, implying that the transistor is operating in the triode region.

In the following example, we repeat the preceding analysis with the same MOS-FET model except that it has a nonzero channel-length modulation coefficient, $\lambda = 0.04 \text{ V}^{-1}$. We then find in the Spice output file the following DC operating-point information:

```
****      SMALL SIGNAL BIAS SOLUTION        TEMPERATURE =   27.000 DEG C
*********************************************************************************

NODE    VOLTAGE     NODE    VOLTAGE    NODE    VOLTAGE    NODE   VOLTAGE

(   1)   10.0000  (    2)    9.8960


VOLTAGE SOURCE CURRENTS
NAME            CURRENT

Vdd          -9.896E-05


TOTAL POWER DISSIPATION   9.90E-04  WATTS


****      OPERATING POINT INFORMATION       TEMPERATURE =   27.000 DEG C
*********************************************************************************
```

```
**** MOSFETS

NAME          M1
MODEL         nmos_depletion_mosfet
ID            9.90E-05
VGS           0.00E+00
VDS           1.04E-01
VBS           0.00E+00
VTH           -1.00E+00
VDSAT         1.00E+00
```

We see that only the voltage at node 2 has changed, to 9.8960 V from the previous value of 9.8956 V. All other values listed appear to be the same as before, given the three or four digits of accuracy. Thus, the inclusion of the term $(1 + \lambda v_{DS})$ does not seem to affect the behavior of the circuit significantly when the device is operated in the triode region.

5.3

Describing JFETs to Spice

Like MOSFETs, JFETs are described to Spice using an element statement and a model statement. Figure 5.16 summarizes the syntax for both n-channel and p-channel JFETs.

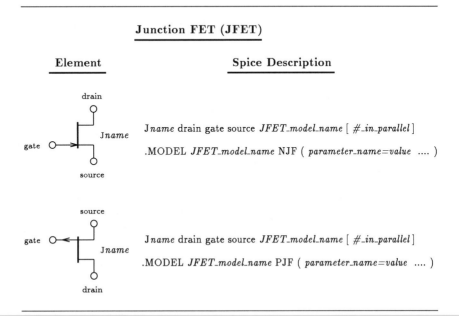

Figure 5.16 Spice element description for n-channel and p-channel JFETs and the general form of the JFET model statement. A partial listing of the parameters applicable to either an n-channel or p-channel JFET is given in Table 5.2.

5.3.1 JFET Element Description

JFETs are described to Spice using an element statement beginning with a unique name prefixed by the letter J, followed by a list of the nodes to which the drain, gate, and source of the JFET are connected. The next field specifies the name of the model that characterizes the transistor's terminal behavior. The final field is optional; its purpose is to allow the size of the device to be scaled by specifying the number of JFETs connected in parallel.

5.3.2 JFET Model Description

As shown in Fig. 5.16, the model statement for either the n-channel or the p-channel JFET transistor begins with the key word .MODEL followed by the name of the model used by a JFET element statement, the nature of the JFET (i.e., NJF or PJF), and a list of the JFET parameter values enclosed in parentheses. The mathematical model of the JFET in Spice is very similar to the level 1 MOSFET model.

Figure 5.17 shows the general schematic form of the DC Spice model for an n-channel JFET. The bulk resistance of the drain and source regions of the JFET are lumped into two linear resistances, r_D and r_S, respectively. The DC characteristic of the intrinsic JFET is determined by the nonlinear, dependent current source i_D, and the two diodes represent the two substrate junctions that define the channel region. A similar model applies to the p-channel device, but the direction of the diodes, the direction of the current source, and the polarities of the terminal voltages are all reversed.

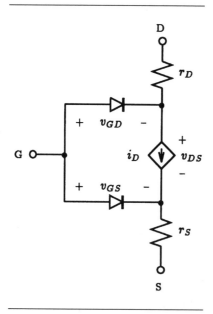

Figure 5.17   Spice large-signal n-channel JFET model under static conditions.

The expression for drain current i_D, assuming that the drain is at a higher potential than the source, is described by the following equation:

$$i_D = \begin{cases} 0 & \text{for } v_{GS} < V_t \\ \beta\left[2(v_{GS} - V_t)v_{DS} - v_{DS}^2\right](1 + \lambda v_{DS}) & \text{for } v_{GS} > V_t \text{ and } v_{DS} \leq v_{GS} - V_t \\ \beta(v_{GS} - V_t)^2(1 + \lambda v_{DS}) & \text{for } v_{GS} > V_t \text{ and } v_{DS} \geq v_{GS} - V_t \end{cases} \quad (5.4)$$

where the device parameters β and V_t are written in terms of I_{DSS} and V_P as

$$\beta = \frac{I_{DSS}}{V_P^2} \quad (5.5)$$

and

$$V_t = V_P \quad (5.6)$$

From these equations, we see that the JFET has three parameters that define its operation: I_{DSS}, V_P, and λ. I_{DSS} is the drain current when $v_{GS} = 0$ V, and V_P corresponds to the pinch-off voltage of the channel. The parameter λ is the channel-length modulation parameter and represents the influence of the drain–source voltage on the drain current i_D when the device is in pinch-off. The sign of this parameter is always positive, regardless of the nature of the device. A note on notation: The transconductance coefficient β is denoted by the parameter K in Sedra and Smith.

Table 5.2 gives a partial listing of the parameters for the Spice JFET model under static conditions and the default value that each parameter assumes if no value is specified. (A complete list appears in Appendix A.) Parameter values are specified by writing, for example, beta = 1m, Vto = −1V, lambda = 0.01, and so on.

5.3.3 An N-Channel JFET Example

Consider the circuit shown in Fig. 5.18. The JFET is n-channel with parameters $V_P = -4$ V, $I_{DSS} = 16$ mA, and $\lambda = 0$, and the model statement in the input file is

```
.model n_jfet NJF (beta=1m Vto=-4V lambda=0)
```

Here we had to convert the device parameters into terms that Spice recognizes. This required computing β from I_{DSS} and V_P from $\beta = I_{DSS}/V_P^2$. V_t is equal to V_P.

As usual, we compute the DC operating point of the circuit shown in Fig. 5.18. The input file, including the appropriate .OP command, is listed in Fig. 5.19.

Table 5.2 Partial listing of the Spice parameters for the JFET model.

Symbol	Spice Name	Model Parameter	Units	Default
β	beta	Transconductance coefficient	A/V^2	100 μ
V_t, V_P	Vto	Threshold voltage	V	−2.0
λ	lambda	Channel-length modulation	V^{-1}	0
r_D	Rd	Drain ohmic resistance	Ω	0
r_S	Rs	Source ohmic resistance	Ω	0

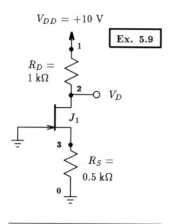

$V_{DD} = +10$ V

Ex. 5.9

$R_D = 1$ kΩ

V_D

J_1

$R_S = 0.5$ kΩ

Figure 5.18   An n-channel JFET circuit.

```
A Simple N-Channel JFET Circuit

** Circuit Description **
* DC supplies
Vdd 1 0 DC +10V
* JFET circuit
J1 2 0 3  n_jfet
Rd 1 2 1k
Rs 3 0 0.5k
* n-channel JFET model statement
.model n_jfet NJF (beta=1m Vto=-4V lambda=0)
** Analysis Requests **
* calculate DC bias point
.OP
** Output Requests **
* none required
.end
```

Figure 5.19   Spice input file for calculating the DC operating point of the n-channel JFET circuit shown in Fig. 5.18.

The following results are found in the output file:

```
****     SMALL SIGNAL BIAS SOLUTION        TEMPERATURE =   27.000 DEG C
********************************************************************************

 NODE    VOLTAGE      NODE    VOLTAGE      NODE    VOLTAGE      NODE   VOLTAGE

(   1)    10.0000  (    2)     6.0000  (    3)     2.0000
```

```
VOLTAGE SOURCE CURRENTS
NAME              CURRENT

Vdd             -4.000E-03

TOTAL POWER DISSIPATION    4.00E-02  WATTS
```

**** OPERATING POINT INFORMATION   TEMPERATURE =   27.000 DEG C
****

**** JFETS

```
NAME           J1
MODEL          n_jfet
ID             4.00E-03
VGS           -2.00E+00
VDS            4.00E+00
```

Here we see that the JFET is biased with a drain current of 4 mA and that the drain is at a voltage of 6 V.

This next example is along the same lines as the previous JFET example except that the JFET is p-channel. The p-channel JFET circuit that we will analyze for its DC bias conditions is displayed in Fig. 5.20. Here the p-channel device has parameters $V_P = +2$ V, $I_{DSS} = 4$ mA, and $\lambda = 0$. This example highlights the fact that the threshold voltage (V_t) of the p-channel JFET is specified on the model statement as a negative value, even though V_P is positive. Spice users must be careful of this convention because it is opposite that used in Sedra and Smith. The originators of Spice adopted this convention so that a depletion-mode device, such as the JFET, would have $V_t < 0$ whether it is n-channel or p-channel.

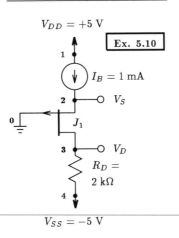

Figure 5.20   A p-channel JFET circuit.

```
A Simple P-Channel JFET Circuit

** Circuit Description **
* DC supplies
Vdd 1  0 DC +5V
Vss 4  0 DC -5V
Ib  1  2 DC 1mA
* JFET circuit
J1 3 0 2 p_jfet
Rd 3 4 2k
* p-channel JFET model statement
.model p_jfet PJF (beta=1m Vto=-2V lambda=0)
** Analysis Requests **
* calculate DC bias point
.OP
** Output Requests **
* none required
.end
```

Figure 5.21 Spice input file for calculating the DC operating point of the p-channel JFET circuit shown in Fig. 5.20.

The Spice input file describing the circuit shown in Fig. 5.20 is listed in Fig. 5.21. After processing, the following results are found in the output file:

```
****      SMALL SIGNAL BIAS SOLUTION      TEMPERATURE =   27.000 DEG C
****************************************************************************

NODE    VOLTAGE      NODE    VOLTAGE      NODE    VOLTAGE      NODE   VOLTAGE

(    1)    5.0000  (    2)    -1.0000  (    3)    -3.0000  (    4)    -5.0000

       VOLTAGE SOURCE CURRENTS
       NAME            CURRENT

       Vdd           -1.000E-03
       Vss            1.000E-03

       TOTAL POWER DISSIPATION    1.00E-02   WATTS

****      OPERATING POINT INFORMATION      TEMPERATURE =   27.000 DEG C
****************************************************************************

**** JFETS

NAME          J1
MODEL         p_jfet
ID            -1.00E-03
VGS            1.00E+00
VDS           -2.00E+00
```

The transistor is biased at 1 mA, and the source and drain are -1 V and -3 V, respectively.

To see how much the DC bias calculation changes with the inclusion of the transistor channel-length modulation effect, we repeat the above analysis with $\lambda = 0.04$ V^{-1}. After modifying the Spice deck shown in Fig. 5.21 to reflect this change, we submit the revised Spice deck. The following results are then obtained:

```
****      SMALL SIGNAL BIAS SOLUTION          TEMPERATURE =   27.000 DEG C
******************************************************************************

   NODE   VOLTAGE      NODE   VOLTAGE      NODE   VOLTAGE      NODE   VOLTAGE

 (    1)    5.0000  (     2)   -1.0371  (     3)   -3.0000  (     4)   -5.0000


     VOLTAGE SOURCE CURRENTS
     NAME            CURRENT

     Vdd            -1.000E-03
     Vss             1.000E-03


     TOTAL POWER DISSIPATION    1.00E-02   WATTS


   ****     OPERATING POINT INFORMATION      TEMPERATURE =    27.000 DEG C
******************************************************************************


**** JFETS

   NAME          J1
   MODEL         p_jfet
   ID            -1.00E-03
   VGS            1.04E+00
   VDS           -1.96E+00
```

Since the JFET is biased externally with a current source, the effect of the channel-length modulation will manifest itself in the voltages that appear across the terminals of the device. For instance, the gate–source voltage increases from 1 V to 1.04 V. Likewise, the drain–source voltage experiences the same change. In both cases, the resulting voltage change is small, which provides further justification that it is reasonable to neglect the presence of channel-length modulation when performing DC bias calculations by hand.

5.4

FET Amplifier Circuits

Field-effect transistors are commonly employed in the design of linear amplifiers. Small-signal linear analysis often is a means of estimating attributes of amplifier behavior when amplifiers are subjected to small input signals. Amplifier attributes include input and output resistances and current and voltage signal gain.

Spice has a small-signal linear model of the MOSFET and another for the JFET. We shall illustrate the effectiveness of small-signal analysis as it applies to a common-source n-channel enhancement-mode MOSFET amplifier by comparing the results calculated by the formulas in Sedra and Smith with those generated by Spice. But before these undertakings, we shall illustrate the importance of biasing a FET around an operating point that is well within the saturation region of the device.

5.4.1 *Effect of Bias Point on Amplifier Conditions*

In this section we consider the effect that a misplaced DC operating point has on the large-signal operation of a MOSFET amplifier. For the enhancement-type n-channel MOSFET amplifier shown in Fig. 5.22 with a +5 V fixed-biasing scheme, the DC operating point of the MOSFET has been set at approximately $I_D = 9$ mA and $V_{DS} = 8$ V. This is because the MOSFET has an assumed threshold voltage V_t of +2 V, a conductance parameter $K = \frac{1}{2}\mu_n C_{OX}(W/L) = 1$ mA/V^2, and a channel-length modulation factor $\lambda = 0.01$ V^{-1}. Figure 5.23 shows the i_D–v_{DS} characteristics for this device obtained using the curve tracer method described in Section 5.1. Also shown superimposed on this graph is the 1.33 kΩ load line. We see that the resulting operating point Q_1 is located slightly to the left of the midway point between the triode and cut-off regions. If the drain resistance, R_D, is increased to 1.78 kΩ, then the DC operating point, Q_2, located at $I_D = 9.3$ mA and $V_{DS} = 3.6$ V, moves closer to the edge of the triode region.

To illustrate the effect that the location of the DC operating point has on the large-signal behavior of the amplifier, consider the application of a 1 V peak-to-peak triangular waveform input signal of 1 kHz frequency in series with the 5 V DC biasing voltage as shown in Fig. 5.22. This process will be repeated for the same amplifier but

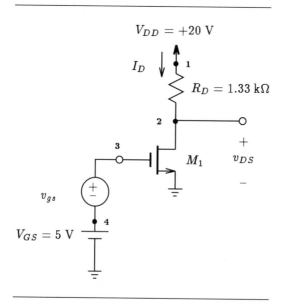

Figure 5.22   A fixed-bias n-channel MOSFET amplifier configuration.

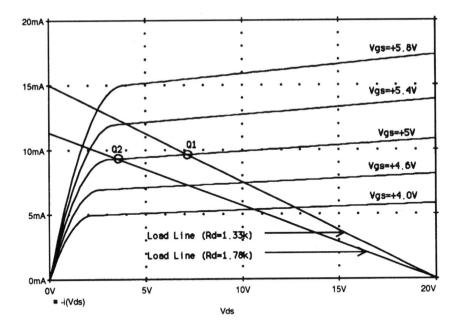

Figure 5.23 The i_D–v_{DS} characteristics of an n-channel MOSFET having model parameters $V_t = +2$ V, $\frac{1}{2}\mu_n C_{OX}(W/L) = 1$ mA/V^2, and $\lambda = 0.01$ V^{-1} and load lines for the MOSFET amplifier shown in Fig. 5.22 for drain resistances of 1.33 kΩ and 1.78 kΩ. Q_1 and Q_2 indicate the DC operating point of each amplifier with a gate–source bias voltage of 5 V.

with a drain resistance of 1.78 kΩ, and the resulting output signals from each amplifier will be compared. The Spice input file describing the first amplifier is provided in Fig. 5.24. Here the triangular input is described using the piecewise linear source statement for a time duration of 3 ms. A transient analysis command is included to compute the behavior of this amplifier over the same time period.

The results of the two Spice analyses are shown in Fig. 5.25. The top graph in part (a) of this figure illustrates both the input and output waveforms for a drain resistance of 1.33 kΩ. The output signal is reasonably linear with a peak-to-peak level about eight times that of the input signal (more precisely, 7.53 times). In contrast, when the drain resistance is increased to 1.78 kΩ, we see in Fig. 5.25(b) that the output signal has become quite distorted in the lower portion of its waveform. This example helps to illustrate the importance of placing the DC operating point of an amplifier well within the saturation region of the MOSFET in order to maximize the output voltage swing of the amplifier while maintaining linear operation.

5.4.2 Small-Signal Model of the FET

The linearized small-signal model for the MOSFET is shown in Fig. 5.26(a). It consists of two voltage-controlled current sources with transconductances g_m and g_{mb}.

```
An Enhancement-Type NMOS Amplifier

** Circuit Description **
* DC supplies
Vdd 1 0 DC +20V
Vgs 4 0 DC +5V
* small-signal input
vi 3 4 PWL (0,0 0.25ms,+0.5V 0.75ms,-0.5V 1.25ms,+0.5V 1.75ms,-0.5V
+           2.25ms,+0.5V 2.75ms,-0.5V 3ms,0V)
* amplifier circuit
M1 2 3 0 0 nmos_enhancement_mosfet L=10u W=10u
Rd 1 2 1.33k
* MOSFET model statement (level 1 by default)
.model nmos_enhancement_mosfet nmos (kp=2m Vto=+2V)
** Analysis Requests **
.OP
.TRAN 10us 3ms 0ms 10us
** Output Requests **
.PLOT TRAN V(2) V(3)
.probe
.end
```

Figure 5.24   Spice input file for calculating the large-signal transient behavior of the MOSFET amplifier shown in Fig. 5.22 with a drain resistance of 1.33 kΩ.

The MOSFET transconductance, g_m, is related to the DC bias current I_D and the device parameters (ignoring the channel-length modulation effect) as follows:

$$g_m = \left.\frac{\partial i_D}{\partial v_{GS}}\right|_{OP} \approx \sqrt{2\mu C_{OX}}\,\sqrt{W/L}\,\sqrt{I_D} \tag{5.7}$$

Here $|_{OP}$ indicates that the derivative is obtained at the DC operating point of the device. The next term, g_{mb}, is known as the body transconductance and is related to several MOSFET bias conditions and device parameters according to the following equation:

$$g_{mb} = \left.\frac{\partial i_D}{\partial v_{BS}}\right|_{OP} \approx \frac{\gamma}{2\sqrt{2\phi_f + V_{SB}}}g_m \tag{5.8}$$

To account for the presence of channel-length modulation, the output conductance g_{ds}, which is also equal to $1/r_o$, is given approximately by the expression:

$$g_{ds} = \left.\frac{\partial i_D}{\partial v_{DS}}\right|_{OP} \approx \lambda I_D \tag{5.9}$$

Finally, the resistances r_D and r_S represent the ohmic resistance of drain and source regions, respectively.

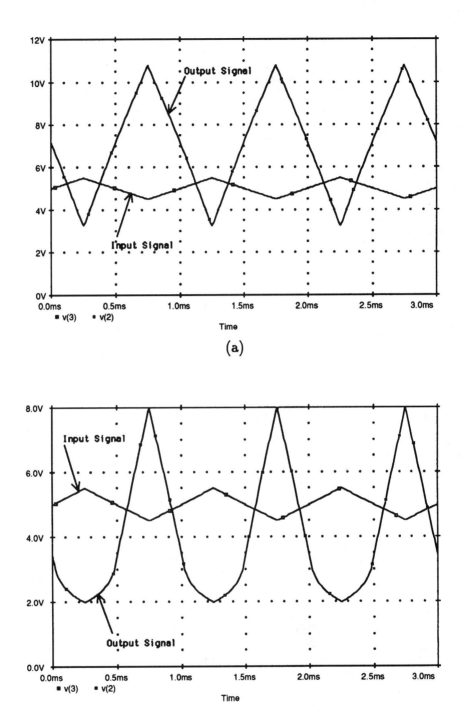

Figure 5.25 The input and output large-signal transient behavior of the MOSFET amplifier shown in Fig. 5.22 for a drain resistance of (a) 1.33 kΩ and (b) 1.78 kΩ.

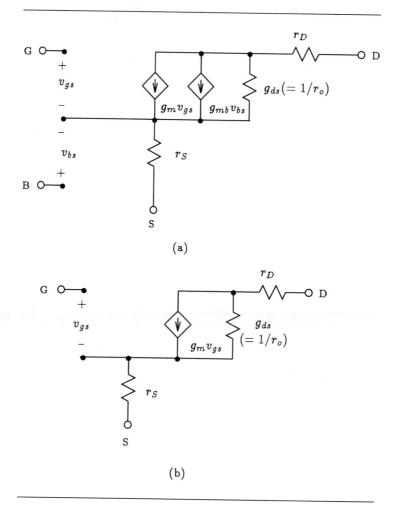

Figure 5.26   Static Spice small-signal model of (a) a MOSFET and (b) a JFET.

The small-signal linear model of a JFET is shown in Fig. 5.26(b). It consists of a single voltage-controlled current source having a transconductance g_m that is related to the DC bias current I_D and to the device parameters (ignoring the channel-length modulation effect) according to the equation:

$$g_m = \left.\frac{\partial i_D}{\partial v_{GS}}\right|_{OP} \approx \frac{2I_{DSS}}{|V_P|}\sqrt{\frac{I_D}{I_{DSS}}} \tag{5.10}$$

Taking into account the presence of the channel-length modulation, the output conductance g_{ds}, which is also equal to $1/r_o$, is given approximately by the expression:

$$g_{ds} = \left. \frac{\partial i_D}{\partial v_{DS}} \right|_{OP} \approx \lambda I_D \tag{5.11}$$

Last, the resistances r_D and r_S represent the ohmic resistance of the drain and source regions, respectively.

Many of the Spice analyses performed on a transistor circuit use its linear small-signal equivalent circuit. The parameters of the small-signal model of each transistor in a given circuit as computed by Spice are available to the user through the operating point (.OP) command. Consider a single NMOS transistor with its drain biased at +5 V, its gate biased at +3 V, and its source connected to ground. We shall assume that the substrate is biased at −5 V. The transistor is assumed to have the following parameters: $V_{t0} = +1$ V, $\mu_n C_{OX} = 20\ \mu\text{A/V}^2$, $L = 10\ \mu\text{m}$, and $W = 400\ \mu\text{m}$. Furthermore, $\lambda = 0.05$ V^{-1} and $\gamma = 0.9$ V$^{1/2}$. The input file for this circuit is

```
Small-Signal Model of an N-Channel MOSFET

** Circuit Description **
* MOSFET terminal bias
Vd 1 0 DC +5V
Vg 2 0 DC +3V
Vb 3 0 DC -5V
* MOSFET under test
M1 1 2 0 3 nmos_enhancement_mosfet L=10u W=400u
* MOSFET model statement (level 1 by default)
.model nmos_enhancement_mosfet nmos (kp=20u Vto=1V lambda=0.05 gamma=0.9)
** Analysis Requests **
.OP
** Output Requests **
* none required
.end
```

The output file contains the following complete list of parameters pertaining to the small-signal MOSFET model:

```
****      OPERATING POINT INFORMATION      TEMPERATURE =   27.000 DEG C

**** MOSFETS

NAME          M1
MODEL         nmos_enhancement_mosfet
ID            1.61E-04
VGS           3.00E+00
VDS           5.00E+00
VBS          -5.00E+00
VTH           2.43E+00
VDSAT         5.67E-01
GM            5.67E-04
GDS           6.44E-06
GMB           1.08E-04
CBD           0.00E+00
CBS           0.00E+00
CGSOV         0.00E+00
```

CGDOV	0.00E+00
CGBOV	0.00E+00
CGS	0.00E+00
CGD	0.00E+00
CGB	0.00E+00

This list of operating-point information includes DC bias conditions (which include drain current and various terminal voltages), device transconductances g_m and g_{mb}, output conductance g_{ds}, and device capacitances accounting for MOSFET dynamic effects. All these parameters, except for the capacitances, have been discussed previously, and their meanings should be clear. A discussion of MOSFET capacitances (all zero for now) will be deferred until Chapter 7. The reader can confirm that similar results extend to the JFET.

5.4.3 A Basic FET Amplifier Circuit

Figure 5.27 displays an enhancement-mode NMOS amplifier in which the input signal v_I is coupled to the gate of the MOSFET through a large capacitor, and the output signal at the drain is coupled to the load resistance R_L via another large capacitor. The MOSFET is assumed to have device parameters $V_t = +1.5$ V, $\mu C_{OX} = 0.5$ mA/V^2, and $\lambda = 0.02$ V^{-1}. A small-signal hand analysis of this same circuit in Example 5.11 of Sedra and Smith found through a two-step process that this amplifier has a voltage gain of -3.3 V/V and an input resistance of 2.33 MΩ. The first step was to obtain the DC bias conditions of the circuit, specifically the drain current, and from that to obtain the small-signal model for the transistor. The second step was to analyze the linear small-signal equivalent circuit of the amplifier to obtain the voltage gain and the input resistance. In each step of the analysis some simplifying assumptions

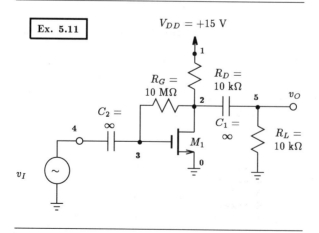

Figure 5.27 Common-source amplifier circuit.

are made. For instance, during the DC bias calculation, the transistor channel-length modulation effect was ignored, and in the second step a simplifying assumption was made regarding the current flowing through the feedback resistor R_G. We saw earlier that ignoring the channel-length modulation effect results in small variations in the drain current of a MOSFET and thus produces only small variations in the small-signal model parameters. We would like to use Spice to investigate the practicality of these two simplifications as they apply to the common-source amplifier in Fig. 5.27.

According to the hand analysis performed in Example 5.11 of Sedra and Smith (which ignores the channel-length modulation effect) the drain current of the MOS-FET is found to be 1.06 mA. From Eqs. (5.7) and (5.9) the MOSFET transconductance g_m equals 0.725 mA/V and the output resistance r_o is equal to 47 kΩ.

To compare these results with those calculated by Spice, we have created the input file listed in Fig. 5.28. The infinite-valued coupling capacitors are represented by 1 GF to ensure that the capacitors behave as short circuits at the signal frequencies of interest. The results generated by Spice are

```
**** MOSFETS

NAME         M1
MODEL        nmos_enhancement_mosfet
ID           1.07E-03
```

```
An Enhancement-Mode NMOS Amplifier

** Circuit Description **
* DC supplies
Vdd 1 0 DC +15V
* input AC signal
Vi 4 0 DC 0V
* amplifier circuit
M1 2 3 0 0 nmos_enhancement_mosfet L=10u W=10u
Rd 1 2 10k
Rg 2 3 10Meg
C1 2 5 1GF
C2 3 4 1GF
* load
Rl 5 0 10k
* MOSFET model statement (level 1 by default)
.model nmos_enhancement_mosfet nmos (kp=0.25m Vto=+1.5V lambda=0.02)
** Analysis Requests **
.OP
** Output Requests **
* none required
.end
```

Figure 5.28 Spice input file for calculating the DC operating point and small-signal model parameters of the amplifier circuit shown in Fig. 5.27.

```
VGS          4.31E+00
VDS          4.31E+00
VBS          0.00E+00
VTH          1.50E+00
VDSAT        2.81E+00
GM           7.62E-04
GDS          1.97E-05
GMB          0.00E+00
```

Here we see that the MOSFET is biased at a drain current of 1.07 mA, has a transconductance g_m equal to 0.762 mA/V, and an output conductance of 19.7 μS, or an output resistance r_o of 50.8 kΩ. We can see that these results generated by Spice are quite close to the hand-calculated values of Sedra and Smith, with at most a 7.5% error. In Table 5.3 we show the results and the error between them.

Once the small-signal equivalent circuit of the amplifier is obtained, we proceed to use standard circuit analysis techniques to obtain pertinent amplifier parameters, such as voltage gain and input resistance. To gain insight into circuit behavior, closed-form expressions are usually derived from the equivalent circuit, often resulting in expressions that are complicated, large, and not very insightful. One solution is to make simplifying assumptions based on practical considerations that lead to simpler but more insightful expressions. In Example 5.11 of Sedra and Smith, the linear small-signal equivalent circuit of the common-source amplifier shown in Fig. 5.27 was analyzed, several practical assumptions were made, and the following expressions for its voltage gain and input resistance were obtained:

$$A_V = -g_m(R_D \parallel R_L \parallel r_o) \tag{5.12}$$

and

$$R_{\text{in}} = \frac{R_G}{1 - A_V} \tag{5.13}$$

The simplicity of these two formulas makes them useful in circuit design, but how accurate are they?

To verify their accuracy, we substitute into these formulas the appropriate circuit parameters and the small-signal parameters of the MOSFET generated by Spice and evaluate and compare these results with those computed directly by Spice. This hand analysis finds $A_V = -3.468$ V/V and $R_{\text{in}} = 2.238$ kΩ.

For this Spice analysis, we might be tempted to request a transfer function (.TF) analysis and directly obtain both the small-signal voltage gain and the amplifier

Table 5.3 Comparing the transistor drain current I_D and its corresponding small-signal model parameters g_m and r_o as computed by straightforward hand analysis and Spice.

Parameter	Hand analysis	Spice	% Error
I_D	1.06 mA	1.07 mA	0.93
g_m	0.725 mA/V	0.762 mA/V	4.9
r_o	47 kΩ	50.8 kΩ	7.5

input resistance. Unfortunately, the amplifier is AC coupled and is intended to amplify signals containing frequencies other than DC. Since the .TF analysis calculates the small-signal input–output behavior of a circuit only at DC, in this situation the results produced would not be very useful. Instead, we apply a 1 V AC voltage signal to the input of the amplifier and compute the voltage appearing at the amplifier output using the .AC analysis command of Spice at a single midband frequency of, say, 1 Hz. A 1 V input level is usually chosen because in this way the output voltage would be directly equal to the input–output transfer function. The input signal level of 1 V is not considered to be above the small-signal limit of the amplifier because the .AC analysis by Spice is performed directly with the small-signal equivalent circuit of the amplifier, so that any input level would work. The Spice statements necessary to invoke this analysis are

```
Vi 4 0 AC 1V
.AC LIN 1 1Hz 1Hz
```

In a similar way, we can compute the input resistance of the amplifier by determining the current supplied by the input voltage source and computing the input resistance as the ratio of the input voltage to input current. To obtain this information, we ask Spice to print both the magnitude and phase of the output voltage signal (node 5), followed by the magnitude and phase of the current supplied by Vi, using the following statement:

```
.PRINT AC Vm(5) Vp(5) Im(Vi) Ip(Vi)
```

Recall that Spice uses complex variables to evaluate the AC small-signal response of a circuit, so we must indicate the form of the complex variable (i.e., magnitude, whether real or imaginary, etc.) we want Spice to print.

The AC analysis results computed by Spice at 1 Hz then would be found in the output file as follows:

```
****      AC ANALYSIS                    TEMPERATURE =    27.000 DEG C

   FREQ         VM(5)         VP(5)       IM(Vi)       IP(Vi)
   1.000E+00    3.467E+00   -1.800E+02    4.467E-07    1.800E+02
```

The small-signal voltage gain of the amplifier (A_V) is therefore -3.476 V/V and the input resistance R_{in} is 2.238 MΩ ($= 1/4.467 \times 10^{-7}$MΩ). The phase of the input current IP(Vi) indicates that the sign of the current supplied by the input voltage source Vi is negative, which is consistent with Spice's convention that current supplied by a source to a circuit is always negative.

When we compare results derived using hand analysis and Spice-generated small-signal model parameters (-3.468 V/V, 2.238 MΩ) with results directly from Spice (-3.476 V/V, 2.238 MΩ), we see that they are either identical or are quite close, with a relative error of only 0.23%. We can therefore conclude that the simplifying assumptions used to derive the formula for small-signal voltage gain and input resistance in Sedra and Smith are reasonable and introduce little error. When small-signal parameters computed by hand are used instead of the Spice-generated model parameters, the accuracy of the results decreases but remains within practical limits (i.e., $A_V = -3.3$ V/V and $R_{in} = 2.33$ MΩ). The relative errors of the two calculations are about 5%

and are largely due to the error incurred when the DC hand analysis ignores the transistor channel-length modulation effect. For easy reference, Table 5.4 summarizes the preceding discussion.

5.5

Investigating Bias Stability with Spice

To obtain a stable DC operating point in discrete transistor circuits, a biasing scheme utilizing some form of negative feedback is usually employed. This ensures that the DC bias currents through the transistors in the circuit remain relatively constant under the influence of normal manufacturing or environmental variations.

To illustrate the effectiveness of incorporating negative feedback in the biasing network of a transistor amplifier, let us compare the DC sensitivities of the two amplifiers shown in Fig. 5.29 with the resistive biasing components. Each MOSFET will be assumed to have the following parameters: $V_t = +2$ V and $\mu_n C_{OX} = 2$ mA/V^2. For a fair comparison, each amplifier is biased at approximately the same current level of 3 mA. The n-channel enhancement MOSFET amplifier of Fig. 5.29(a) is simply biased by a voltage appearing at its gate terminal. Although the MOSFET in the amplifier of Fig. 5.29(b) is also biased with a voltage at its gate through a voltage divider circuit, a resistor is included in the source lead of the MOSFET, which provides a feedback action that acts to stabilize the drain current when the MOSFET is subjected to change. This means that if one of the biasing elements undergoes a small change, the resulting drain current of the MOSFET will experience less change than when no feedback action is present. To see this, we created the two Spice input files listed in Figs. 5.30 and 5.31 with the following sensitivity analysis request:

```
.SENS I(Vdrain)
```

This command will invoke Spice to compute the DC sensitivities of the drain current of each MOSFET as monitored by the zero-valued voltage source that appears in series with its drain terminal. The sensitivity analysis command of Spice was first introduced to the reader in Section 4.5.

The results of this analysis found in the Spice output file are shown below for the amplifier circuit of Fig. 5.29(a). First we show the bias conditions of the amplifier, followed by a list of the circuit DC sensitivities.

```
    ****      OPERATING POINT INFORMATION      TEMPERATURE =    27.000 DEG C

    **** MOSFETS

    NAME        M1
    MODEL       nmos
    ID          3.06E-03
    VGS         3.75E+00
    VDS         8.88E+00
    VBS         0.00E+00
    VTH         2.00E+00
    VDSAT       1.75E+00
```

Table 5.4 Comparing the voltage gain and the input resistance of the amplifier shown in Fig. 5.27 as calculated by three different methods: closed-form expression using hand estimates of the small-signal model parameters, closed-form expression using Spice-calculated small-signal model parameters, and direct calculation using Spice.

| Parameter | Hand Analysis | | | % Error | |
	Hand Estimate of Small-Signal Model Parameters	Spice Small-Signal Model Parameters	Spice (Direct Calculation)	Hand Estimate of Small-Signal Model Parameters	Spice Small-Signal Model Parameters
A_V	−3.3 V/V	−3.468 V/V	−3.476 V/V	5.1	0.23
R_{in}	2.33 MΩ	2.238 MΩ	2.238 MΩ	−4.1	0

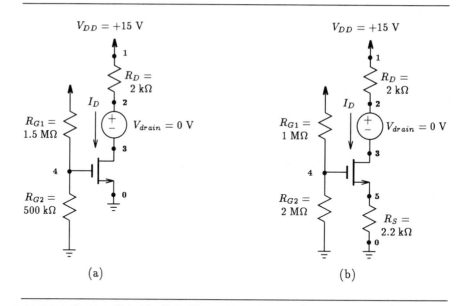

Figure 5.29 Two different MOSFET biasing arrangements: (a) fixed biasing and (b) biasing with source-resistance feedback. The zero-valued voltage source in series with the drain terminal of each MOSFET is used to directly monitor the drain current.

```
Simple MOSFET Bias Network (No Feedback Mechanism)

** Circuit Description **
* DC supply
Vdd 1 0 DC +15V
* amplifier circuit
M1 3 4 0 0 nmos L=100u W=100u
Rg1 1 4 1.5Meg
Rg2 4 0 500k
Rd 1 2 2k
* drain current monitor
Vdrain 2 3 0
* MOSFET model statement (level 1 by default)
.model nmos nmos (kp=2m Vto=+2V lambda=0)
** Analysis Requests **
.OP
.SENS I(Vdrain)
** Output Requests **
* none required
.end
```

Figure 5.30 Spice input file for calculating the DC sensitivities of the drain current of the MOSFET amplifier shown in Fig. 5.29(a).

```
MOSFET Bias Network with Feedback

** Circuit Description **
* DC supply
Vdd 1 0 DC +15V
* amplifier circuit
M1 3 4 5 0 nmos L=100u W=100u
Rg1 1 4 1Meg
Rg2 4 0 2Meg
Rd 1 2 2k
Rs 5 0 2.2k
* drain current monitor
Vdrain 2 3 0
* MOSFET model statement (level 1 by default)
.model nmos nmos (kp=2m Vto=+2V lambda=0)
** Analysis Requests **
.OP
.SENS I(Vdrain)
** Output Requests **
* none required
.end
```

Figure 5.31   Spice input file for calculating the DC sensitivities of the drain current of the MOSFET amplifier shown in Fig. 5.29(b).

```
****       DC SENSITIVITY ANALYSIS              TEMPERATURE =    27.000 DEG C

DC SENSITIVITIES OF OUTPUT I(Vdrain)
```

ELEMENT NAME	ELEMENT VALUE	ELEMENT SENSITIVITY (AMPS/UNIT)	NORMALIZED SENSITIVITY (AMPS/PERCENT)
Rg1	1.500E+06	-6.563E-09	-9.844E-05
Rg2	5.000E+05	1.969E-08	9.844E-05
Rd	2.000E+03	-3.063E-15	-6.125E-14
Vdd	1.500E+01	8.750E-04	1.313E-04
Vdrain	0.000E+00	-1.000E-12	0.000E+00

Likewise, the results of the sensitivity analysis for the amplifier circuit of Fig. 5.29(b), together with the DC operating-point information for the MOSFET, are as follows:

```
****       OPERATING POINT INFORMATION        TEMPERATURE =    27.000 DEG C

**** MOSFETS

NAME        M1
MODEL       nmos
ID          2.87E-03
```

VGS	3.69E+00
VDS	2.96E+00
VBS	-6.31E+00
VTH	2.00E+00
VDSAT	1.69E+00

****   DC SENSITIVITY ANALYSIS   TEMPERATURE =   27.000 DEG C

DC SENSITIVITIES OF OUTPUT I(Vdrain)

ELEMENT NAME	ELEMENT VALUE	ELEMENT SENSITIVITY (AMPS/UNIT)	NORMALIZED SENSITIVITY (AMPS/PERCENT)
Rg1	1.000E+06	-1.336E-09	-1.336E-05
Rg2	2.000E+06	6.679E-10	1.336E-05
Rd	2.000E+03	-2.867E-15	-5.734E-14
Rs	2.200E+03	-1.149E-06	-2.527E-05
Vdd	1.500E+01	2.672E-04	4.008E-05
Vdrain	0.000E+00	-1.000E-12	0.000E+00

Reviewing the above sensitivity results, we see that Spice has generated several columns of output. The first column indicates the element with respect to which the sensitivity of the drain current I_D is taken. The second column indicates the nominal value of that element as it appears in the Spice input file. The third column indicates the sensitivity quantity $\partial I_D/\partial x$, where x is the corresponding element from the first column. The units of this sensitivity quantity are the units of the output variable specified on the .SENS statement divided by the units of the element x. For instance, in the case of element R_{G1}, the sensitivity quantity $\partial I_D/\partial R_{G1}$ is expressed in A/Ω. The final column is a normalized sensitivity measure. It simply expresses the sensitivity in more convenient units of A/%. Mathematically, it is written as $\partial I_D/(\partial R_{G1}/R_{G1})_\%$ where $(\partial R_{G1}/R_{G1})_\%$ is the relative accuracy of R_{G1} expressed as a percentage.

Returning to the results of the two sensitivity analyses, let us consider one interpretation of these results: If we were to build the two amplifiers shown in Fig. 5.29 using resistors that have a tolerance of ± 5%, then according to the principle of a total derivative, the total change in the drain current due to variations in the biasing resistors [i.e., R_{G1}, R_{G2}, R_D, and R_S in the case of the amplifier in Fig. 5.29(b)] can be approximated by the following:

$$\Delta I_D \approx \frac{\partial I_D}{(\partial R_{G1}/R_{G1})_\%} \times (\Delta R_{G1}/R_{G1})_\% + \frac{\partial I_D}{(\partial R_{G2}/R_{G2})_\%} \times (\Delta R_{G2}/R_{G2})_\%$$

$$+ \frac{\partial I_D}{(\partial R_S/R_S)_\%} \times (\Delta R_S/R_S)_\% + \frac{\partial I_D}{(\partial R_D/R_D)_\%} \times (\Delta R_D/R_D)_\% \quad (5.14)$$

Now, if we assume the worst-case situation in which each resistor undergoes the same amount of change with a sign that contributes to the total sum (as opposed to reducing the sum), we can write Eq. (5.14) as

$$\Delta I_D \approx \left[\left| \frac{\partial I_D}{(\partial R_{G1}/R_{G1})_\%} \right| + \left| \frac{\partial I_D}{(\partial R_{G2}/R_{G2})_\%} \right| + \left| \frac{\partial I_D}{(\partial R_S/R_S)_\%} \right| + \left| \frac{\partial I_D}{(\partial R_D/R_D)_\%} \right| \right] \times (\Delta R/R)_\%$$

$$(5.15)$$

where we denote the relative magnitude change of each resistor as $(\Delta R/R)_\%$. Under the assumed worst-case condition, with $(\Delta R/R)_\% = 5\%$, we can substitute the sensitivities previously computed by Spice into Eq. (5.15) and determine the expected change in the drain current of the MOSFET in each amplifier of Fig. 5.29. On doing so, we find that for the amplifier with the fixed-gate-voltage biasing scheme [Fig. 5.29(a)] the expected change in the drain current will be about 984.4 μA, whereas in the amplifier with a negative-feedback biasing scheme [Fig. 5.29(b)] the worst-case change in the drain current is expected to be only 260 μA. This is about four times less change than the former case.

This same type of analysis can be repeated for a variation in the supply voltage V_{DD}. If we consider a 1% change in the supply voltage, we can expect that the drain current of the MOSFET in the fixed-bias amplifier will experience a 131 μA change. In contrast, the MOSFET drain current of the amplifier with a negative-feedback biasing scheme will only change by 40 μA.

Unfortunately, the preceding sensitivity analysis does not list the sensitivity relative to the MOSFET. To obtain this information, we use a simple brute-force approach. Let us change the device parameters K and V_t separately and observe their effects on the drain current through an operating-point (.OP) command.

For instance, in the case of the fixed-bias MOSFET amplifier stage shown in Fig. 5.29, let us change the value of $kp = 2$ mA/V^2 in the MOSFET model statement seen in Fig. 5.30 to $kp = 2.2$ mA/V^2, a positive change of +10%. After running the Spice job, we obtain the following operating-point information for the MOSFET:

```
**** MOSFETS

NAME          M1
MODEL         nmos
ID            3.37E-03
VGS           3.75E+00
VDS           8.26E+00
VBS           0.00E+00
VTH           2.00E+00
VDSAT         1.75E+00
```

The drain current is now 3.37 mA. Comparing this with the previous drain current of 3.06 mA prior to the change in kp, we see that a +10% change in the process transconductance kp results in a +10.1% change in the drain current.

Similarly, if we repeat this same experiment with the MOSFET amplifier with a feedback biasing scheme (i.e., if we change kp from 2 mA/V^2 to 2.2 mA/V^2 in the Spice deck listed in Fig. 5.31) and resimulate the circuit, we get the following operating-point results:

```
**** MOSFETS

NAME          M1
MODEL         nmos
ID            2.90E-03
VGS           3.62E+00
VDS           2.83E+00
```

```
VBS              -6.38E+00
VTH               2.00E+00
VDSAT             1.62E+00
```

The drain current becomes 2.90 mA. Comparing this with the previous drain current of 2.87 mA prior to the change in kp, we see that a $+10\%$ change in the process transconductance kp results in a $+1.04\%$ change in the drain current. The feedback biasing scheme is about 10 times less sensitive to a process change in kp than the fixed biasing scheme is.

A similar approach can be taken for changes in the threshold voltage. If we assume that, because of a process variation, the threshold voltage of the MOSFET changes by -5%, we find that the drain current for the fixed biasing scheme changes by $+11.8\%$. In contrast, the drain current in the MOSFET of the amplifier having a negative-feedback biasing scheme would only change by $+1.39\%$.

The benefits of an amplifier biasing scheme that incorporates negative feedback should now be apparent.

5.6

Integrated-Circuit MOS Amplifiers

In this section we shall investigate the behavior of several different types of fully integratable amplifiers that are constructed with MOSFETs only.

5.6.1 Enhancement-Load Amplifier Including the Body Effect

Figure 5.32 shows an enhancement-load NMOS amplifier with the substrate connections. This arrangement would be typical of an amplifier implemented in an NMOS fabrication process. One significant drawback to this amplifier is that its voltage gain

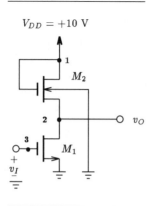

Figure 5.32 Enhancement-load amplifier.

is reduced because of the presence of the MOSFET body effect in transistor M_2. To see this, let us consider that the two MOSFETs in Fig. 5.32 have the following device parameters: a process transconductance coefficient ($\mu_n C_{OX}$) of 0.25 mA/V^2, a zero-bias threshold voltage of 1 V, a channel-length modulation factor λ of 0.02 V^{-1}, and a body-effect coefficient γ of 0.9 V$^{1/2}$. Transistor M_1 will have a length–width dimension of 10 μm by 100 μm, whereas the dimensions of transistor M_2 will reciprocate at 100 μm by 10 μm. The input file for this arrangement is in Fig. 5.33. A DC sweep of the input voltage level between ground and V_{DD} is requested. For comparison, we shall repeat this analysis on a circuit with identical device parameters except that the body-effect coefficient will be set to zero. The input deck for this case is added to the end of the one in Fig. 5.33 and both are submitted to Spice for analysis. Figure 5.34 shows the DC transfer characteristics of the enhancement-load amplifier with and without the transistor body effect. Clearly, transistor body effect alters the transfer characteristics of the enhancement-load amplifier significantly.

When the input level is below 1 V, corresponding to the threshold of M_1, the output voltage is held constant at either 7.2 V or 9 V, depending on the curve. On the curve with the body effect present when input levels increase above the 1 V level, the output voltage decreases linearly from 7.2 V at a rate of about -7.9 V/V until the input exceeds about 1.8 V and the output is 0.75 V. Above this input voltage level, transistor M_1 leaves the saturation region and enters the triode region, resulting in nonlinear amplifier characteristics.

```
An Enhancement-Load Amplifier Including Body Effect

** Circuit Description **
* DC supplies
Vdd 1 0 DC +10V
* input signal
Vi 3 0 DC 0V
* amplifier circuit
M1 2 3 0 0 nmos L=10u W=100u
M2 1 1 2 0 nmos L=100u W=10u
* MOSFET model statement (level 1 by default)
.model nmos nmos (kp=0.25m Vto=+1.0V lambda=0.02 gamma=0.9)
** Analysis Requests **
.OP
.DC Vi 0V 10V 100mV
** Output Requests **
.PLOT DC V(2)
.Probe
.end
```

Figure 5.33 Spice input file for calculating the DC transfer characteristic of the enhancement-load amplifier shown in Fig. 5.32. Each MOSFET is modeled to include the transistor body effect.

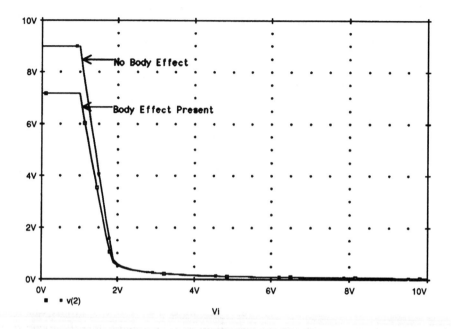

Figure 5.34 DC transfer characteristic of the enhancement-load amplifier shown in Fig. 5.32 with and without the MOSFET body effect.

In the case of the amplifier with the body effect eliminated, at inputs above 1 V the output level (beginning at 9 V) decreases linearly at a rate of -9.2 V/V. As before, when the input voltage level exceeds 1.8 V (and the output is at 0.75 V), transistor M_1 enters the triode region and the amplifier characteristic curve becomes nonlinear. Comparing these curves suggests that the presence of transistor body effect has decreased the effective gain of this enhancement-load amplifier.

To further demonstrate this, let us compute this amplifier's voltage gain with and without the body effect using the transfer function (.TF) analysis command. For illustrative purposes we shall bias the input to the amplifier at 1.5 V because this input level maintains both amplifiers in their linear regions. We modify each of the two input decks used previously by changing the source statement to read

```
Vi 3 0 DC +1.5V
```

and including the following .TF command:

```
.TF V(2) Vi
```

When we look at the output files of the two small-signal transfer function analyses, we find that the enhancement-load amplifier with body effect present shows

```
****    SMALL-SIGNAL CHARACTERISTICS

    V(2)/Vi = -7.316E+00
```

```
INPUT RESISTANCE AT Vi =   1.000E+20

OUTPUT RESISTANCE AT V(2) =   5.508E+03
```

and the amplifier with the body effect eliminated shows

```
****      SMALL-SIGNAL CHARACTERISTICS

V(2)/Vi = -9.027E+00

INPUT RESISTANCE AT Vi =   1.000E+20

OUTPUT RESISTANCE AT V(2) =   6.677E+03
```

This reconfirms our earlier claim that the MOSFET body effect acts to decrease the effective gain of an enhancement-load amplifier.

It also would be instructive to confirm the accuracy of the small-signal formula for the voltage gain of the amplifier that includes the transistor body effect. According to Section 5.9 of Sedra and Smith, the voltage gain of the amplifier shown in Fig. 5.32 is given by

$$A_V = -\frac{g_{m_1}}{g_{m_2} + g_{mb_2} + g_{ds_1} + g_{ds_2}} \tag{5.16}$$

Through an operating point (.OP) analysis command (included in the previous analysis), we found the following values for the small-signal model parameters of each transistor:

```
**** MOSFETS
```

NAME	M1	M2
MODEL	nmos	nmos
ID	3.32E-04	3.32E-04
VGS	1.50E+00	6.87E+00
VDS	3.13E+00	6.87E+00
VBS	0.00E+00	-3.13E+00
VTH	1.00E+00	2.04E+00
VDSAT	5.00E-01	4.83E+00
GM	1.33E-03	1.37E-04
GDS	6.25E-06	5.84E-06
GMB	7.72E-04	3.20E-05

Substituting the appropriate values into Eq. (5.16), we find $A_V = -7.344$ V/V, which is very close to the value predicted directly by Spice (i.e., $A_V = -7.316$ V/V). If the output conductances are neglected in this calculation, we would find $A_V = -7.869$ V/V, a more than adequate result for most practical applications.

5.6.2   CMOS Amplifier

Figure 5.35 shows a CMOS amplifier with an active current-source load, which is an example of an amplifier that is fully integratable using MOS technology. We will compute and plot the amplifier DC transfer characteristic (v_O vs. v_I) with Spice.

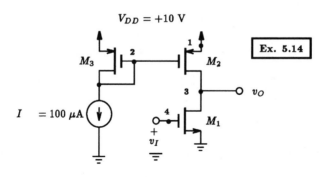

Figure 5.35   CMOS amplifier with current-source biasing.

```
A CMOS Amplifier

** Circuit Description **
* DC supplies
Vdd 1 0 DC +10V
Iref 2 0 DC 100uA
* input signal
Vi 4 0 DC 0V
* amplifier circuit
M1 3 4 0 0 nmos L=10u W=100u
M2 3 2 1 1 pmos L=10u W=100u
M3 2 2 1 1 pmos L=10u W=100u
* MOSFET model statements (level 1 by default)
.model nmos nmos (kp=20u Vto=+1V lambda=0.01)
.model pmos pmos (kp=10u Vto=-1V lambda=0.01)
** Analysis Requests **
* calculate DC transfer characteristics
.DC Vi 0V +10V 10mV
** Output Requests **
.PLOT DC V(3)
.probe
.end
```

Figure 5.36   Spice input file for calculating the DC transfer characteristic of the CMOS amplifier circuit shown in Fig. 5.35.

In Example 5.14 of Sedra and Smith this circuit was assumed to have $V_{tn} = |V_{tp}| = 1$ V, $\mu_n C_{OX} = 2\mu_p C_{OX} = 20 \ \mu A/V^2$, and $\lambda = 0.01 \ V^{-1}$ for n and p devices. Furthermore, for all devices it was assumed that $W = 100 \ \mu m$ and $L = 10 \ \mu m$.

Figure 5.36 shows the input file for this CMOS amplifier. A zero-valued DC voltage source V_I is initially applied to the amplifier input and will be varied over a range beginning at ground potential and increasing to V_{DD} in 10 mV steps. The output voltage (V(3)) will then be plotted as a function of the input voltage V_I.

The Spice-calculated DC transfer characteristic of the CMOS amplifier is shown in Fig. 5.37. We see that for input signals less than about +1.0 V the output voltage starts at about 10 V, at +1.0 V it begins to change gradually, at around 2 V it drops dramatically but linearly to a gradual change, and by +2.5 V it is near ground potential. Thus, the CMOS amplifier experiences large voltage gain in the vicinity of the 2 V input level.

To see the high-gain region of this amplifier more closely, we will repeat the previous DC sweep analysis and evaluate the amplifier transfer characteristic between +1.9 V and +2.1 V. A very small step size of 100 μV is used to obtain a smooth curve. The results are shown in Fig. 5.38, where the linear region of the amplifier is clearly visible, bounded by input voltages of 1.955 V and 2.027 V. The output voltage varies correspondingly between 8.589 V and 0.9966 V. These results suggest that the gain of this amplifier in this linear region is approximately (8.589 − 0.9966)/(1.955 − 2.027) = −118.9 V/V.

A similar result is obtained when the small-signal gain is evaluated around a single operating point inside this linear region. A point midway between the extremes of the linear region would correspond to an input bias level of about 2 V. If we modify the input deck shown in Fig. 5.36 so that the input is biased at 2 V and replace the DC sweep command by a .TF analysis command, we find the following small-signal DC transfer function information in the output file:

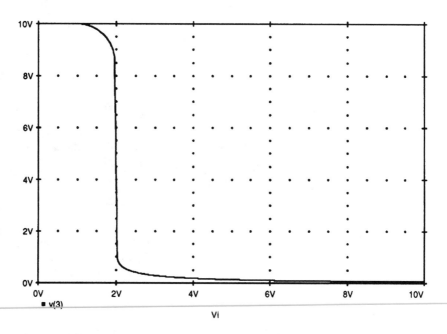

Figure 5.37 DC transfer characteristic of the CMOS amplifier shown in Fig. 5.35 as calculated by Spice.

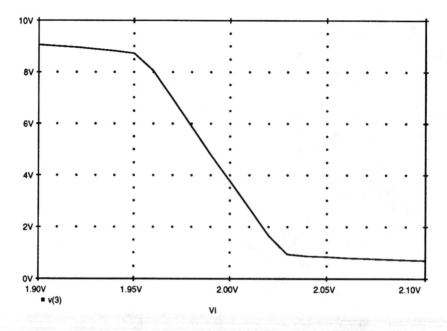

Figure 5.38 An expanded view of the large-signal transfer characteristic of the CMOS amplifier shown in Fig. 5.35 in its high-gain region.

```
****        SMALL-SIGNAL CHARACTERISTICS

     V(3)/Vi = -1.050E+02

     INPUT RESISTANCE AT Vi =  1.000E+20

     OUTPUT RESISTANCE AT V(3)  =  5.059E+05
```

The small-signal gain of the CMOS amplifier is -105 V/V at an input DC voltage level of 2 V.

In Section 5.9 of Sedra and Smith, a simple formula is derived for the voltage gain in the linear region of the CMOS amplifier in Fig. 5.35:

$$A_V = -\frac{\sqrt{\frac{1}{2}\mu_n C_{OX}(W/L)_1}\,|V_A|}{\sqrt{I_{\text{ref}}}} \tag{5.17}$$

This formula results in a gain that is quite close to that computed by Spice. Substituting the appropriate device parameter values into Eq. (5.17), we get $A_V = -100$ V/V.

5.7

MOSFET Switches

MOSFETs are commonly used as switches for both analog and digital signals. In analog circuit applications a switch is used to control the passage of an analog current

signal between two nodes in a circuit in either direction without distortion or attenuation. In digital applications a switch is commonly used as a transmission gate to
realize specific logic functions when combined with other logic gates.

   An ideal mechanical switch is depicted in Fig. 5.39(a). In the on state (i.e., switch
closed) a direct connection is made between nodes 1 and 2. Thus, a signal applied
to node 1 will also appear at node 2. The reverse is also true if nodes 1 and 2 are
interchanged. In practice, a signal passing through a switch will experience a signal
attenuation due to the electrical resistance of the switch. Thus, we can model this
behavior by adding a resistance R_{ON} in series with the ideal switch, as illustrated in
Fig. 5.39(b). Clearly, the larger R_{ON} is, the more loss a signal will experience as it
passes through the switch.

   To judge the on-resistance of a single n-channel MOSFET switch, let us create a
Spice input file that represents the situation shown in Fig. 5.40(a). In this circuit, we
shall sweep the input voltage v_I from $V_{SS} = -5$ V to $V_{DD} = +5$ V and compute the
current that is supplied to the switch by v_I. The other terminal of the MOSFET switch
is connected directly to ground, thus the resistance of the switch R_{ON} is given by the
ratio of the input voltage to the current supplied by this voltage source. The gate of the
MOSFET, being the control terminal of the switch, will be held at $V_{DD} = +5$ V to
ensure that the switch is turned on. The resulting Spice input file is shown in Fig. 5.41.
The NMOS transistor will be assumed to have a process transconductance parameter
$\mu_n C_{OX}$ equal to 0.25 mA/V^2, a zero-bias threshold voltage of 1 V, a channel-length
modulation factor λ of 0.02 V^{-1}, and a body-effect coefficient γ of 0.9 V$^{1/2}$. The
dimensions of the MOSFET will be 100 μm by 100 μm.

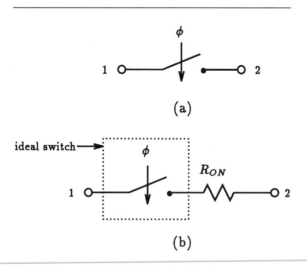

Figure 5.39 (a) An ideal mechanical switch. (b) Electrical equivalent of a real
switch with on-resistance R_{ON}.

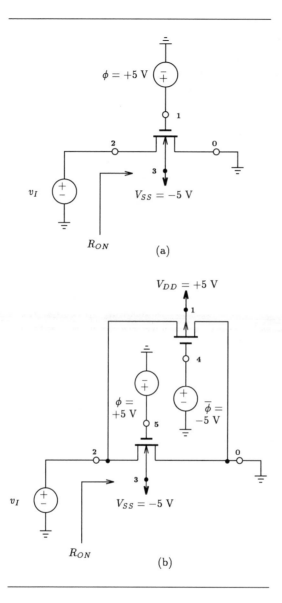

Figure 5.40   MOSFET switch realizations: (a) a single n-channel MOSFET and (b) a CMOS transmission gate.

It should be pointed out that the DC sweep was selected to have a voltage step of 10.01 mV instead of a more even 10 mV to ensure that when we compute the ratio of input voltage to input current, we don't try to divide 0 by 0. Moreover, our choice of which terminal of the MOSFET is the source and which the drain was arbitrary. The results should be the same regardless of our choice. (If there is any doubt, the reader should repeat the simulation with the source and drain terminals interchanged.)

```
Computing Switch On-Resistance of an NMOS Switch

** Circuit Description **
* input signal
Vi 2 0 DC 0V
* substrate bias
Vss 3 0 DC -5V
* control signal
Vphi 1 0 DC +5V
* switch: NMOS transistor
M1 2 1 0 3 nmos_enhancement_mosfet L=100u W=100u
* MOSFET model statement (level 1 by default)
.model nmos_enhancement_mosfet nmos (kp=0.25m Vto=+1.0V lambda=0.02 gamma=0.9)
** Analysis Requests **
* sweep the input voltage from Vss to Vdd (skip over 0,0 point)
.DC Vi -5V +5V 10.01mV
** Output Requests **
.PLOT DC I(Vi) V(2)
.Probe
.end
```

Figure 5.41 Spice input file for calculating the input current as a function of input voltage for the n-channel MOSFET switch of Fig. 5.40(a). Postprocessing will be used to compute the on-resistance of the switch.

On completion of the Spice job, we divide the input voltage by the input current and plot the results as a function of the input voltage. The curve representing this result is shown in Fig. 5.42 and is labeled the *NMOS switch.* As evident, the on-resistance of the switch is lowest at $R_{ON} = 560 \ \Omega$ when the input equals $V_{SS} = -5$ V. The switch on-resistance steadily increases as the input voltage increases towards $V_{DD} = +5$ V. In fact, as the input voltage approaches V_{DD}, the switch on-resistance increases substantially, for example, at $v_I = +5$ V, $R_{ON} = 5.6 \ \text{k}\Omega$. We can conclude that the on-resistance of this switch is dependent on the input signal level, which introduces undesirable nonlinearity.

If we repeat the preceding experiment on a p-channel MOSFET, similar outcomes would result. These are also shown in Fig. 5.42 as the curve labeled *PMOS switch.* Clearly, the PMOS switch has similar but complementary behavior. Observing these results, we can conclude that a single MOSFET does not make a very effective switch.

A better approach, and one that is extensively used in analog IC design when a high-quality switch is required, is to connect both the n-channel and the p-channel MOSFETs in parallel as illustrated in Fig. 5.40(b). The gate control of each MOSFET is driven by complementary signals. To see why this transistor arrangement makes an effective switch, let us compute the on-resistance of this switch. To turn on the switch, control signal ϕ is set to $V_{DD} = +5$ V and $\overline{\phi}$ is set to $V_{SS} = -5$ V. The Spice input file describing the situation is listed in Fig. 5.43. The MOSFETs are assumed to have the same device parameters used in the previous example. For easy comparison, the

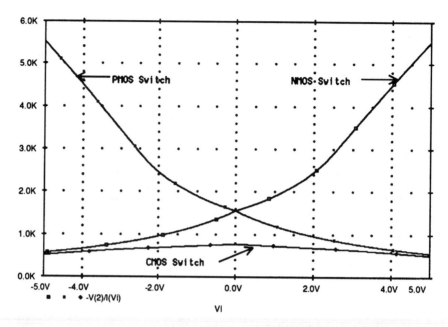

Figure 5.42 The on-resistance R_{ON} as a function of the input signal level of an analog switch realized as (a) a single n-channel MOSFET, (b) a single p-channel MOSFET, or (c) a parallel combination of an n-channel and a p-channel MOSFET.

Spice results for this CMOS switch are shown superimposed on the same graph of on-resistances of the NMOS and PMOS switches in Fig. 5.42. Clearly, the on-resistance of the CMOS transmission gate is much more constant and also much lower than that of either single-MOSFET switch. The on-resistance is seen to vary between 500 Ω and 800 Ω, with the maximum on-resistance occurring when the input signal is zero.

This example did not show another limitation of the single transistor switch, namely, its limited range of operation, which is caused by its nonzero threshold voltage. To illustrate this limitation, consider the switch arrangement shown in Fig. 5.44. Here a single n-channel MOSFET is connected between the voltage source v_I and a load consisting of a 100 kΩ resistor and a 10 pF capacitor. The n-channel MOSFET is assumed to be identical to the one previously described. Moreover, the gate of this MOSFET is biased at +5 V to turn the switch fully on. The Spice deck for this case is provided in Fig. 5.45. Now if we simulate the action of this switch using Spice with a 10 V step input beginning at −5 V, instead of the voltage at the output following the input in its entirety, we see that the output voltage actually clamps at a level of about 1.85 V as seen in Fig. 5.46. It is easily shown that this clamping limit depends on the threshold voltage of the MOSFET (see Sedra and Smith), but it may not be quite so obvious that this clamping limit also depends on the i–v characteristics of the switch. To prove this, let us increase the size of the transistor so that its i–v characteristics change. For example, let us increase the width of the MOSFET to 500 μm. On rerunning the Spice simulation we find that the output voltage has behavior similar to the previous case except that the output voltage clamps at a higher voltage of 2.0 V (also

Computing Switch On-Resistance of a CMOS Switch

```
** Circuit Description **
* input signal
Vi 2 0 DC 0V
* substrate bias
Vdd 1 0 DC +5V
Vss 3 0 DC -5V
* control signal
Vphi    4 0 DC -5V
Vphibar 5 0 DC +5V
* switch: NMOS + PMOS transistors
M1 2 5 0 3 nmos_enhancement_mosfet L=100u W=100u
M2 2 4 0 1 pmos_enhancement_mosfet L=100u W=100u
* MOSFET model statement (level 1 by default)
.model nmos_enhancement_mosfet nmos (kp=0.25m Vto=+1.0V lambda=0.02 gamma=0.9)
.model pmos_enhancement_mosfet pmos (kp=0.25m Vto=-1.0V lambda=0.02 gamma=0.9)
** Analysis Requests **
* sweep the input voltage from Vss to Vdd (skip over 0,0 point)
.DC Vi -5V +5V 10.01mV
** Output Requests **
.PLOT DC I(Vi) V(2)
.Probe
.end
```

Figure 5.43   Spice input file for calculating the input current as a function of input voltage for the CMOS transmission gate of Fig. 5.40(b). Postprocessing will be used to compute the on-resistance of the switch.

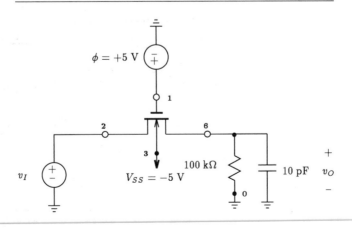

Figure 5.44   Circuit arrangement illustrating the effect of MOSFET threshold voltage on the switch range of operation.

The Effect of the Threshold Voltage on Switch Operation

```
** Circuit Description **
* input signal
Vi 2 0 DC 0V PWL (0 -5V 100ns -5V 200ns 5V 10ms 5V)
* substrate bias
Vss 3 0 DC -5V
* control signal
Vphi 1 0 DC +5V
* switch: NMOS transistor
M1 2 1 6 3 nmos_enhancement_mosfet L=100u W=100u
* load resistor
Rload 6 0 100k
Cload 6 0 10pF
* MOSFET model statement (level 1 by default)
.model nmos_enhancement_mosfet nmos (kp=0.25m Vto=+1.0V lambda=0.02 gamma=0.9)
** Analysis Requests **
.TRAN 1ns 500ns 0ns 1ns
** Output Requests **
.PLOT TRAN V(6)
.Probe
.end
```

Figure 5.45   Spice input file for calculating the step response of the NMOS switch arrangement shown in Fig. 5.44.

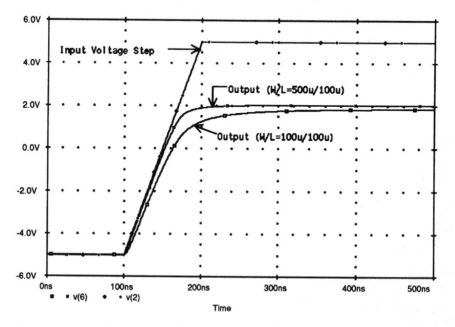

Figure 5.46   Step response of the NMOS switch arrangement shown in Fig. 5.44 for two differently sized MOSFETs.

shown in Fig. 5.46). The reason for this behavior is that the MOSFET must supply a continuous current to the RC load in order to sustain the output voltage.

5.8

Describing MESFETs to PSpice

Built-in models for metal-semiconductor FETs (MESFETs) do not exist in Spice (versions 2G6 and earlier). Instead, we shall rely on the built-in MESFET model of PSpice for our simulations.

MESFETs are described to PSpice in exactly the same way as any other semiconductor device, that is, using an element statement and a model statement. A summary of the syntax for these two statements as they apply to the n-channel MESFET is provided in Fig. 5.47.

5.8.1 MESFET Element Description

The element statement describing MESFETs in a PSpice listing begins with a unique name prefixed with the letter B, followed by a list of nodes to which the drain, gate, and source of the MESFET are connected. In the next field the name of a model characterizing the particular MESFET is given—more on this in a moment. The final field of this statement is optional and allows us to scale the size of the device by declaring the number of MESFETs connected in parallel.

5.8.2 MESFET Model Description

The model statement for the n-channel MESFET begins with the key word .MODEL followed by the name of the model used by a MESFET element statement, the nature

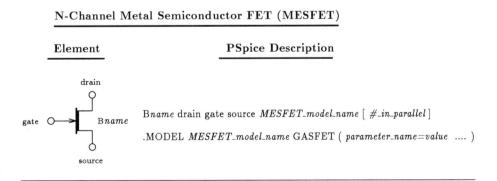

Figure 5.47 PSpice element description for the n-channel depletion-mode GaAs MESFET and the general form of the associated MESFET model statement. A partial listing of the parameters applicable to the n-channel MESFET is given in Table 5.5.

of the MESFET (i.e., GASFET), and a list of the parameters characterizing the terminal behavior of the MESFET enclosed in parentheses. Parameters not specified in the model statements are assigned default values by PSpice. The parameters describing the terminal characteristics of the MESFET are similar to those used to describe the JFET. An additional parameter (α) has been added to properly account for the early saturation phenomenon of the MESFET [Hodges and Jackson, 1983].

The general form of the DC PSpice model for an n-channel MESFET is illustrated schematically in Fig. 5.48. The bulk resistance of the drain, gate, and source regions of the MESFET are lumped into three linear resistances, r_D, r_G, and r_S, respectively. The DC characteristic of the intrinsic MESFET is determined by the nonlinear dependent current source i_D and the two Schottky-barrier diodes. These two Schottky diodes represent the metal-semiconductor junctions that define the channel region. The functional description of the drain current i_D can take on three different forms, depending on the value assigned to the parameter *level* that appears in the MESFET model statement. PSpice defaults to a level 1 model when none is specified. The level 1 MESFET model is probably the most widely used of the three and will be the only one discussed here. Details on the other models can be found in the *PSpice Users' Manual.*

The equation describing the drain current for the level 1 MESFET model is

$$i_D = \begin{cases} 0 & \text{for } v_{GS} < V_t \\ \beta(v_{GS} - V_t)^2(1 + \lambda v_{DS})\tanh(\alpha v_{DS}) & \text{for } v_{GS} > V_t \end{cases} \tag{5.18}$$

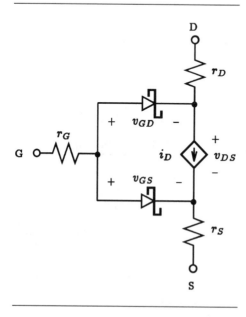

Figure 5.48   PSpice large-signal MESFET model under static conditions.

Here the same equation is used to describe both the triode and saturation regions of the device; the distinction between them is provided by the factor $\tanh(\alpha v_{DS})$. This factor also accounts for the early saturation phenomenon observed in MESFETs. See the i–v characteristics in Fig. 5.49 for a typical MESFET having device parameters $V_t = -1\,\text{V}$, $\beta = 10^{-4}\,\text{A/V}^2$, $W = 100\,\mu\text{m}$, $L = 1\,\mu\text{m}$, $\lambda = 0.05\,\text{V}^{-1}$, and α taking on values of 0.5, 1.0, and 2.0. As an extreme limit, the i–v characteristic for $\alpha = 10^6$ is also included. In these equations β is the device transconductance parameter, and λ is the channel-length modulation coefficient, just as in the Spice JFET model.

A partial listing of the parameters associated with the PSpice MESFET model under static conditions is given in Table 5.5 along with the associated default value that each parameter assumes if a value is not specified for it in the .MODEL statement. (A complete listing appears in Appendix A.) To specify a parameter value we write, for example, level=1, beta=20u, Vto=−1V, and so on.

5.8.3   Small-Signal MESFET Model

The linear small-signal model of the n-channel MESFET is identical to that of the JFET in Fig. 5.26(b). It consists of a single voltage-controlled current source having transconductance g_m. Here g_m is the MESFET transconductance and is related to the DC bias according to the following equation:

$$g_m = \left.\frac{\partial i_D}{\partial v_{GS}}\right|_{OP} = 2\beta(V_{GS} - V_t)(1 + \lambda V_{DS})\tanh(\alpha V_{DS}) \qquad (5.19)$$

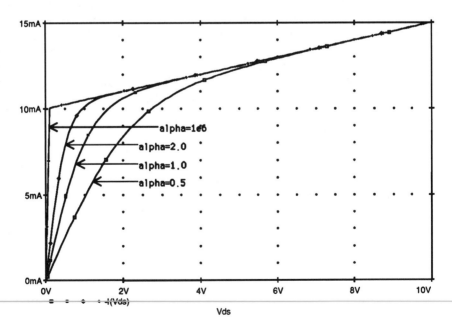

Figure 5.49   Illustrating the dependence of the i–v characteristic of a single MESFET on the value of α.

Table 5.5 Partial listing of the PSpice parameters for the level 1 MESFET model.

Symbol	PSpice Name	Model Parameter	Units	Default
	Level	Model type		1
β	beta	Transconductance coefficient	A/V^2	0.1
α	alpha	Saturation voltage parameter	V^{-1}	2.0
V_t, V_P	Vto	Pinch-off voltage	V	-2.5
λ	lambda	Channel-length modulation	V^{-1}	0
r_G	Rg	Gate ohmic resistance	Ω	0
r_D	Rd	Drain ohmic resistance	Ω	0
r_S	Rs	Source ohmic resistance	Ω	0

The output conductance g_{ds}, which is also equal to $1/r_o$, is computed as follows:

$$g_{ds} = \left. \frac{\partial i_D}{\partial v_{DS}} \right|_{OP} \tag{5.20}$$

5.8.4 *A MESFET Biasing Example*

As an example of an application of GaAs MESFETs, in Fig. 5.50 we display a simple MESFET amplifier circuit. The lengths of the two devices are assumed to be equal to the minimum value for a particular technology (1 μm in this case), and the widths of B_1 and B_2 are $W_1 = 100$ μm and $W_2 = 50$ μm, respectively. The characteristics of the GaAs process are specified in terms of the electrical parameters of minimum-sized devices (i.e., $L = 1$ μm and $W = 1$ μm), which will be assumed to be $V_t = -1$ V, $\beta = 10^{-4}$ A/V^2, and $\lambda = 0.1$ V^{-1}. To determine the parameters of a particular device, it is a simple matter to scale the appropriate parameters of the unit-sized device, or we can let PSpice perform this scaling operation for us. We shall illustrate the latter, since PSpice is less likely to make mistakes.

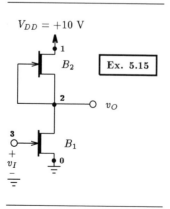

Figure 5.50   MESFET amplifier with MESFET load.

Using the device parameters for the minimum-sized device, we can establish the following PSpice model statement for the unit-sized n-channel MESFET:

```
.model n_mesfet gasfet (beta=0.1m Vto=-1.0V lambda=0.1)
```

Parameters not specified will, as usual, assume default values.

To account for the different widths of the two transistors, we specify in the element statement of each MESFET the ratio of the transistor width to the unit-sized width. In this case the unit-sized width is 1 μm, so for B_1 with a device width of 100 μm we specify a scale factor of 100, which is the same as connecting 100 unit-sized transistors in parallel. The element statement for transistor B_1 is

```
B1 2 3 0 n_mesfet 100
```

We specify a scale factor of 50 for the other transistor.

Using the PSpice listing given in Fig. 5.51, we would like to determine the DC transfer characteristics of this amplifier. Here we are sweeping the input voltage level between −10 V and +1 V in 10 mV steps. The input voltage normally should remain negative, otherwise the input metal-semiconductor junction of B_1 (the gate–source Schottky diode) becomes forward biased. We are allowing the input voltage to go slightly positive simply to observe the effect of forward biasing this junction.

The resulting DC transfer characteristic of this MESFET amplifier is displayed in Fig. 5.52. Here we see that the amplifier has an inverterlike characteristic, with its high-gain region in the vicinity of a −0.3 V input level. When the input signal level goes more positive than 0.7 V we see that the output voltage begins to rise, because the input metal-semiconductor junction of the MESFET has become forward biased and the gate voltage no longer controls the drain-to-source current. Using the transfer function (.TF) command we can determine the voltage gain in the high-gain region of the amplifier. This requires that we bias the input to the amplifier at −0.3 V, as shown in the PSpice deck of Fig. 5.51. Further, we need to change the analysis request from a DC sweep to the following small-signal transfer function command:

```
.TF V(2) Vi
```

We also include an operating-point command (.OP) to obtain the small-signal model parameters. On completion of the job, we find the following small-signal characteristics associated with the MESFET amplifier:

```
**** SMALL - SIGNAL CHARACTERISTICS

V(2) / Vi = -2.143E+01

INPUT RESISTANCE AT Vi = 4.269E+10

OUTPUT RESISTANCE AT V(2) = 1.010E+03
```

Here we see that the gain in the linear region of the MESFET amplifier is −21.43 V/V, about five times less than the gain available with the simple CMOS amplifier discussed previously in Section 5.6. This result seems to correspond directly to the result that the output resistance for the MESFET amplifier is about five times lower than the output resistance of the CMOS amplifier.

```
A MESFET Amplifier

** Circuit Description **
* DC supplies
Vdd 1 0 DC +10V
* input signal
Vi 3 0 DC -0.3V
* amplifier circuit
B1 2 3 0 n_mesfet 100
B2 1 2 2 n_mesfet 50
* MESFET model statements (level 1 by default)
.model n_mesfet gasfet (beta=0.1m Vto=-1.0V lambda=0.1)
** Analysis Requests **
* calculate DC transfer characteristics
.DC Vi -10V +1V 0.01V
** Output Requests **
.PLOT DC V(2)
.probe
.end
```

Figure 5.51   Spice input file for calculating the DC transfer characteristic of the MESFET amplifier circuit shown in Fig. 5.50.

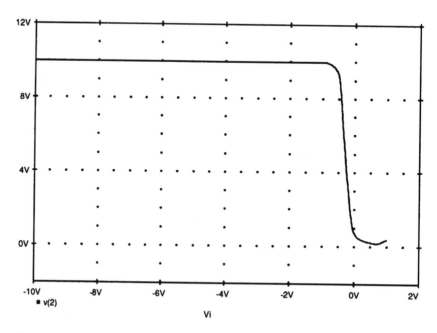

Figure 5.52   DC transfer characteristic of the MESFET amplifier shown in Fig. 5.50 as calculated by PSpice.

The small-signal model parameters of the two MESFETs as calculated by PSpice are found in the output file:

```
**** GASFETS

NAME            B1              B2
MODEL           n_mesfet        n_mesfet
ID              7.42E-03        7.42E-03
VGS            -3.00E-01        0.00E+00
VDS             5.15E+00        4.85E+00
GM              2.12E-02        1.48E-02
GDS             4.90E-04        5.00E-04
```

It is reassuring that when these values are substituted into the formula for the small-signal voltage gain given by $A_V = -g_m(r_{o1} \parallel r_{o2})$, we get results very similar to the gain computed using the .TF command (i.e., $A_V = -21.41$ V/V).

5.9

Spice Tips

- FETs are described to Spice using an element statement and a model statement. The element statement describes the connections that the FET makes to the rest of the network and the name of the model that specifies its terminal behavior. The model statement assigns particular values to internal parameters of the built-in FET model of Spice. Parameters whose values are not specified in the model statement are assigned default values.
- Spice has built-in models for MOSFETs and JFETs.
- PSpice has built-in models for MOSFETs, JFETs, and MESFETs.
- V_{t0} is positive for enhancement-mode n-channel MOSFETs and for depletion-mode p-channel MOSFETs.
- V_{t0} is negative for depletion-mode n-channel MOSFETs and for enhancement-mode p-channel MOSFETs.
- V_t is negative for both n-channel and p-channel JFETs.
- V_t is negative for n-channel depletion MESFETs and is positive for n-channel enhancement MESFETs.
- The sensitivity analysis command of Spice (.SENS) does not compute the sensitivities of the circuit variables to the parameters of the MOSFET model. These must be computed directly by perturbing each parameter of the MOSFET model and observing the effect.
- A small-signal analysis of an amplifier should always be computed around a known operating point that is normally within the linear region of the amplifier. The best way to determine the appropriate operating point is by first performing a DC sweep of the input voltage over the range defined by the power supplies in order to locate the linear region of the amplifier. The input to the amplifier is then biased inside this region where Spice can compute the small-signal behavior.

5.10

Bibliography

R.L. Geiger, P. E. Allen, and N. R. Strader, *VLSI Design Techniques for Analog and Digital Circuits,* New York: McGraw-Hill, 1990.

D.A. Hodges and H. Jackson, *Analysis and Design of Digital Integrated Circuits,* New York: McGraw-Hill, 1983.

PSpice Users' Manual, MicroSim Corporation, Irvine, CA, Jan. 1991.

A. Vladimirescu and S. Liu, *The Simulation of MOS Integrated Circuits Using SPICE2,* Memorandum No. M80/7, Electronics Research Laboratory, University of California, Berkeley, CA, Feb. 1980.

5.11

Problems

5.1 An enhancement PMOS transistor has $\mu C_{OX} = 40\ \mu A/V^2$, $L = 25\ \mu m$, $W = 50\ \mu m$, $V_t = -1.5$ V, and $\lambda = 0.02\ V^{-1}$. The gate is connected to ground and the source to $+5$ V. Find the drain current using Spice for (a) $v_D = 4$ V, (b) $v_D = +1.5$ V, (c) $v_D = 0$ V, and (d) $v_D = -5$ V.

5.2 Using Spice as a curve tracer, plot the i_D–v_{DS} characteristics of a depletion-mode p-channel MOSFET having $\mu C_{OX} = 50\ \mu A/V^2$, $L = 25\ \mu m$, $W = 50\ \mu m$, $V_t = +1.2$ V, and $\lambda = 0.04\ V^{-1}$. Provide curves for $v_{GS} = -1, -2, -3, -4$, and -5 V. Show the characteristics for v_{DS} from 0 to -10 V.

5.3 Consider an n-channel JFET with $I_{DSS} = 4$ mA and $V_P = -2$ V. If the source is grounded and a -1 V DC voltage source is applied to the gate, using Spice find the drain current that corresponds to the minimum drain voltage that results in pinch-off operation.

5.4 For a JFET having $V_P = -2$ V and $I_{DSS} = 8$ mA operating at $v_{GS} = -1$ V and a very small v_{DS}, with the aid of Spice determine the value of r_{DS}.

5.5 Using Spice as a curve tracer, plot the i_D–v_{DS} characteristics of an n-channel JFET with $I_{DSS} = 8$ mA, $V_P = -4$ V, and $\lambda = 0.01 V^{-1}$. Provide curves for $v_{GS} = -5, -4, -3, -2$, and 0 V. Show the characteristics for v_{DS} up to 10 V.

5.6 The NMOS transistor in the circuit of Fig. P5.6 has $V_t = 1$ V, $\mu C_{OX} = 1$ mA/V^2, and $\lambda = 0.02\ V^{-1}$. If v_G is a pulse with 0- and 5-V levels and a 1 ms pulse width, simulate the circuit using Spice, and determine the pulse signal that appears at the output.

5.7 Simulate the circuit in Fig. P5.7 to determine the drain current and the drain voltage. Assume that the depletion MOSFET has $V_t = -1$ V, $\mu C_{OX} = 1$ mA/V^2, and $\lambda = 0.03\ V^{-1}$. If K of the MOSFET increases by a factor of 2, determine the new level of drain current and drain voltage. Compare this to the previous case and note whether the biasing scheme is very effective.

5.8 A MOSFET having $\mu C_{OX} = 2$ mA/V^2 and $V_t = 1$ V operates in a feedback bias arrangement such as that shown in Fig. P5.8 from a 10 V supply with $R_D = 8\ k\Omega$ and $R_G = 10\ M\Omega$. What value of I_D results? If the FET is replaced by another with (a) $\mu C_{OX} = 1$ mA/V^2 and $V_t = 1$ V and (b) $\mu C_{OX} = 2$ mA/V^2 and $V_t = 2$ V, what percentage change in I_D results?

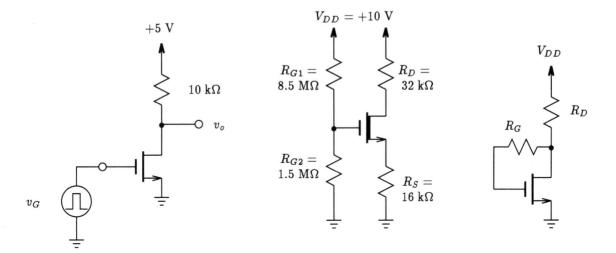

Figure P5.6   **Figure P5.7**   **Figure P5.8**

5.9 The JFET in the amplifier circuit in Fig. P5.9 has $V_P = -4$ V, $I_{DSS} = 12$ mA, and $\lambda = 0.003$ V^{-1}. Using Spice, answer the following:
  (a) Determine the DC bias quantities V_G, I_D, V_{GS}, and V_D associated with the JFET.
  (b) Determine the overall voltage gain v_o/v_i and input resistance R_{in}.

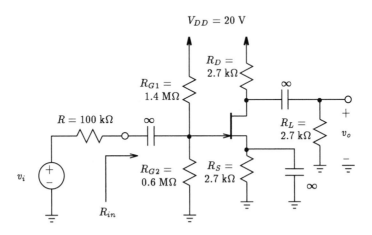

Figure P5.9

5.10 In the circuit of Fig. P5.10 all devices are matched and assumed to have the following parameters: $\mu C_{OX} = 50$ μA/V^2, $W = L = 25$ μm, and $V_t = 2$ V. Using Spice, plot the transfer characteristics v_O versus v_I between V_{DD} and ground. What is the small-signal voltage gain in the linear region of this amplifier? What is the corresponding output resistance?

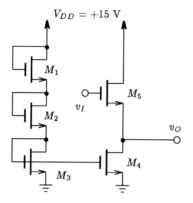

Figure P5.10

5.11 A CMOS switch for which $\mu C_{OX} = 50\ \mu A/V^2$, $\gamma = 0.3\ V^{1/2}$, and $|V_{t0}| = 2$ V is placed in the circuit setup shown in Fig. P5.11 for measuring its on-resistance. Sweep the input voltage signal v_I from -5 V to $+5$ V using Spice, and determine the resistance of the switch over this range of signals. Compare these results to those obtained from a switch consisting of a single n-channel MOSFET.

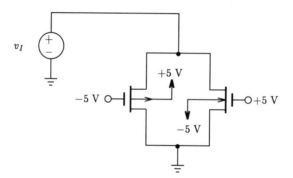

Figure P5.11

5.12 A CMOS switch is used to connect a sinusoidal source of $0.1 \sin(2\pi 10^3 t)$ to a load consisting of a 10 kΩ resistor and a 5 pF capacitor. If the control terminals of the switch are driven by complementary signals of ± 5 V at a frequency of 10 kHz, simulate the transient behavior of the switch using Spice. Assume that the model parameters of the NMOS and PMOS devices are $\mu_n C_{OX} = 20\ \mu A/V^2$, $\mu_p C_{OX} = 10\ \mu A/V^2$, $|V_{t0}| = 1$ V, $\gamma = 0.3\ V^{1/2}$, $L_n = L_p = 5\ \mu m$, $W_n = 15\ \mu m$, and $W_p = 30\ \mu m$.

5.13 Using Spice as a curve tracer, plot the i_D–v_{DS} characteristics of an n-channel MESFET with $\beta = 10^{-4}\ A/V^2$, $V_t = -1$ V, $I_S = 10^{-15}$ A, and $\lambda = 0.1\ V^{-1}$. Provide curves for $v_{GS} = -5, -4, -3, -2,$ and 0 V. Show the characteristics for v_{DS} up to 10 V.

5.14 The n-channel MESFET in the circuit of Fig. 5.50 has $V_t = -1$ V, $\beta = 0.1\ mA/V^2$, and $\lambda = 0.1\ V^{-1}$. If v_I is a pulse with -5 and 0 V levels of 1 μs pulse width, simulate the circuit using Spice, and plot the pulse signal that appears at the output.

Differential and Multistage Amplifiers

Up to this point we have been looking mainly at transistor circuits driven by unbalanced inputs (i.e., one terminal of the signal source is grounded). A more important transistor arrangement is one that receives differential or balanced input signals. The differential pair is an example of such an arrangement used extensively in present-day monolithic IC operational amplifiers. In this chapter we demonstrate how to generate the appropriate input signals for investigating the behavior of differential amplifiers, and we investigate several circuits involving differential and multistage amplifiers and a source.

6.1

Input Excitation for the Differential Pair

The differential pair, shown in Fig. 6.1, is the most widely used circuit building block in analog integrated circuits. It operates by amplifying only the difference between the signals at the two inputs. The signals appearing as common-mode at the amplifier inputs ideally are not amplified. We would like to use Spice to investigate the effect of varying the input common-mode and differential-mode signal levels on the collector currents. The focus of this problem is more on how we generate the Spice input signals for differential amplifiers, rather than on the behavior of the differential pair.

For example, to analyze the behavior of a differential pair subject to a differential input signal, first-time users of Spice are often tempted to apply a single ungrounded source to the amplifier input as illustrated in Fig. 6.1. Unfortunately, that causes Spice to generate erroneous data because there is no defined common-mode input level. To see this, we perform a DC analysis of the differential pair shown in Fig. 6.1, assuming the BJTs have the following parameters: $I_S = 14$ fA, $\beta_F = 100$, and $V_{AF} = 100$ V. The input file is in Fig. 6.2.

Differential and Multistage Amplifiers

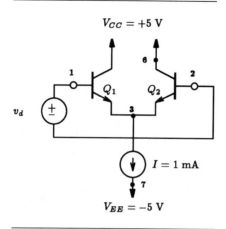

Figure 6.1 A BJT differential pair driven by a differential input signal without a defined input common-mode level. This approach is not recommended.

```
A BJT Differential Pair

** Circuit Description **
* power supply
Vdd 6 0 DC +5V
Vee 7 0 DC -5V
* input differential signal source
Vd 1 2 DC 0V
* differential pair
Q1 6 1 3 npn_transistor
Q2 6 2 3 npn_transistor
* bias source
I 3 7 DC 1mA
* transistor model statements
.model npn_transistor npn ( Is=14fA Bf=100 VAf=100V )
** Analysis Requests **
.OP
** Output Requests **
* none required
.end
```

Figure 6.2 Spice input file for illustrating the problem of applying a single ungrounded voltage between the input terminals of the differential amplifier in Fig. 6.1.

The Spice results of the DC analysis are then found in this output file:

```
****      SMALL SIGNAL BIAS SOLUTION          TEMPERATURE =    27.000 DEG C
***********************************************************************************

NODE    VOLTAGE      NODE    VOLTAGE      NODE    VOLTAGE      NODE  VOLTAGE

(    1)-22.19E+03  (    2)-22.19E+03  (    3)-22.19E+03  (    6)    5.0000
(    7)   -5.0000
```

We see that the DC voltage appearing at the two inputs to the differential pair (nodes 1 and 2) and at the emitters of the two transistors (node 3) are very large negative levels, outside the limits of the supply voltages—certainly not levels observed in any real circuit. The lack of a common-mode input level, which causes this, can be corrected by revising the input excitation to include a common-mode level among the differential components.

The circuit shown in Fig. 6.3 accomplishes this in a convenient way. It allows the input common-mode level to be adjusted by varying V_{CM} independently of the value of the differential-mode component being established by the two VCVSs connected across the input terminals of the differential pair. The level of each VCVS is one-half the voltage set by the isolated voltage source V_d. This voltage source is loaded

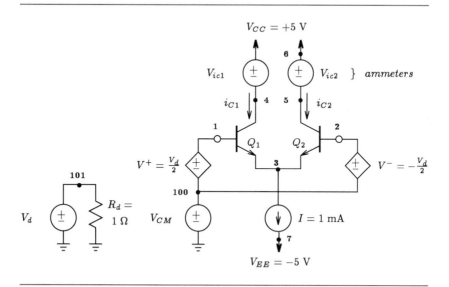

Figure 6.3 A BJT differential pair driven by a set of input voltage sources arranged so that the common-mode and differential voltage components can be independently varied. More important, both components will always exist and are clearly defined. Voltage sources V_{iC1} and V_{iC2} are used to monitor the collector current of each transistor.

arbitrarily with a 1 Ω resistor in order to satisfy the Spice requirement that every node in a circuit have at least two connections.

To demonstrate the versatility of this input–source arrangement, we look at the effect of separately varying the input common-mode and differential signal components on the collector currents of the differential pair of Fig. 6.3. The input file describing this circuit is shown in Fig. 6.4. The parameters of the *npn* BJTs are assumed to be the same as before (i.e., $I_S = 14$ fA, $\beta_F = 100$, and $V_{AF} = 100$ V). As our first analysis request, we ask Spice to perform a DC sweep of the input common-mode voltage level, V_{CM}, beginning at -5 V and ending at $+6$ V in increments of 100 mV. The differential input component V_d is set to zero during this analysis, also making V^+ and V^- equal zero. The results in Fig. 6.5 show the behavior of the two collector currents as a function of V_{CM}. Both Q_1 and Q_2 are conducting equal currents of 0.5 mA for most input common-mode levels, but when V_{CM} exceeds 5.6 V the collector currents of both Q_1 and Q_2 fall off very rapidly to 0 mA. Any further increase in V_{CM}

```
A BJT Differential Pair

** Circuit Description **
* power supply
Vdd 6 0 DC +5V
Vee 7 0 DC -5V
* differential-mode signal level
Vd 101 0 DC 0V
Rd 101 0 1
EV+ 1 100 101 0 +0.5
EV- 2 100 101 0 -0.5
* common-mode signal level
Vcm 100 0 DC 0V
* monitor collector currents of Q1 and Q2
Vic1 6 4 0
Vic2 6 5 0
* differential pair
Q1 4 1 3 npn_transistor
Q2 5 2 3 npn_transistor
* bias source
I 3 7 DC 1mA
* transistor model statements
.model npn_transistor npn ( Is=14fA Bf=100 VAf=100V )
** Analysis Requests **
.DC Vcm -5V +6V 100mV
** Output Requests **
.PLOT DC I(Vic1) I(Vic2)
.probe
.end
```

Figure 6.4   Spice input file for analyzing the effect of common-mode input signals on the collector currents of the BJT differential pair of Fig. 6.3.

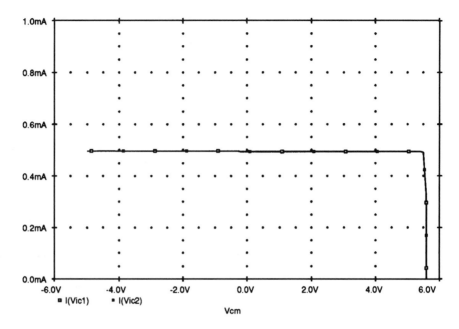

Figure 6.5 Collector currents of the BJT differential pair for a range of input common-mode levels.

causes the collector currents to go negative because the base–collector junctions of both Q_1 and Q_2 become forward biased, and they conduct appreciable currents in the direction opposite the normal flow of collector current.

To investigate the effect of a differential signal at the input terminals of the differential pair, we simply revise the input file in Fig. 6.4 by replacing the DC sweep command given there by the following one:

```
.DC Vd -500mV +500mV 10mV
```

We are requesting that Spice vary the level of V_d between -500 mV and $+500$ mV in 10 mV increments. The common-mode input level V_{CM} is to remain at 0 V throughout this analysis. Spice produces the display of the collector currents in Fig. 6.6. The results are as expected: Depending on the level of input differential signal, the biasing current of 1 mA is *steered* between the two transistors.

6.2

Small-Signal Analysis of the Differential Amplifier: Symmetric Conditions

This section investigates the small-signal behavior of the BJT differential-pair amplifier configuration shown in Fig. 6.7. We assume throughout this section that the circuit remains symmetric, that is, that the resistances in collectors are equal and that transistors Q_1 and Q_2 are matched. Our purpose here is twofold: to illustrate how one

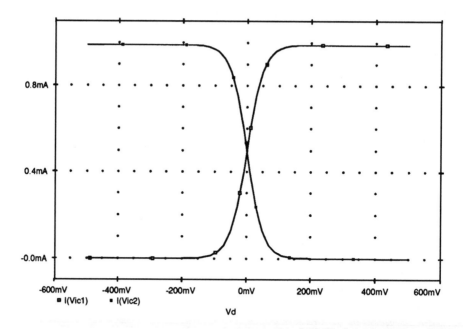

Figure 6.6 Collector currents of the BJT differential pair versus the level of differential input signal.

uses Spice to determine the small-signal behavior of a differential amplifier using the multiple-source arrangement discussed in the previous section and to determine the accuracy of the expressions derived for differential amplifiers in Sections 6.2 and 6.3 of Sedra and Smith.

It is important to point out that when we compare results computed by Spice with those computed with closed-form expressions derived by hand analysis from a small-signal circuit model of the transistor circuit, the accuracy of the results depends on two factors: (1) the precision of the estimate of the values of the parameters of the small-signal model and (2) the accuracy of the small-signal expressions, given that certain simplifying assumptions are made in their development. Within this section we address the second issue, particularly the accuracy of Sedra and Smith's expressions for the BJT differential amplifier under various circuit conditions. To do this, we will use the small-signal model parameters computed by Spice directly through a DC operating-point analysis command, rather than estimate them ourselves from the DC circuit conditions.

According to Sedra and Smith, the small-signal analysis of the differential pair shown in Fig. 6.7 is relatively straightforward. Table 6.1 summarizes the expressions used in its development and includes expressions for both the differential-mode and common-mode voltage gain (A_d and A_{CM}) and the input resistances (R_{id} and R_{iCM}), as well as expressions for the input-referred offset voltage, the input bias current, and the input offset current (V_{OS}, I_B, and I_{OS}, respectively) under asymmetric circuit conditions.

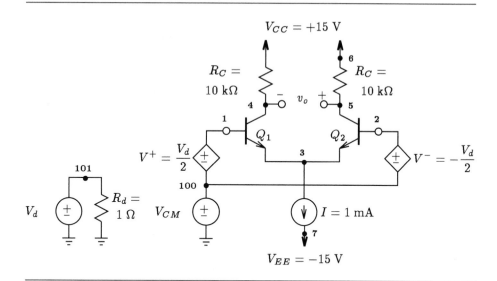

Figure 6.7 Basic BJT differential-pair amplifier configuration driven with the multiple-source arrangement described in Section 6.1.

Table 6.1 General expressions for estimating the small-signal behavior of the differential amplifier shown in Fig. 6.7 (derived in Sections 6.2 and 6.3 in Sedra and Smith). Note that R is the output resistance of the current source I.

Parameter	Formula
A_d	$-g_m(R_C\|r_o)$
R_{id}	$2r_\pi$
A_{CM}	$\dfrac{R_C}{2R}\dfrac{\Delta R_C}{R_C}$
R_{iCM}	$\left(\dfrac{r_\mu}{2}\right)\|[(\beta+1)R]\|\left[(\beta+1)\dfrac{r_o}{2}\right]$
$\|V_{os}\|$	$V_T\sqrt{\left(\dfrac{\Delta R_C}{R_C}\right)^2+\left(\dfrac{\Delta I_S}{I_S}\right)^2}$
I_B	$\dfrac{I/2}{\beta+1}$
$\|I_{os}\|$	$I_B\left(\dfrac{\Delta\beta}{\beta}\right)$

We wish to use Spice to calculate the differential-mode voltage gain and input resistance of the differential amplifier shown in Fig. 6.7. We assume that each transistor has device parameters $I_S = 14$ fA and $\beta = 100$. For the moment, we shall neglect the effect of the transistor's Early voltage (i.e., $V_A = \infty$), which is represented by VAf = 0 in the transistor model statement. The input file describing the differential

amplifier of Fig. 6.7 is listed in Fig. 6.8. The input to the differential amplifier is arranged using the multiple-source setup described in the last section. It consists of two
voltage-controlled sources (EV+ and EV−) and two independent input voltage signals (Vd and Vcm). The DC levels of both of these voltage sources are set equal to 0
V because we already know that the two transistors of the differential pair will remain
in the active region under these bias conditions. That is, with a transistor quiescent
current of 0.5 mA the voltage at the collector and emitter of each transistor will be approximately 10 V and −0.7 V, respectively. The differential-input voltage source Vd
also includes an AC component of 1 V—the reason for it will be made clear in a moment.

Our first analysis request is a .TF command written as

```
.TF V(5,4) Vd
```

```
A BJT Differential Pair

** Circuit Description **
* power supply
Vdd 6 0 DC +15V
Vee 7 0 DC -15V
* differential-mode signal level
Vd 101 0 DC 0V AC 1V
Rd 101 0 1
EV+ 1 100 101 0 +0.5
EV- 2 100 101 0 -0.5
* common-mode signal level
Vcm 100 0 DC 0V
* differential pair
Q1 4 1 3 npn_transistor
Q2 5 2 3 npn_transistor
* load resistors
Rc1 6 4 10k
Rc2 6 5 10k
* bias source
I 3 7 DC 1mA
* transistor model statements
.model npn_transistor npn ( Is=14fA Bf=100 VAf=0 )
** Analysis Requests **
.TF V(5,4) Vd
.AC LIN 1 1Hz 1Hz
.OP
** Output Requests **
.PRINT AC Im(EV+) Im(EV-) Vm(1,2)
.probe
.end
```

Figure 6.8   Spice input file for calculating the two-port equivalent of the differential amplifier shown in Fig. 6.7.

which will provide us with an equivalent circuit representation of the differential amplifier evaluated around 0 V DC as seen looking into the port made by the differential-input signal source V_d and the output port denoted between nodes 5 and 4 in Fig. 6.7. Unfortunately, with this multiple-source arrangement the resistance seen by the input source V_d is just the 1 Ω resistor connected across V_d—a resistor added just to increase the number of connections at node 101 and serving no circuit function. We require another means of obtaining the resistance seen looking into this differential amplifier.

A 1 V AC signal applied across R_d also will appear across the input terminals of the differential pair. By calculating the current that flows into the terminals of the differential pair as a result of this AC signal, we can obtain the input resistance as the reciprocal of the current.

In the input deck listed in Fig. 6.8, we have already indicated this 1 V AC component in the statement describing the differential-input voltage source. The appropriate AC current then can be monitored with the two dependent voltage sources in series with the input terminals (i.e., EV+ and EV−). To access the data, we add the following two command statements to the Spice deck:

```
.AC LIN 1 1Hz 1Hz
.PRINT AC Im(EV+) Im(EV-) Vm(1,2)
```

The first command tells Spice to perform an AC analysis at a single frequency—1 Hz. The second statement instructs Spice to send the magnitude of the AC current to the output file. As a means of ensuring that the correct AC voltage appears at the input terminals of the amplifier, we are also requesting that Spice print out the voltage there.

Finally, our last analysis request is an operating-point command (.OP), which will provide us with the parameters of the small-signal model of each transistor.

On completion of Spice, we obtain the following parameters of the two-port equivalent circuit of the differential amplifier shown in Fig. 6.7:

```
****      SMALL-SIGNAL CHARACTERISTICS

        V(5,4)/Vd =  1.914E+02

        INPUT RESISTANCE AT Vd =  1.000E+00

        OUTPUT RESISTANCE AT V(5,4) =  2.000E+04
```

This particular differential amplifier has a small-signal voltage gain of 191.4 V/V and an output resistance of 20 kΩ. The input resistance indicated in the output file is the 1 Ω resistance shunting the input signal generator V_d, rather than the input resistance of the amplifier. To determine the input resistance of the amplifier we use the results of the AC analysis found in the output file:

```
****      AC ANALYSIS

  FREQ          IM(EV+)       IM(EV-)       VM(1,2)

  1.000E+00   9.570E-05   9.570E-05   1.000E+00
```

The input differential resistance to the amplifier is then found to be 10.44 kΩ.

It would be interesting to check the results obtained by hand analysis to see whether they agree with those computed by Spice. Consider the parameters of the small-signal model of each transistor as computed by Spice:

```
**** BIPOLAR JUNCTION TRANSISTORS

NAME            Q1                      Q2
MODEL           npn_transistor  npn_transistor
IB              4.95E-06                4.95E-06
IC              4.95E-04                4.95E-04
VBE             6.28E-01                6.28E-01
VBC            -1.01E+01               -1.01E+01
VCE             1.07E+01                1.07E+01
BETADC          1.00E+02                1.00E+02
GM              1.91E-02                1.91E-02
RPI             5.22E+03                5.22E+03
RX              0.00E+00                0.00E+00
```

Using the expressions given in Table 6.1, we can expect that this amplifier will have a differential gain of 191.0 V/V and an input resistance of 10.44 kΩ, values that are very close to those computed by Spice.

To further analyze the differential amplifier of Fig. 6.7, let us compute the common-mode voltage gain and common-mode input resistance using Spice. We proceed in exactly the same way as for the prior differential-mode analysis using the .TF command, except that we change the source reference from Vd to Vcm as shown here:

```
.TF V(5,4) Vcm
```

The results of this analysis are

```
****        SMALL-SIGNAL CHARACTERISTICS

    V(5,4)/Vcm =  0.000E+00

    INPUT RESISTANCE AT Vcm =  4.927E+11

    OUTPUT RESISTANCE AT V(5,4) =  2.000E+04
```

Here we see that the common-mode voltage gain of this amplifier is exactly zero, which is what we would expect given that the resistances in the two collectors are equal. Note that the input resistance is 492.7 GΩ rather than the theoretically expected value of infinity. This is a numerical artifact of Spice. Also, the output resistance computed during this analysis is identical to that found during the previous analysis for differential-mode behavior.

In the development of most of the expressions in Table 6.1—the expressions we used for our hand analysis—the current-source resistance was assumed to be infinite. In practice, such a current source cannot be achieved. In our analysis we assumed that the differential pair was biased by an infinite-output-resistance current source I. It is imperative to repeat the preceding analysis including a current-source resistance, such as that shown in Fig. 6.9, to verify the accuracy of the expressions in Table 6.1.

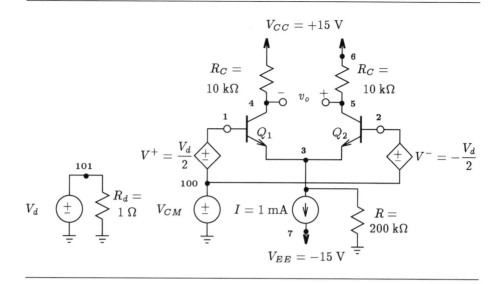

Figure 6.9 Including a 200 kΩ current-source resistance in the differential amplifier of Fig. 6.7.

We begin by assuming a current-source resistance of 200 kΩ and modifying the input file in Fig. 6.8 to include a Spice statement for the current-source resistance:

```
R 3 0 200k
```

Spice produces the following pertinent small-signal differential-mode information:

```
****        SMALL-SIGNAL CHARACTERISTICS

      V(5,4)/Vd =  1.908E+02

      INPUT RESISTANCE AT Vd =  1.000E+00

      OUTPUT RESISTANCE AT V(5,4) =  2.000E+04

****        AC ANALYSIS

   FREQ           IM(EV+)       IM(EV-)       VM(1,2)

     1.000E+00    9.540E-05    9.540E-05    1.000E+00
```

We find here that the differential-mode voltage gain is 190.8 V/V and that the input differential resistance is 10.48 kΩ. These results are very close to the previous case when the resistance of the current source was assumed infinite. The small differences in these small-signal results are a result of altering the DC bias conditions in the two transistors.

If we modify the Spice deck to compute the common-mode input equivalent circuit, as was done before, we arrive at these results:

```
****        SMALL-SIGNAL CHARACTERISTICS

    V(5,4)/Vcm =   0.000E+00

    INPUT RESISTANCE AT Vcm =   2.020E+07

    OUTPUT RESISTANCE AT V(5,4) =   2.000E+04
```

Using the formulas given in Table 6.1 together with the small-signal model parameters for Q_1 and Q_2, we would obtain exactly the same results as those computed by Spice.

The Early effect was neglected in each of the preceding analyses performed. Let us now consider an Early voltage of 100 V for each transistor of the differential pair. We maintain the current-source resistance at 200 kΩ. This Spice input file is similar to that shown in Fig. 6.8, but with the following modified model statement for each transistor:

```
.model npn_transistor npn ( Is=14fA Bf=100 VAf=100V )
```

The results of both the differential-mode and common-mode analyses computed by Spice are

```
Differential-Mode Analysis:

    ****        SMALL-SIGNAL CHARACTERISTICS

        V(5,4)/Vd =   1.827E+02

        INPUT RESISTANCE AT Vd =   1.000E+00

        OUTPUT RESISTANCE AT V(5,4) =   1.914E+04

    ****        AC ANALYSIS

        FREQ          IM(EV+)        IM(EV-)        VM(1,2)

        1.000E+00    8.676E-05    8.676E-05    1.000E+00

Common-Mode Analysis:

    ****        SMALL-SIGNAL CHARACTERISTICS

        V(5,4)/Vcm =  -7.733E-15

        INPUT RESISTANCE AT Vcm =   7.825E+06

        OUTPUT RESISTANCE AT V(5,4) =   1.914E+04
```

We see the differential-mode voltage gain is 182.7 V/V. The input differential resistance is calculated from the input current of 86.76 μA to be 11.52 kΩ. The common-mode voltage gain is zero for all practical purposes, and the input common-mode resistance is 7.825 MΩ.

As a means of verifying the formulas in Table 6.1 under the conditions of finite Early voltage, we also list the parameters of small-signal model computed by Spice:

```
**** BIPOLAR JUNCTION TRANSISTORS

NAME            Q1                       Q2
MODEL           npn_transistor    npn_transistor
IB              4.49E-06                 4.49E-06
IC              4.94E-04                 4.94E-04
VBE             6.26E-01                 6.26E-01
VBC            -1.01E+01                -1.01E+01
VCE             1.07E+01                 1.07E+01
BETADC          1.10E+02                 1.10E+02
GM              1.91E-02                 1.91E-02
RPI             5.76E+03                 5.76E+03
RX              0.00E+00                 0.00E+00
RO              2.23E+05                 2.23E+05
```

Substituting these parameter values into the expressions for A_d, R_{id}, A_{CM}, and R_{iCM} in Table 6.1 and approximating r_μ by $10r_o\beta$, we obtain the following estimates of the amplifier's small-signal behavior: $A_d = 182.8$ V/V, $R_{id} = 11.52$ kΩ, $A_{CM} = 0$, and $R_{iCM} = 6.75$ MΩ. These results are in good agreement with those computed directly by Spice, except for R_{iCM}. The formula for R_{iCM} in Table 6.1 seems to underestimate the actual input common-mode resistance by about 14%, perhaps because of our approximation of r_μ by $10r_o\beta$.

To summarize the results of this section, we have compiled a list in Table 6.2 that compares the results computed directly by Spice with those computed using the formulas presented in Table 6.1 and shows the percentage error. The results predicted by the formulas of Table 6.1 agree well with the results computed by Spice. We can

Table 6.2 Comparing the small-signal parameters of the differential amplifier shown in Figs. 6.7 and 6.9 as calculated by hand analysis and as computed by Spice.

Condition	Amplifier Parameter	Hand Analysis	Spice	% Error
$V_A = \infty$ $R = \infty$	A_d	191.0 V/V	191.4 V/V	0.21
	R_{id}	10.44 kΩ	10.45 kΩ	0.1
	R_o	20 kΩ	20 kΩ	0
	A_{CM}	0	0	0
	R_{iCM}	∞	4.927×10^{11} Ω	—
$V_A = \infty$ $R = 200$ kΩ	A_d	191.0 V/V	190.8 V/V	−0.1
	R_{id}	10.48 kΩ	10.48 kΩ	0
	R_o	20 kΩ	20 kΩ	0
	A_{CM}	0	0	0
	R_{iCM}	20.20 MΩ	20.20 MΩ	0
$V_A = 100$ V $R = 200$ kΩ	A_d	182.8 V/V	182.7 V/V	−0.05
	R_{id}	11.52 kΩ	11.53 kΩ	0.09
	R_o	19.14 kΩ	19.14 kΩ	0
	A_{CM}	0	0	0
	R_{iCM}	6.75 MΩ	7.825 MΩ	13.7

conclude that when given good estimates of the small-signal model parameters, the formulas of Table 6.1 will predict quite accurately the small-signal differential and common-mode voltage gain of a differential amplifier and its corresponding input resistances.

6.3

Small-Signal Analysis of the Differential Amplifier: Asymmetric Conditions

The previous section assumed that many of the components in the differential amplifier of Figs. 6.7 and 6.9 were matched. In practice this is rarely the case. In this section we investigate the effect of asymmetric circuit conditions on amplifier behavior, specifically the effect of variations, or mismatches, on collector resistances, transistor saturation (scale) currents, and transistor βs on circuit behavior.

6.3.1 Input Offset Voltage

Let us use Spice to determine the input offset voltage V_{OS} for the differential amplifier in Fig. 6.9, assuming that each collector resistor undergoes a change of 5%—one positive, the other negative—resulting in a net change of 10%. The input deck is shown in Fig. 6.10. We request a DC sweep of the input differential voltage V_d with a range of -10 mV to $+10$ mV. The range was determined by considering the formula for input offset voltage V_{OS} in Table 6.1. Given that $\Delta R_C/R_C = 10\%$, we can calculate an input offset voltage having a magnitude of 2.6 mV. We decided that our sweep should not exceed this by very much so that details, such as zero crossings, can be easily seen. In the input deck is a .TF command to compute the common-mode voltage gain. As we can see from Table 6.1, the magnitude of the common-mode voltage gain A_{CM} will no longer be zero but approximately 2.5 mV/V.

The results of this DC sweep of the input differential voltage are shown in Fig. 6.11. We see that the transfer function curve (v_o vs. v_d) for this change in resistance no longer passes through the origin. Instead, careful probing using the cursor feature of the Probe facility of PSpice indicates that the output offset voltage for zero input is -472.7 mV. Conversely, the input voltage that corresponds to zero output voltage is $+2.6$ mV. This, then, is the negative of the input offset voltage (i.e., $V_{OS} = -2.6$ mV). Interestingly enough, this corresponds exactly with the voltage predicted above by the formula given in Table 6.1.

If we use Spice to find the common-mode voltage gain (v_o/v_{cm}) in this same situation, we see

```
****       SMALL-SIGNAL CHARACTERISTICS

   V(5,4)/Vcm = -2.042E-03

   INPUT RESISTANCE AT Vcm =   7.825E+06

   OUTPUT RESISTANCE AT V(5,4)  =   1.914E+04
```

```
Differential Amplifier: Asymmetric Collector Resistance

** Circuit Description **
* power supply
Vdd 6 0 DC +15V
Vee 7 0 DC -15V
* differential-mode signal level
Vd 101 0 DC 0V AC 1V
Rd 101 0 1
EV+ 1 100 101 0 +0.5
EV- 2 100 101 0 -0.5
* common-mode signal level
Vcm 100 0 DC 0V
* differential pair
Q1 4 1 3 npn_transistor
Q2 5 2 3 npn_transistor
* unequal collector resistors (10Rc1 6 4 9.50k
Rc2 6 5 10.5k
* bias source
I 3 7 DC 1mA
R 3 0 200k
* transistor model statements
.model npn_transistor npn ( Is=14fA Bf=100 VAf=100V )
** Analysis Requests **
.OP
.DC Vd -10mV +10mV 100uV
.TF V(5,4) Vcm
** Output Requests **
.PLOT DC V(5,4)
.probe
.end
```

Figure 6.10 Spice input file for calculating the input offset voltage V_{OS} of the differential amplifier shown in Fig. 6.9 when the collector resistors undergo a 10% relative variation. Included is a .TF command to compute the common-mode voltage gain.

The common-mode voltage gain is -2.042 mV/V, which is reasonably close to the 2.5 mV/V that was predicted by the formula given in Table 6.1.

We can repeat the forgoing analysis and determine the effect of a 5% difference between the saturation currents I_S of the two transistors on the amplifier's input offset voltage. According to the formula in Table 6.1, we can expect an input offset voltage of 1.3 mV. The input file is shown in Fig. 6.12, and the results are superimposed on Fig. 6.11. Using Probe, we determine that the output offset voltage is -236.4 mV and the input offset voltage is -1.3 mV. Our estimate of the input offset voltage agrees with that obtained with Spice.

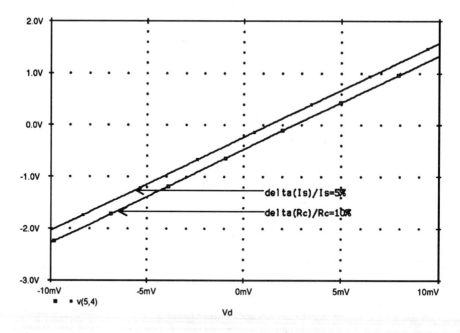

Figure 6.11 Input and output offset voltages of the differential amplifier shown in Fig. 6.9 subject to two different mismatches: $\Delta R_C/R_C = 10\%$ and $\Delta I_S/I_S = 5\%$.

6.3.2 *Input Bias and Offset Currents*

Mismatches in transistor βs result in different base currents, which in turn give rise to differences in the input bias currents to the amplifier. To demonstrate this, let us alter by -5% the β of Q_1 in the differential amplifier shown in Fig. 6.9 and the β of Q_2 by $+5\%$. The following model statements would be used to describe this situation to Spice:

```
.model npn_transistor1 npn ( Is=14fA Bf=95  VAf=100V )
.model npn_transistor2 npn ( Is=14fA Bf=105 VAf=100V )
```

These two statements can be used to replace the transistor model statements shown in the input file in Fig. 6.12. Only the .OP command is necessary to see the effect of β variation on input bias currents. The input bias current to the amplifier can be seen directly from the bias information generated by Spice for the two transistors.

Spice produces this DC operating-point information for the two transistors:

```
**** BIPOLAR JUNCTION TRANSISTORS
```

NAME	Q1	Q2
MODEL	npn_transistor1	npn_transistor2
IB	4.72E-06	4.27E-06
IC	4.94E-04	4.94E-04
VBE	6.26E-01	6.26E-01
VBC	-1.01E+01	-1.01E+01

```
Differential Amplifier: Asymmetric Saturation Current

** Circuit Description **
* power supply
Vdd 6 0 DC +15V
Vee 7 0 DC -15V
* differential-mode signal level
Vd 101 0 DC 0V AC 1V
Rd 101 0 1
EV+ 1 100 101 0 +0.5
EV- 2 100 101 0 -0.5
* common-mode signal level
Vcm 100 0 DC 0V
* differential pair
Q1 4 1 3 npn_transistor1
Q2 5 2 3 npn_transistor2
* load resistors
Rc1 6 4 10k
Rc2 6 5 10k
* bias source
I 3 7 DC 1mA
R 3 0 200k
* transistor model statements
.model npn_transistor1 npn ( Is=13.65fA Bf=100 VAf=100V )
.model npn_transistor2 npn ( Is=14.35fA Bf=100 VAf=100V )
** Analysis Requests **
.DC Vd -10mV +10mV 100uV
.OP
.TF V(5,4) Vcm
** Output Requests **
.PLOT DC V(5,4)
.probe
.end
```

Figure 6.12 Spice input file for calculating the input offset voltage V_{OS} of the differential amplifier shown in Fig. 6.9 when the saturation currents of the two transistors differ by 5%.

VCE	1.07E+01	1.07E+01
BETADC	1.05E+02	1.16E+02
GM	1.91E-02	1.91E-02
RPI	5.48E+03	6.05E+03
RX	0.00E+00	0.00E+00
RO	2.23E+05	2.23E+05

We see that the base currents of Q_1 and Q_2 are not equal but differ by 450 nA. The input bias current to the differential amplifier is the average of these two currents, 4.50 μA. Correspondingly, the offset current for the amplifier is 450 nA. It is interesting to note that the formula provided in Table 6.1 generates the same value.

It is also noteworthy that according to the preceding list of small-signal model parameters, the BETADC of Q_1 is 105, and for Q_2 it is 116. This is quite different from the β that was assigned on the transistor model statement. This difference arises from the definition Spice uses to compute BETADC, which is the ratio of the collector current to base current and includes the effect of transistor output resistance. This is not what β represents in the model statement, where it represents short-circuit current gain.

6.4

Current-Source Biasing in Integrated Circuits

A typical approach for implementing a current source in IC technology is with current-mirror circuits. Figure 6.13 illustrates a simple current mirror arranged as a current source. The input terminal of the current mirror is fed with a 942 kΩ resistor connected to the positive power supply. This results in an input bias current of about 10 μA. The output terminal is connected to a variable DC voltage source V_{out} to simulate the action of a load. With the aid of Spice, we would like to determine the range of output voltages that can appear across the output port of the current mirror before the circuit ceases to operate as a current source. We would also like to determine the Norton equivalent circuit representation of the output port. We shall assume that the transistors are all matched and are typical of the small npn integrated transistor variety manufactured by the Gennum Corporation [Gennum IC Data Book, 1991]. Typically these transistors have a forward β_{ac} of approximately 90 at a bias current of 1 mA and an Early voltage in the neighborhood of 80 V. Furthermore, as all transistors in this technology share a common p-type substrate, we must specify the substrate connection in its element description statement.[†] In this case it would be the ground node (node 0) because this

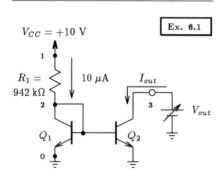

Figure 6.13   Simple current-mirror circuit set up as a current source.

[†] Whenever an IC transistor is employed and its model is obtained from a library, the substrate connection must be defined.

```
A Simple Bipolar Current Source

** Circuit Description **
Vcc 1 0 DC +10V
R1 1 2 942k
Q1 2 2 0 0 npn
Q2 3 2 0 0 npn
Vout 3 0 DC 5V
* transistor model statement for integrated npn transistor by Gennum Corp.
.MODEL npn NPN (IS=5E-17 BF=147 VAF=80 IKF=4.3E-3 ISE=8E-18 NE=1.233
+               BR=1.9 VAR=11 IKR=6E-4 ISC=5E-16 NC=1.08 RE=12 RB=1200 RBM=200 RC=25
+               CJE=58E-15 VJE=0.83 MJE=0.35 CJC=133E-15 VJC=0.6 MJC=0.44 XCJC=1
+               CJS=830E-15 VJS=0.6 MJS=0.4 ISS=1E-16 FC=0.85 TF=60P XTF=48 ITF=3E-2
+               TR=10N EG=1.16 XTI=3 XTB=1.6)
** Analysis Requests **
.OP
* sweep the output voltage Vout between 0 V and +5 V
.DC Vout 0V 5V 10mV
** Output Requests **
.PLOT DC I(Vout)
.Probe
.end
```

Figure 6.14   Spice input file for computing the I_{out}–V_{out} characteristic of the simple current-mirror circuit shown in Fig. 6.13.

will ensure that the *pn* junction formed by the substrate is reverse biased by connecting the substrate to the lowest potential in the circuit.

The input file describing this circuit is given in Fig. 6.14, in which we have requested a DC sweep of the output voltage V_{out} beginning at 0 V and ending at 5 V. The output collector current of Q_2 will then be plotted as a function of V_{out}. The plot is shown in Fig. 6.15. For output voltages larger than 220 mV, we see that the output current remains relatively constant around 10 μA, increasing slightly at a rate of 0.13 μA per volt. When the output voltage drops below 220 mV the output current drops quickly to zero. The circuit operates as an effective current source provided the voltage across the output remains above 220 mV.

We can go further and characterize the output port of this circuit (assuming $V_{out} >$ 220 mV) using the Norton equivalent circuit shown in Fig. 6.16. The current-source value of 9.64 μA is the y-axis intersection of the extrapolation of the *i–v* curve in Fig. 6.15 for $V_{out} >$ 220 mV.

The output resistance R_o = 7.8 MΩ is the inverse of the slope of this curve and is found by the Probe facility of PSpice. Similar resistance is found through a .TF command with the output voltage biased to a value inside the linear region of the current source, and hand analysis generates almost the same output resistance of V_A/I = 80 V/10 μA = 8 MΩ.

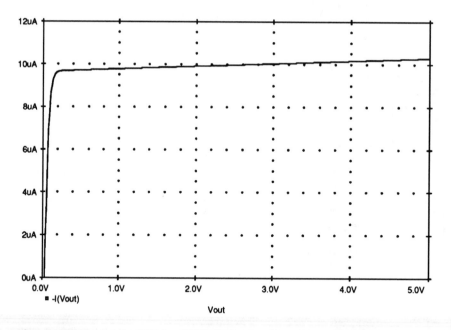

Figure 6.15 I_{out} versus V_{out} for the current-source implementation shown in Fig. 6.13.

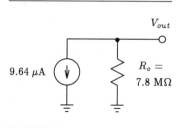

Figure 6.16 Norton equivalent representation of the current source in Fig. 6.13 for $V_{out} > 220$ mV.

6.5

A BJT Multistage Amplifier Circuit

Figure 6.17 illustrates the circuit of a simple operational amplifier. It consists of a cascade of several gain stages, two of which are made from differential pairs, and an output buffer. The positive and negative input terminals to the amplifier are labeled as V^+ and V^-, respectively, and the output terminal is denoted as V_o. Sedra and Smith analyze this operational amplifier in their Examples 6.2 and 6.3, and at the end of this discussion we will summarize their results and compare them with ours.

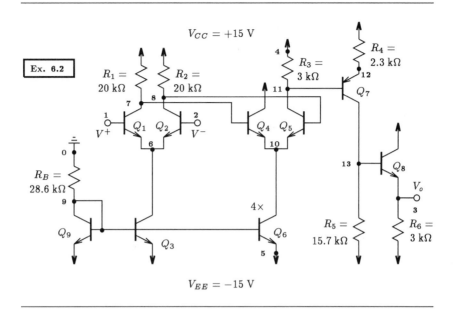

Figure 6.17   Simple operational amplifier consisting of three stages.

For our analysis we assume that both the *npn* and *pnp* transistors have the following parameters: $I_S = 18$ fA, $\beta_F = 100$, and $V_{AF} = 100$ V. In addition, Q_6 is considered to have four times the area of Q_9 and Q_3. Using Spice we would like to determine the operational amplifier's DC operating point, which includes output offset voltage, input bias currents, and quiescent power dissipation. We also would like to compute the small-signal differential voltage gain (in the high-gain region of the amplifier) and the input common-mode range.

The Spice input file describing the makeup of the operational amplifier shown in Fig. 6.17 and the multiple input voltage source arrangement discussed in Section 6.1 is given in Fig. 6.18. Notice how Q_6 is specified to have four times the area of Q_3 and Q_9. Two analysis request commands are listed here: a DC operating-point command and a DC sweep command. In a moment we shall explain the role of a .NODESET command in the analysis request section of the Spice deck.

The DC operating-point command will determine the power dissipated by the amplifier and the input bias currents in addition to calculating the DC operating point and the transistor small-signal model parameters. The DC sweep command will be used to compute the large-signal transfer characteristic of the amplifier by varying the DC level of the input differential source V_d between -15 V and $+15$ V in 100 mV increments. The common-mode input V_{CM} will be held at zero throughout this DC sweep. We may be tempted to add a small-signal transfer function analysis request here, but this should be deferred until we see the large-signal differential transfer characteristic of the amplifier and can determine the DC input conditions so that the Spice small-signal analysis is performed around a known operating point.

```
A Simple Operational Amplifier

** Circuit Description **
* power supplies
Vcc 4 0 DC +15V
Vee 5 0 DC -15V
* differential-mode signal level
Vd 101 0 DC 0V
Rd 101 0 1
EV+ 1 100 101 0 +0.5
EV- 2 100 101 0 -0.5
* common-mode signal level
Vcm 100 0 DC 0V
* 1st stage
R1 4 7 20k
R2 4 8 20k
Q1 7 1 6 npn_transistor
Q2 8 2 6 npn_transistor
Q3 6 9 5 npn_transistor
* 2nd stage
R3 4 11 3k
Q4 4 7 10 npn_transistor
Q5 11 8 10 npn_transistor
Q6 10 9 5 npn_transistor 4
* 3rd or output stage
R4 4 12 2.3k
Q7 13 11 12 pnp_transistor
R5 13 5 15.7k
Q8 4 13 3 npn_transistor
R6 3 5 3k
* biasing stage
Rb 0 9 28.6k
Q9 9 9 5 npn_transistor
* transistor model statements
.model npn_transistor npn ( Is=18fA Bf=100 VAf=100V )
.model pnp_transistor pnp ( Is=18fA Bf=100 VAf=100V )
** Analysis Requests **
* compute DC operating point using the following initial guesses
.OP
.NODESET V(3)=0V V(6)=-0.7V V(7)=+10V V(8)=+10V V(9)=-14.3V V(10)=+9.3V
+        V(11)=+12V V(12)=+12.7V V(13)=+0.7V
* compute large-signal differential-input transfer characteristics of amplifier
.DC Vd -15V +15V 100mV
** Output Requests **
.PLOT DC V(3)
.probe
.end
```

Figure 6.18 Spice input file for computing the DC operating point and the large-signal differential transfer characteristic of the amplifier shown in Fig. 6.17.

To assist in the DC bias calculation, the input file includes a list of initial guesses of some node voltages in the circuit on the statement beginning with the keyword .NODESET. If we provide good estimates of the circuit's DC bias point, the time required by Spice to compute it is reduced, but more important, including a .NODESET statement prevents Spice from terminating the analysis without obtaining a DC solution because of nonconvergence. In older versions of Spice (version 2G6 and less) this step is usually required, but newer versions, such as PSpice, have improved algorithms that have more robust convergence capabilities and rarely require guesses of the node voltages.

Part of the Spice data from the input of Fig. 6.18 is shown in the following DC analysis output:

```
****      SMALL SIGNAL BIAS SOLUTION         TEMPERATURE =   27.000 DEG C
**********************************************************************************

   NODE    VOLTAGE     NODE    VOLTAGE     NODE    VOLTAGE     NODE    VOLTAGE

(    1)     0.0000  (    2)     0.0000  (    3)     2.0677  (    4)    15.0000
(    5)   -15.0000  (    6)    -.6035  (    7)     9.4295  (    8)     9.4295
(    9)   -14.3790  (   10)     8.7868  (   11)    11.6170  (   12)    12.2590
(   13)     2.7493  (  100)     0.0000  (  101)     0.0000


      VOLTAGE SOURCE CURRENTS
      NAME            CURRENT

      Vcc           -9.691E-03
      Vee            1.020E-02
      Vd             0.000E+00
      Vcm           -4.887E-06


      TOTAL POWER DISSIPATION    2.98E-01   WATTS

**** VOLTAGE-CONTROLLED VOLTAGE SOURCES

NAME            EV+             EV-
V-SOURCE        0.000E+00       0.000E+00
I-SOURCE       -2.443E-06      -2.443E-06
```

We see that this amplifier has an output DC offset of $+2.0677$ V and input bias currents of 2.443 μA. The static power dissipated by the amplifier is 0.298 W.

The collector currents of each device have also been computed by Spice and are listed in Table 6.3 with the currents computed by hand analysis in Example 6.2 of Sedra and Smith (assuming that $\beta \gg 1$ and ignoring the effect of transistor Early voltage). The third column shows the percentage error between these two currents. As we can see, the hand estimates are reasonably close to the Spice results; the largest error is 15.3%. It is reassuring that reasonably good estimates of the DC collector currents of a complicated transistor circuit can be obtained by assuming ideal transistor behavior (i.e., $\beta \gg 1$ and $V_A = \infty$).

As a further check, in Table 6.4 we compiled the collector currents that were computed by Spice for various combinations of β and V_A values (I_S remains at 18 fA). As the table shows, when β and V_A approach infinity (i.e., the transistors become more ideal), the results approach those computed by the simplified hand analysis.

Table 6.3 DC collector currents of the operational amplifier shown in Fig. 6.17 expressed in mA as computed by hand analysis and by Spice.

Transistor	Hand Analysis	Spice	% Error
Q_1	0.25	0.267	6.4
Q_2	0.25	0.267	6.4
Q_3	0.5	0.540	7.4
Q_4	1.0	1.18	15.3
Q_5	1.0	1.14	12.3
Q_6	2.0	2.34	14.5
Q_7	1.0	1.18	15.3
Q_8	5.0	5.64	11.3
Q_9	0.5	0.47	−5.5

Table 6.4 Variation in the DC collector currents (in mA) of the operational amplifier for different βs and V_As computed by Spice, compared with those computed by simple hand analysis.

	Hand Analysis	Spice			
Transistor	$\beta \gg 1$ $V_A = \infty$	$\beta = 10^6$ $V_A = \infty$	$\beta = 100$ $V_A = \infty$	$\beta = 100$ $V_A = 100$ V	$\beta = 100$ $V_A = 35$ V
Q_1	0.25	0.251	0.235	0.267	0.328
Q_2	0.25	0.251	0.235	0.267	0.328
Q_3	0.5	0.503	0.474	0.540	0.661
Q_4	1.0	1.01	0.939	1.18	1.61
Q_5	1.0	1.01	0.939	1.14	1.45
Q_6	2.0	2.01	1.90	2.34	3.08
Q_7	1.0	1.03	0.926	1.18	1.57
Q_8	5.0	5.18	4.35	5.64	7.58
Q_9	0.5	0.503	0.474	0.474	0.474

The large-signal differential transfer characteristic of this amplifier is displayed in Fig. 6.19, which illustrates a view of the operational amplifier differential-input transfer characteristics between −15 V and +15 V. We can see that the high-gain region of the amplifier is in the vicinity of 0 V, but the resolution of the input voltage axis does not enable us to be certain of the boundaries of this high-gain region. We shall rerun the input file with the DC sweep command modified to the following so that it will produce an expanded view of the high-gain region between −5 mV and +5 mV:

```
.DC Vd -5mV +5mV 10uV
```

The results of this analysis are displayed in Fig. 6.20. For inputs less than −2 mV the output remains saturated at −15 V. In the region between −2 mV and +1 mV the output level changes from −15 V to +10 V in a linear manner. Thus, the output voltage swing for this amplifier is bounded between −15 V and +10 V, a

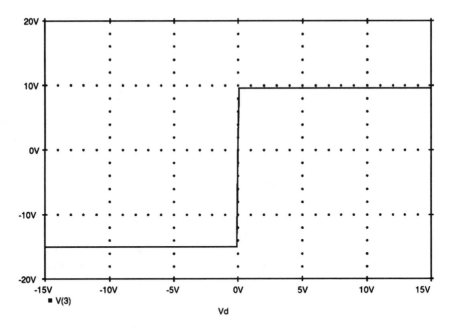

Figure 6.19 Large-signal differential transfer characteristic of the operational amplifier shown in Fig. 6.17. The input common-mode voltage V_{CM} is set to zero.

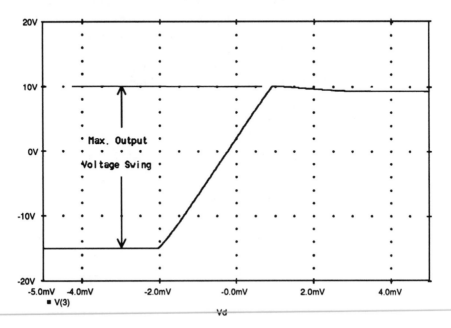

Figure 6.20 Expanded view of the high-gain differential region of the operational amplifier shown in Fig. 6.17.

somewhat asymmetrical voltage swing. The gain experienced by input signals in this linear region is approximately $[10 - (-15)]/3$ mV $= 8.33$ kV/V. For inputs greater than $+2$ mV the output levels off at $+8.5$ V. We also see that the input offset voltage V_{OS} for this amplifier is $+230.0$ μV (by convention, the negative value of the x-axis intercept). This is, of course, the systematic offset of the amplifier and does not include the components due to various imbalances in the circuit (see Section 6.3).

Now that we know the boundaries of the high-gain region of the amplifier, we can use the transfer function command of Spice to compute the small-signal equivalent circuit parameters of the amplifier in its linear region. First we must decide on the point to use inside the linear region of the amplifier. For most applications a high-gain amplifier such as that shown in Fig. 6.17 usually is used in conjunction with negative feedback, and as a result the output potential of the amplifier is held close to ground potential (when no input signal is applied). Thus, the small-signal parameters of the amplifier should be obtained around the bias point that has the amplifier output voltage close to 0 V. This is obtained by applying the negative of the amplifier input-referred offset voltage (i.e., $-V_{OS}$) across the input terminals of the amplifier. For the multiple-source arrangement suggested in Fig. 6.3 this is achieved by setting V_d equal to $-V_{OS}$.

In this example, we shall modify the Spice statement for V_d:

```
Vd 101 0 DC -230.0uV AC 1V
```

A 1 V AC voltage has also been appended to this input voltage source statement, which will enable us to compute the AC current circulating around the input terminals of the op amp when a 1 V AC voltage signal is placed across its input terminals. From this we can calculate the input differential resistance. For more details on this approach, the reader should refer back to Section 6.2.

Two analysis requests and a .PRINT command will be added to the input file to compute the small-signal parameters of the amplifier:

```
.TF V(3) Vd
.AC LIN 1 1Hz 1Hz
.PRINT AC Im(EV+) Im(EV-) Vm(1,2)
```

In the output file, some of the small-signal circuit parameters of this amplifier are found as follows:

```
****      SMALL-SIGNAL CHARACTERISTICS

     V(3)/Vd =   8.834E+03

     INPUT RESISTANCE AT Vd =   1.000E+00

     OUTPUT RESISTANCE AT V(3) =   1.337E+02
```

Here we see that the actual voltage gain in the amplifier linear region is 8.834 kV/V, very close to the 8.83 kV/V we estimated from the amplifier large-signal differential transfer characteristic in Fig. 6.20. The input resistance listed here is not the input resistance for the amplifier, but rather the resistance seen by the voltage source V_d—in this case, the 1 Ω resistor connected in series with V_d. The output resistance of 133.7

Ω listed above is the actual output resistance for this amplifier. As an estimate of the input differential resistance of the amplifier we use the results of the AC analysis given here:

```
****      AC ANALYSIS

   FREQ          IM(EV+)       IM(EV-)       VM(1,2)

   1.000E+00    4.729E-05     4.718E-05     1.000E+00
```

We see that current $IM(EV+)$ is not quite the same as the current $IM(EV-)$ but is very close. The reason for this difference lies in the amplifier's systematic offset because of transistor Early voltage. To determine the input differential resistance of the amplifier under such asymmetric conditions, we shall work with the average of the two base currents so that we can eliminate the presence of the offset current in the resistance calculation. Thus, we compute the average input base current to be 47.235 μA, and we obtain the input differential resistance at 21.17 kΩ.

The final analysis that we would like to perform on the operational amplifier shown in Fig. 6.17 is to determine its input common-mode range. Let us perform a DC sweep of the input common-mode voltage V_{CM} between the rails of the two power supplies. This requires that we replace the DC sweep command in Fig. 6.18 by

```
.DC Vcm -15V +15V 0.1V
```

We also maintain an input differential offset voltage of $V_d = -230\,\mu$V to ensure that the amplifier is biased inside its linear region.

Resubmitting the revised input file to Spice and observing the results in the output file, we obtain the amplifier large-signal common-mode transfer characteristic given in Fig. 6.21. As we can see from these results, the transfer characteristic has linear behavior over the range of V_{CM} between -14.3 V and $+9.6$ V. Outside these limits the characteristic becomes nonlinear. Thus, the input common-mode range for this amplifier is between -14.3 V and $+9.6$ V. We should also note that our large-signal differential transfer characteristic computed earlier is valid, because it was obtained with an input common-mode voltage that falls within the input common-mode range of the amplifier.

It is interesting to correlate the limits of the amplifier common-mode range with the mode of operation of the transistors in the front-end stage. Specifically, Sedra and Smith mention in their text that the upper limit to the common-mode range is determined by Q_1 or Q_2 saturating and that the lower limit is determined by Q_3 saturating. We can determine when these transistors saturate by observing the voltage across the base–collector junction of each transistor and recalling that a transistor enters its saturation region when the base–collector junction becomes forward biased. For instance, in Fig. 6.22 we display the voltage across the base–collector junctions of Q_1 and Q_3 and observe that Q_1 saturates when V_{CM} exceeds $+9.6$ V. In contrast, transistor Q_3 saturates when V_{CM} goes below -14.3 V.

Table 6.5 is a summary of what we have learned about this op amp through the application of Spice and compares those results with the simplified hand analysis performed by Sedra and Smith in their Examples 6.2 and 6.3.

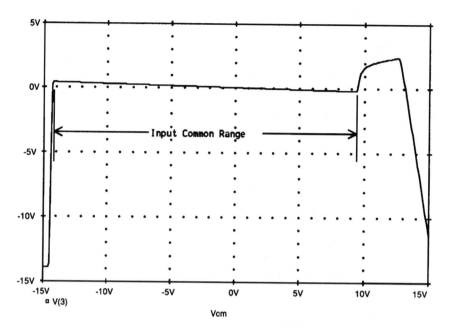

Figure 6.21 Large-signal common-mode DC transfer characteristic of the BJT amplifier shown in Fig. 6.17. An input differential offset voltage of $-230\ \mu V$ is applied to the amplifier input to prevent premature saturation.

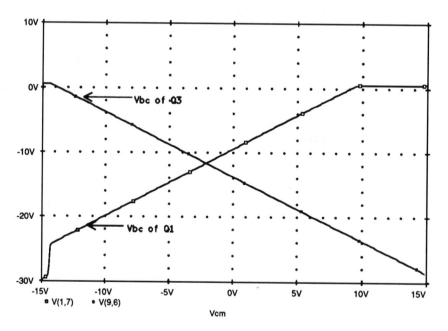

Figure 6.22 Effect of a common-mode input voltage V_{CM} on the linearity of the input stage of the operational amplifier shown in Fig. 6.17. Here we illustrate the base–collector voltage of Q_1 and Q_3 as a function of V_{CM}. The first stage of the amplifier leaves the active region when the base–collector junction of either Q_1 or Q_3 becomes forward bias.

Table 6.5 Comparison of the results of the analysis of the operational amplifier shown in Fig. 6.17 by hand (in Sedra and Smith) and using Spice.

Parameter	Units	Hand Calculation	Spice
Output-referred offset voltage	V	0	2.07
Input-referred offset voltage	μV	0	230.0
Input bias currents	μA	2.5	2.443
Quiescent power dissipation	mW	262.5	298
Differential voltage gain	kV/V	8.513	8.834
Input differential resistance	kΩ	20.1	21.17
Output resistance	Ω	152	133.7
Input common-mode range	V	−13.6 to +10.0	−14.3 to +9.6
Output voltage swing	V	—	−15.0 to +10.0

6.6

Spice Tips

- Spice can be conveniently used to compute both the large- and small-signal characteristics of amplifier circuits.
- Inputs to differential amplifiers should consist of both a differential and common-mode level. An interesting arrangement of several voltage sources was given in this chapter (see Fig. 6.3), illustrating how the differential and common-mode levels can be independently adjusted.
- The high-gain linear region of an amplifier is located by first sweeping the input differential voltage V_d between the limits of the power supplies with the common-mode voltage V_{CM} equal to 0 V. This analysis is repeated with a reduced sweep range centered more closely around the high-gain region of the amplifier until a smooth transition through the high-gain region is achieved. Following this, one must check to see whether the input common-mode voltage of 0 V keeps the amplifier in its linear region.
- Care must be exercised when computing the small-signal characteristics of an amplifier using Spice. One should first decide what DC input conditions are required so that the amplifier is linearized around an appropriate DC operating point. Generally, selecting an input differential offset voltage that forces the output voltage to 0 V will bias the amplifier at an operating point that is quite close to the operating point that results when some external negative-feedback connection is made around the amplifier.
- Any time an IC transistor is used in a circuit and its model is obtained from a library, the substrate connection must be defined.
- At times Spice will not be able to compute the DC node voltages in a nonlinear circuit because of a DC convergence problem. Sometimes this convergence problem can be alleviated by providing Spice with a set of initial estimates of

the DC node voltages of the circuit. These are entered into the Spice deck using the .NODESET command of Spice.

■ The small-signal input resistance to a differential amplifier is computed using Spice by applying a known AC voltage across the input terminals of the differential amplifier and computing the AC currents that flow into the amplifier terminals. In many practical amplifier situations these currents will not be equal, so the average of these two currents is used in the input resistance calculation.

6.7

Bibliography

1990–1991 IC Data Book, Gennum Corporation, Burlington, ON, Canada.

6.8

Problems

6.1 A BJT differential amplifier is biased from a 2 mA constant-current source and includes a 100 Ω resistor in each emitter. The collectors are connected to +10 V via 5 kΩ resistors. A differential input signal of 0.1 V is applied between the two bases. Assume that the transistors are matched and have $\beta = 100$ and $I_S = 14$ fA.
(a) With the aid of Spice, determine the signal current in the emitters (i_e) and the base–emitter voltage v_{be} for each BJT.
(b) What is the total emitter current in each BJT?
(c) What is the signal voltage at each collector?
(d) What is the voltage gain realized when the output is taken between the two collectors?

6.2 The differential amplifier circuit of Fig. P6.2 utilizes a resistor connected to the negative power supply to establish the bias current I.

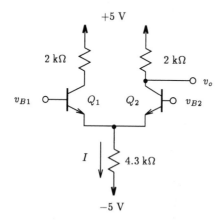

Figure P6.2

(a) For $v_{B1} = v_d/2$ and $v_{B2} = -v_d/2$, where v_d is a small signal with zero average, find the magnitude of the differential gain, $|v_o/v_d|$, using Spice.
(b) For $v_{B1} = v_{B2} = v_{CM}$, find the magnitude of the common-mode gain, $|v_o/v_{CM}|$, using Spice.

(c) Calculate the common-mode rejection ratio.

(d) If $v_{B1} = 0.1 \sin(2\pi t) + 0.005 \sin(2\pi t)$ volts, $v_{B2} = 0.1 \sin(2\pi t) - 0.005 \sin(2\pi t)$ volts, plot the output voltage v_O using the .PLOT command of Spice.

6.3 A BJT differential amplifier is biased from a 300 μA constant-current source, and the collectors are connected to +7.5 V via 50 kΩ resistors. If the scale currents I_S of the two transistors have a nominal value of 10 fA but differ by 10%, what is the resulting input offset voltage?

6.4 A BJT differential amplifier is biased from a 1 mA constant-current source, and the collectors are connected to +15 V via 10 kΩ resistors. If the βs of the two transistors are 100 and 200, what is the resulting input offset voltage?

6.5 Compute the voltages at all nodes and the currents through all branches in the circuit of Fig. P6.6, assuming β is very large, using Spice. Compare this to the case when $\beta = 100$.

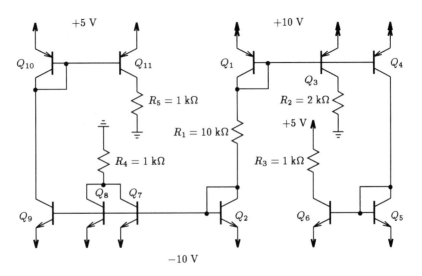

Figure P6.6

6.6 Compute the voltages at all nodes and the currents through all branches in the circuit of Fig. P6.6, assuming V_A is very large, using Spice. (Setting $V_A = 0$ in the model statement for the BJT is equivalent to setting $V_A = \infty$.) Compare this to the case when $|V_A| = 100$ V.

6.7 Consider the effect of power-supply variation on the DC bias of the op amp circuit of Fig. 6.17: If +15 V is lowered to +14 V, what is the effect on the amplifier parameters listed in Table 6.5? If −15 V is raised to −14 V separately, what is the effect on the amplifier parameters in Table 6.5?

6.8 In the multistage amplifier of Fig. 6.17, 100 Ω resistors are introduced into the emitter lead of each transistor in the differential pair of the first stage (Q_1 and Q_2), and 25 Ω resistors for each of the second-stage transistors (Q_4 and Q_5). With the aid of Spice, find the effect that these additions have on the input resistance, the voltage gain of the first stage, and the overall voltage gain.

6.9 A JFET differential amplifier is loaded with the basic BJT current mirror. The JFETs have $V_P = -2$ V, $I_{DSS} = 4$ mA, and $V_A = 100$ V. The BJTs have $|V_A| = 100$ V, and β

is large. The bias current $I = 2$ mA. Find R_i, G_m, R_o, and the open-circuit voltage gain.

6.10 For the amplifier circuit in Fig. P6.10 let $W_1 = 100$ μm, $W_2 = 500$ μm, $W_3 = 50$ μm, and $W_4 = 250$ μm. With the aid of Spice compute the voltage gain of this circuit. Assume that the MESFETs are modeled with parameters $V_t = -1$ V, $\beta = 0.1$ mA/V^2, and $\lambda = 0.1$ V^{-1}. Assume that the length of each device is 1 μm.

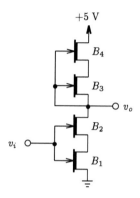

Figure P6.10

7
Frequency Response

In this chapter we use Spice to investigate the frequency response behavior of various amplifier circuits and to verify the accuracy of the gain and bandwidth expressions derived by Sedra and Smith in their Chapter 7.

7.1

Investigating Transfer Function Behavior Using PSpice

A useful feature of PSpice, but not of Spice, is the ability to specify the frequency-dependent gain of a current- or voltage-dependent source as a Laplace transform function. Although we can envision many new applications for this feature, we will limit ourselves to investigating transfer function behavior as a function of frequency.

This PSpice feature is restricted to the VCVS and the VCCS (designated with the letters E and G, respectively). Figure 7.1 illustrates the element statement's general form for specifying gain in terms of the Laplace variable s. The first part of this statement is identical to that given previously for the VCVS or VCCS (i.e., element type with a unique name and the nodes to which the dependent source is connected). After the key word *Laplace,* the variable on which the controlled source depends is

$$\left\{ \begin{array}{c} \text{E}name \\ \text{G}name \end{array} \right\} \quad \text{n+} \quad \text{n-} \quad \text{Laplace} \quad \{\, Controlling\_Variable \,\} \quad = \quad \{\, Laplace\_Transform \,\}$$

Figure 7.1 General form of the PSpice element statement for a voltage- or current-dependent source with its gain written as a function of the complex frequency variable.

specified between braces (e.g., {V(1)}), followed by an equal sign and an expression written in terms of the Laplace variable enclosed between braces.

If we can specify gain as an arbitrary expression in terms of the Laplace variable, we have a simple means for investigating transfer-function frequency behavior. In the simple circuit in Fig. 7.2 a VCVS has a voltage gain expressed in the form of a transfer function $T(s)$ and is excited by an input AC voltage source of 1 V with an arbitrary frequency. The voltage then appearing across the output port is directly equal to the transfer function, that is, $V_o(s) = T(s)$. Using the AC analysis command of PSpice, we can investigate the behavior of the transfer function $T(s)$ for physical frequencies defined by $s = j\omega$.[†] We leave it to the reader to adopt a similar approach for a VCCS.

To illustrate the idea, let us consider having PSpice compute the magnitude and phase behavior of the following transfer function:

$$T(s) = \frac{10s}{(1 + s/10^2)(1 + s/10^5)} \tag{7.1}$$

This transfer function is discussed in Example 7.1 of Sedra and Smith. Figure 7.3 lists a PSpice description of the circuit shown in Fig. 7.2 with the $T(s)$ given in Eq. (7.1). An AC analysis command requests that PSpice evaluate the transfer function using a logarithmic frequency sweep of 10 points per decade beginning at 0.01 Hz and ending at 100 MHz. We ask that PSpice plot the magnitude and phase of the output voltage V(2) because it equals the transfer function specified as the gain of the VCVS. The graph from this analysis is in Fig. 7.4 and can be compared with Sedra and Smith's Bode plot.

7.2

Modeling Dynamic Effects in Semiconductor Devices

Spice performs frequency domain analysis on diode and transistor circuits in much the same way that this analysis is performed by hand. Each diode or transistor in the circuit

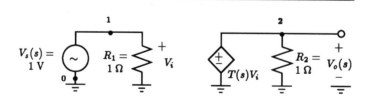

Figure 7.2 Simple circuit arrangement for investigating transfer function behavior.

[†] Resistors R_1 and R_2 are included in the circuit of Fig. 7.2 to satisfy the PSpice requirement of two or more element connections at each node. They do not affect the circuit's overall transfer function.

```
Bode Plot

** Circuit Description **
* input signal
Vin 1 0 AC 1V
* VCVS with transfer function:
*                              10s
*               T(s)  = --------------------------
*                       ( 1+s/10^2 ) ( 1 + s/10^5 )
*
E1 2 0 Laplace V(1)  = (10*s) / ( (1+s/1e2) * (1+s/1e5) )
R1 1 0 1Ohm
R2 2 0 1Ohm
** Analysis Requests **
.AC DEC 10 1e-2 1e8
** Output Requests **
.PLOT AC VdB(2) Vp(2)
.probe
.end
```

Figure 7.3 PSpice input file for computing the magnitude and phase of the transfer function given in Eq. (7.1).

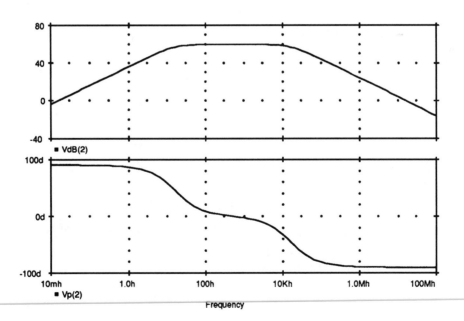

Figure 7.4 Magnitude and phase response of the transfer function given in Eq. (7.1) as calculated by PSpice.

is replaced by its small-signal frequency-dependent equivalent circuit, all DC voltage sources are shorted to ground, and all DC current sources are open-circuited. Standard frequency-domain circuit analysis techniques are then applied to the equivalent small-signal circuit to determine the signal levels at various nodes or branches of the circuit.

The small-signal frequency-domain model of either a diode or a transistor is similar to its small-signal model under static conditions except that capacitors are included to account for the frequency effects. In Fig. 7.5 we illustrate the generally used Spice small-signal dynamic models of the semiconductor junction diode, the BJT (applies to both *npn* and *pnp* devices), and the MOSFET (both n- and p-channel devices). Except for the additional capacitors, these models are identical to those introduced in previous chapters. For each model Table 7.1 provides the description of each capacitor and its Spice name.

To gain insight into circuit frequency-domain behavior, it is sometimes useful to know the values of the parameters of the small-signal model used by Spice in an AC analysis. An .OP (operating point) command in the input file makes these values accessible in the output file.

Consider, for example, biasing a 2N2222A transistor with a 10 μA base current and a collector–emitter voltage of 5 V, as illustrated in Fig. 7.6. The input file for this arrangement with only an .OP analysis request would appear as follows:

```
Small-Signal Model For Q2N2222A (Ib=10uA, Vce=5V)

** Circuit Description **
* bias conditions
Vce 1 0 DC +5V
Ib 0 2 10uA
* transistor under test
Q1 1 2 0 Q2N2222A
* transistor model statement for the 2N2222A
.model Q2N2222A NPN (Is=14.34f Xti=3 Eg=1.11 Vaf=74.03 Bf=255.9 Ne=1.307
+                    Ise=14.34f Ikf=.2847 Xtb=1.5 Br=6.092 Nc=2 Isc=0 Ikr=0 Rc=1
+                    Cjc=7.306p Mjc=.3416 Vjc=.75 Fc=.5 Cje=22.01p Mje=.377 Vje=.75
+                    Tr=46.91n Tf=411.1p Itf=.6 Vtf=1.7 Xtf=3 Rb=10)
** Analysis Requests **
.OP
** Output Requests **
* none required
.end
```

Spice produces the following list of parameters:

```
****        OPERATING POINT INFORMATION        TEMPERATURE =   27.000 DEG C

**********************************************************************************
**** BIPOLAR JUNCTION TRANSISTORS

NAME          Q1
MODEL         Q2N2222A
IB            1.00E-05
IC            1.63E-03
VBE           6.57E-01
```

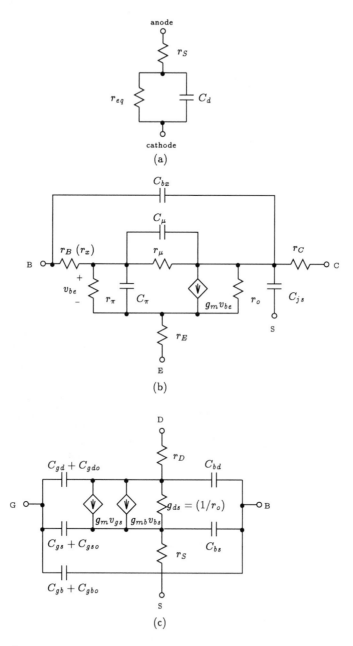

Figure 7.5 Dynamic Spice models of the small-signal behavior of (a) junction diode, (b) BJT, and (c) MOSFET.

```
VBC          -4.34E+00
VCE           5.00E+00
BETADC        1.63E+02
GM            6.26E-02
RPI           2.85E+03
RX            1.00E+01
RO            4.82E+04
CBE           6.24E-11
CBC           3.80E-12
CBX           0.00E+00
CJS           0.00E+00
BETAAC        1.78E+02
FT            1.50E+08
```

Table 7.1 Capacitances associated with the small-signal models of the diode, BJT, and MOSFET shown in Fig. 7.5.

Device	Symbol	Spice Name	Capacitance Description
Diode	C_d	CAP	Junction capacitance
BJT	C_π	CBE	Base–emitter junction capacitance
	C_μ	CBC	Base–collector junction capacitance
	C_{bx}	CBX	Extrinsic base–intrinsic collector capacitance
	C_{js}	CJS	Substrate junction capacitance
MOSFET	C_{bd}	CBD	Bulk–drain junction capacitance
	C_{bs}	CBS	Bulk–source junction capacitance
	C_{gso}	CGSOV	Gate–source overlap capacitance
	C_{gdo}	CGDOV	Gate–drain overlap capacitance
	C_{gbo}	CGBOV	Gate–bulk overlap capacitance
	C_{gs}	CGS	Gate–source intrinsic capacitance
	C_{gd}	CGD	Gate–drain intrinsic capacitance
	C_{gb}	CGB	Gate–bulk intrinsic capacitance

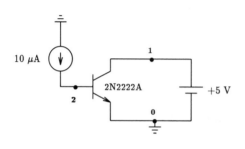

Figure 7.6 One bias arrangement for the commercial transistor 2N2222A.

Most of the transistor DC operating-point and small-signal model parameters should be evident in this list, but some are not. Capacitors C_{BE} and C_{BC} are equivalent to capacitances C_π and C_μ, respectively, of the hybrid-pi model described in Sedra and Smith. Capacitances C_{Bx} and C_{jS} are zero because they represent capacitive coupling to the device substrate. Since the 2N2222A is a discrete transistor, constructed from a sandwich of three separate layers of semiconductor material, it has no substrate layer. Spice assumes that these substrate capacitances do not exist and sets them at zero.

In addition to printing the hybrid-pi model parameters, Spice computes and prints the values β_{dc}, β_{ac}, and f_T, which are related to the transistor's operating point according to

$$\beta_{dc} = \left.\frac{I_C}{I_B}\right|_{OP} \tag{7.2}$$

$$\beta_{ac} = \left.\frac{i_c}{i_b}\right|_{OP} = g_m r_\pi \tag{7.3}$$

and

$$f_T = \frac{1}{2\pi}\frac{g_m}{C_{BE} + C_{BC}} \tag{7.4}$$

This transistor has a DC current gain of 163 and an AC current gain of 178. It is important to note that in general these values are not alike but, more important, that they can be quite different from the value of β_F specified on the model statement in the input deck (i.e., 255.9). This arises because the dependence of the transistor current gain on collector current is captured by the Spice model for the 2N2222A. The β_F value specified on a BJT model statement represents the peak value of β_{dc}. The unity-gain bandwidth f_T for this transistor is estimated to be about 150 MHz.

Similar ideas extend to JFETs, MOSFETs, MESFETs, and junction diodes.

7.3

The Low-Frequency Response of the Common-Source Amplifier

In this section we investigate the low-frequency response of the classical common-source JFET amplifier in Fig. 7.7. The component values for this circuit were selected in Example 7.6 of Sedra and Smith and are such that the amplifier's low-frequency response is dominated by a pole at 100 Hz and the nearest pole or zero will be at least a decade away at a lower frequency. We shall use Spice to verify whether this is accurate.

The input file describing this circuit is shown in Fig. 7.8. The device parameters of the n-channel JFET are set as follows: $V_{t0} = -2$ V, $\beta = 2$mA/V^2 (β is the parameter K in Sedra and Smith such that $K = I_{DSS}/V_P^2$), and $\lambda = 0$ V^{-1}. These device parameters correspond to those used by Sedra and Smith in Example 7.6, that is, $V_P = -2$ V, $I_{DSS} = 8$ mA, and $r_o = \infty$. The amplifier input is excited by a 1 V AC input signal. In order to compute the amplifier's small-signal frequency response, an AC frequency sweep command orders a logarithmic sweep of 10 points per decade, varying the input

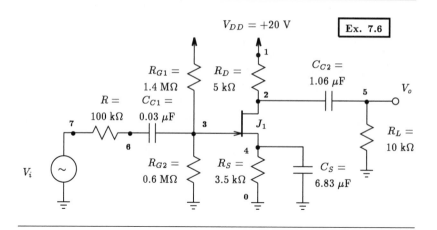

Figure 7.7 Capacitively coupled common-source JFET amplifier.

```
Frequency Response of a JFET Amplifier

** Circuit Description **
* power supply
Vdd 1 0 DC +20V
* input signal and source resistance
Vi 7 0 AC 1V
R 7 6 100k
* JFET amplifier circuit
J1 2 3 4 n_jfet
Rg1 1 3 1.4Meg
Rg2 3 0 0.6Meg
Rd 1 2 5k
Rs 4 0 3.5k
* by pass and coupling capacitors
Cs 4 0 6.83uF
Cc1 6 3 0.03uF
Cc2 2 5 1.06uF
* load
Rl 5 0 10k
* n-channel JFET model statement
.model n_jfet NJF (beta=2m Vto=-2V lambda=0)
** Analysis Requests **
.OP
.AC DEC 10 1Hz 1kHz
** Output Requests **
.PLOT AC VdB(5)
.probe
.end
```

Figure 7.8 Spice input file for computing the low-frequency response behavior of the capacitively coupled common-source JFET amplifier shown in Fig. 7.7.

frequency of excitation from 1 Hz to 1 kHz. A corresponding plot command will then plot the magnitude of the output voltage as a function of frequency. Also included in this input file is an .OP command. We leave it for the reader to confirm that the amplifier is operating correctly in its linear region (i.e., J_1 is in pinch-off).

The frequency response behavior of the JFET amplifier, as calculated by Spice, is shown in Fig. 7.9. The midband voltage gain is found to be +20.64 dB, and the 3 dB frequency is located very near 100 Hz. Interestingly enough, these values correspond almost exactly to the desired design values in Example 7.6 of Sedra and Smith.

The magnitude response of the JFET amplifier shown in Fig. 7.9 does not have a simple one-pole response; instead, it increases at a rate of +40 dB/dec for low frequencies, much like a two-pole response. When the frequency reaches about 10 Hz, the rate of rise in magnitude decreases to +20 dB/dec, which continues until 100 Hz. Then the magnitude response levels off at a constant +20.64 dB for increasing frequencies. This is clarified by the three straight lines superimposed on the graph to highlight the response changes.

It appears that the dominant pole is located at about 100 Hz, and the nondominant pole is located at 10 Hz. The effect of the zero (putting aside the two zeros at DC) is nullified by the presence of another pole located quite close to it. We thus confirm that the dominant pole and the next closest pole or zero are approximately one decade apart.

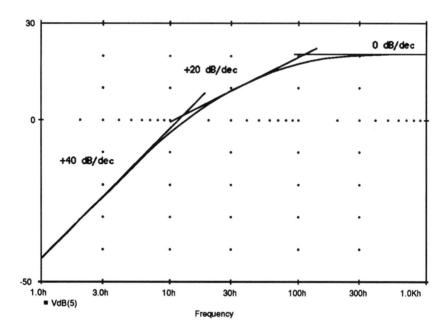

Figure 7.9 The low-frequency magnitude response of the capacitively coupled common-source JFET amplifier shown in Fig. 7.7. Three straight lines having slopes of +40 dB/dec, +20 dB/dec, and 0 dB/dec are superimposed on the magnitude response of the amplifier to illustrate the approximate locations of the different break frequencies.

7.4

High-Frequency Response Comparison of the Common-Emitter and Cascode Amplifiers

Figure 7.10 illustrates two different styles of high-frequency amplifier configuration: the common-emitter amplifier and the cascode amplifier. We use Spice to analyze the high-frequency operation of these two amplifier configurations, to confirm the formulas used to estimate amplifier midband gain and 3 dB bandwidth, and in the process to determine which amplifier configuration is more suited for high-frequency operation.

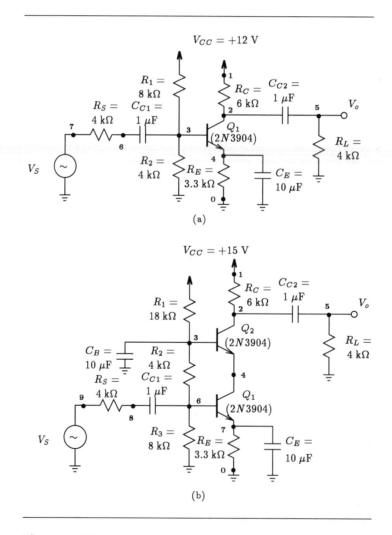

(a)

(b)

Figure 7.10 Two high-frequency amplifier configurations: (a) common-emitter and (b) cascode.

For true comparison we use identical transistors in both amplifiers and model them after the commercially available 2N3904 *npn* transistor. Furthermore, we note that reasonable comparisons can be made between the two amplifier types because their load and source resistances are identical.

The particulars of the two amplifier designs shown in Fig. 7.10, specifically the midband gain and 3 dB frequencies, were presented by Sedra and Smith in their Sections 7.6 and 7.7. In Table 7.2 we summarize the hand-analysis formulas; we plan to verify their accuracy by comparing their results with those computed by Spice.

Figures 7.11 and 7.12 show the input listings for the two amplifier configurations in Fig. 7.10, with the appropriate model parameters for the 2N3904 *npn* transistor. The operating point command (.OP) is included to obtain the small-signal model parameters for the 2N3904 transistor. The AC analysis command (.AC) is necessary to obtain the frequency response behavior of the amplifier. To compare results, we concatenate the two input files before submitting them to Spice so that the Probe facility of PSpice can display the frequency behavior of the two amplifiers on the same graph.

The magnitude responses from the Spice AC analysis are displayed in Fig. 7.13. We can see that both amplifiers have very similar low-frequency responses, but quite different high-frequency behavior. Each amplifier has a lower 3 dB frequency of about 436 Hz and a midband gain of approximately 28 dB. The common-emitter amplifier's upper 3 dB frequency is limited to 575 kHz, while the cascode amplifier has an upper 3 dB frequency of more than 5.8 MHz. Clearly then, the cascode amplifier has a 3 dB bandwidth that is more than 10 times that of the common-emitter amplifier and would be preferred as a wideband amplifier.

Table 7.3 compares the midband gain and 3 dB frequencies of each amplifier as calculated using the equations from Table 7.2 and as computed by exact analysis using Spice. In the hand calculations, we use the small-signal model parameters generated by Spice for each transistor of each amplifier. The actual operating-point information found in the Spice output file is as follows:

Common-Emitter Amplifier		Cascode Amplifier		
NAME	Q1	NAME	Q1	Q2
MODEL	Q2N3904	MODEL	Q2N3904	Q2N3904
IB	7.34E-06	IB	7.29E-06	7.08E-06
IC	9.97E-04	IC	9.80E-04	9.73E-04
VBE	6.65E-01	VBE	6.65E-01	6.64E-01
VBC	-2.04E+00	VBC	-1.33E+00	-3.25E+00
VCE	2.70E+00	VCE	1.99E+00	3.91E+00
BETADC	1.36E+02	BETADC	1.35E+02	1.37E+02
GM	3.80E-02	GM	3.74E-02	3.71E-02
RPI	4.10E+03	RPI	4.13E+03	4.25E+03
RX	1.00E+01	RX	1.00E+01	1.00E+01
RO	7.63E+04	RO	7.69E+04	7.94E+04
CBE	1.79E-11	CBE	1.77E-11	1.76E-11
CBC	2.43E-12	CBC	2.66E-12	2.17E-12
CBX	0.00E+00	CBX	0.00E+00	0.00E+00
CJS	0.00E+00	CJS	0.00E+00	0.00E+00
BETAAC	1.56E+02	BETAAC	1.54E+02	1.58E+02
FT	2.97E+08	FT	2.92E+08	2.98E+08

Table 7.2 General expressions for estimating the midband gain (A_M) and 3 dB frequencies (ω_L and ω_H) of the common-emitter and cascode amplifiers as derived by Sedra and Smith in Sections 7.6 and 7.7 of their text.

Common-Emitter Amplifier

Midband Gain

$$A_M = -\frac{R_{in}}{R_{in} + R_S} g_m R_L'$$

where

$R_{in} = R_1 \parallel R_2 \parallel (r_\pi + r_x)$

$R_L' = R_C \parallel R_L \parallel r_o$

3 dB Frequencies

Lower

$$\omega_L \approx \frac{1}{C_{C_1} R_{C_1}} + \frac{1}{C_E R_E'} + \frac{1}{C_{C_2} R_{C_2}}$$

where

$R_{C_1} = R_S + [R_1 \parallel R_2 \parallel (r_\pi + r_x)]$

$R_E' = R_E \parallel \left(\frac{(R_1 \parallel R_2 \parallel R_S) + r_\pi + r_x}{\beta_{AC} + 1} \right)$

$R_{C_2} = R_L + (R_C \parallel r_o)$

Upper

$$\omega_H \approx \frac{1}{(R_S \parallel R_{in})[C_\pi + C_\mu(1 + g_m R_L')]}$$

where

$R_{in} = R_1 \parallel R_2 \parallel (r_\pi + r_x)$

$R_L' = R_C \parallel R_L \parallel r_o$

Cascode Amplifier

Midband Gain

$$A_M = -\frac{R}{R + R_S} g_m R_L' \times \frac{r_\pi}{r_\pi + r_x + R_2 \parallel R_3 \parallel R_S}$$

where

$R = R_2 \parallel R_3$

$R_L' = R_C \parallel R_L$

3 dB Frequencies

Lower

$$\omega_L \approx \frac{1}{C_{C_1} R_{C_1}} + \frac{1}{C_E R_E'} + \frac{1}{C_{C_2} R_{C_2}}$$

where

$R_{C_1} = R_S + [R_2 \parallel R_3 \parallel (r_\pi + r_x)]$

$R_E' = R_E \parallel \left(\frac{(R_2 \parallel R_3 \parallel R_S) + r_\pi + r_x}{\beta_{AC} + 1} \right)$

$R_{C_2} = R_L + R_C$

Upper

$$\omega_H \approx \frac{1}{R_S'(C_{\pi_1} + 2C_{\mu_1})}$$

where

$R_S' = r_{\pi_1} \parallel [r_{x_1} + R_2 \parallel R_3 \parallel R_S]$

```
A Common-Emitter Amplifier

** Circuit Description **
* power supplies
Vcc 1 0 DC +12V
* input signal source
Vs 7 0 AC 1V
Rs 7 6 4k
* CE stage
Ccl 6 3 1uF
R1 1 3 8k
R2 3 0 4k
Q1 2 3 4 Q2N3904
Re 4 0 3.3k
Rc 1 2 6k
Ce 4 0 10uF
Cc2 2 5 1uF
* output load
Rl 5 0 4k
*
* transistor model statement for 2N3904
.model Q2N3904   NPN (Is=6.734f Xti=3 Eg=1.11 Vaf=74.03 Bf=416.4 Ne=1.259
+                     Ise=6.734f Ikf=66.78m Xtb=1.5 Br=.7371 Nc=2 Isc=0 Ikr=0 Rc=1
+                     Cjc=3.638p Mjc=.3085 Vjc=.75 Fc=.5 Cje=4.493p Mje=.2593 Vje=.75
+                     Tr=239.5n Tf=301.2p Itf=.4 Vtf=4 Xtf=2 Rb=10)
** Analysis Requests **
.OP
.AC DEC 10 1Hz 100MegHz
** Output Requests **
.PLOT AC VdB(5)
.probe
.end
```

Figure 7.11   Spice input file for computing the frequency response of the common-emitter amplifier circuit shown in Fig. 7.10(a).

The last column in Table 7.3 expresses the percentage error in the hand calculation as compared with the Spice results. Clearly, hand analysis provides reasonable estimates of the amplifier's midband gain and 3 dB frequencies with at most an 11.8% error.

Although the preceding hand analysis resulted in very reasonable numbers compared with Spice results, it should be pointed out that Spice is especially useful in several ways. Spice can provide more accurate results than are possible by hand

```
Cascode Amplifier

** Circuit Description **
* power supplies
Vcc 1 0 DC +15V
* input signal source
Vs 9 0 AC 1V
Rs 9 8 4k
* CE stage (input stage)
Cc1 6 8 1uF
R1 1 3 18k
R2 3 6 4k
R3 6 0 8k
Q1 4 6 7 Q2N3904
Re 7 0 3.3k
Ce 7 0 10uF
* CB stage (upper stage)
Q2 2 3 4 Q2N3904
Rc 1 2 6k
Cb 3 0 10uF
Cc2 2 5 1uF
* output load
Rl 5 0 4k
*
* transistor model statement for 2N3904
.model Q2N3904  NPN (Is=6.734f Xti=3 Eg=1.11 Vaf=74.03 Bf=416.4 Ne=1.259
+                    Ise=6.734f Ikf=66.78m Xtb=1.5 Br=.7371 Nc=2 Isc=0 Ikr=0 Rc=1
+                    Cjc=3.638p Mjc=.3085 Vjc=.75 Fc=.5 Cje=4.493p Mje=.2593 Vje=.75
+                    Tr=239.5n Tf=301.2p Itf=.4 Vtf=4 Xtf=2 Rb=10)
** Analysis Requests **
.OP
.AC DEC 10 1Hz 100MegHz
** Output Requests **
.PLOT AC VdB(5)
.probe
.end
```

Figure 7.12 Spice input file for computing the frequency response of the cascode
amplifier circuit shown in Fig. 7.10(b).

analysis. Spice provides an easy way to investigate component trade-offs by allowing the designer to simply change component values and observe the effect on the
frequency response. Unlike hand analysis, Spice is not limited by the complexity of
the circuit, although we must be sensible or we end up swamped in results that are
difficult to interpret or to derive any design insight from.

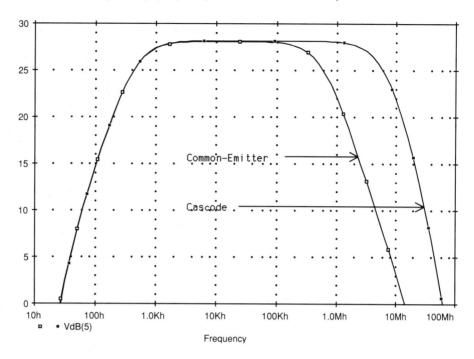

Figure 7.13   Comparison of the frequency response of the two amplifier configurations presented in Fig. 7.10.

Table 7.3   The midband gain and 3 dB frequencies of the common-emitter and cascode amplifiers as computed by hand and as obtained by exact analysis using Spice. The rightmost column presents the relative error (in percentage) between the value predicted by hand and by Spice.

Amplifier Type	Amplifier Parameter	Hand Analysis	Spice	% Error
Common-emitter	A_M	−25.5 V/V	−25.4 V/V	−0.2
	f_L	487.4 Hz	436 Hz	−11.8
	f_H	587.6 kHz	575 kHz	−2.2
Cascode	A_M	−25.8 V/V	−25.6 V/V	−0.9
	f_L	478.8 Hz	436 Hz	−9.8
	f_H	5.97 MHz	5.84 MHz	−2.2

7.5

High-Frequency Response of the CC-CE Amplifier

An example of another important high-frequency amplifier configuration, the common-collector common-emitter (CC-CE) cascade, is shown in Fig. 7.14. Here, a

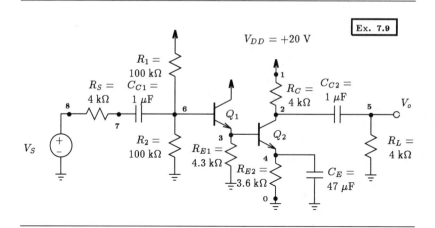

Figure 7.14 Common-collector common-emitter cascade amplifier configuration.

common-collector stage isolates the source resistance R_S from the Miller capacitance of the common-emitter stage, thus extending the high-frequency operation. With Spice, we shall analyze the high-frequency operation of this amplifier configuration and check Sedra and Smith's formulas used for hand estimates of the amplifier's midband gain and 3 dB frequencies. For easy reference, we summarize these hand calculations in Table 7.4.

The input file describing Fig. 7.14's amplifier is listed in Fig. 7.15. Each transistor is modeled after the commercially available 2N3904 transistor. Both a DC operating point and an AC analysis command are included in this Spice deck.

The frequency response of the CC-CE amplifier is shown in Fig. 7.16. Using PSpice's Probe facility, we find that the midband gain is 35.8 dB or 61.9 V/V and that the lower and upper 3 dB frequencies of this amplifier are 122 Hz and 9.2 MHz, respectively. This magnitude plot also shows signs of high-frequency resonance (i.e., magnitude peaking). As we shall see when we compare our hand analysis results with those obtained using Spice, this resonant peaking extends the upper 3 dB frequency of this amplifier somewhat. The extension is not accounted for by the simple theory from Sedra and Smith. The gain-bandwidth product of this amplifier is about 570 MHz, which is the largest of any of the amplifiers seen so far. (The CE and cascode stage presented in the last section had gain-bandwidth products of 14.6 MHz and 149 MHz, respectively.)

To investigate the accuracy of the expressions in Table 7.4 for the CC-CE amplifier stage, we calculate the midband gain and 3 dB frequencies and compare them

Table 7.4 General expressions for estimating the midband gain (A_M) and 3 dB frequencies (ω_L and ω_H) of the common-collector common-emitter cascade amplifier as derived by Sedra and Smith in Sections 7.3 and 7.9 of their text.

Common-Collector Common-Emitter Cascade Amplifier

Midband Gain

$$A_M = -\frac{R_{\text{in}}}{R_{\text{in}} + R_S} g_{m_2} R_L' \times \frac{R_{E_1} \| r_{\pi_2}}{(R_{E_1} \| r_{\pi_2}) + r_{e_1}}$$

where

$$R_{\text{in}} = R_1 \| R_2 \| R_{B_1}'$$

$$R_{B_1}' = r_{\pi_1} + (\beta_1 + 1)(R_{E_1} \| r_{\pi_2})$$

$$R_L' = R_C \| R_L \| r_o$$

3 dB Frequencies

Lower

$$\omega_L \approx \frac{1}{C_{C_1} R_{C_1}} + \frac{1}{C_E R_E'} + \frac{1}{C_{C_2} R_{C_2}}$$

where

$$R_{C_1} = R_S + (R_1 \| R_2 \| R_{B_1}')$$

$$R_{B_1}' = r_{\pi_1} + (\beta_1 + 1)(R_{E_1} \| r_{\pi_2})$$

$$R_E' = R_{E_2} \| \left(\frac{R_{E_1}'}{\beta_2 + 1} \right)$$

$$R_{E_1}' = \left(\frac{(R_1 \| R_2 \| R_S) + r_{\pi_1}}{\beta_1 + 1} \right) \| R_{E_1}$$
$$+ r_{\pi_2}$$

$$R_{C_2} = R_L + (R_C \| r_{o_2})$$

Upper

$$\omega_H \approx \frac{1}{C_{\mu_1} R_{\mu_1} + C_{\pi_1} R_{\pi_1} + C_T R_T + C_{\mu_2} R_L'}$$

where

$$R_{\mu_1} = R_S \| R_{\text{in}}$$

$$R_{\text{in}} = R_1 \| R_2 \| R_{B_1}'$$

$$R_{B_1}' = r_{\pi_1} + (\beta_1 + 1)(R_{E_1} \| r_{\pi_2})$$

$$R_{E_1}' = R_{E_1} \| r_{\pi_2}$$

$$R_{\pi_1} = r_{\pi_1} \| \left(\frac{R_S' + R_{E_1}'}{1 + g_{m_1} R_{E_1}'} \right)$$

$$R_S' = R_S \| R_1 \| R_2$$

$$R_L' = R_C \| R_L$$

$$R_T = R_{E_1}' \| \left(\frac{R_S' + r_{\pi_1}}{\beta_1 + 1} \right)$$

$$C_T = C_{\pi_2} + C_{\mu_2}(1 + g_{m_2} R_L')$$

```
Common-Collector Common-Emitter Cascade Amplifier

** Circuit Description **
* power supplies
Vcc 1 0 DC +10V
* input signal source
Vs 8 0 AC 1V
Rs 8 7 4k
* CE stage (input stage)
Cc1 6 7 1uF
R1 1 6 100k
R2 6 0 100k
Q1 1 6 3 Q2N3904
Re1 3 0 4.3k
* CC stage (output stage)
Q2 2 3 4 Q2N3904
Re2 4 0 3.6k
Rc 1 2 4k
Ce 4 0 47uF
Cc2 2 5 1uF
* output load
Rl 5 0 4k
*
* transistor model statement for 2N3904
.model Q2N3904  NPN (Is=6.734f Xti=3 Eg=1.11 Vaf=74.03 Bf=416.4 Ne=1.259
+                    Ise=6.734f Ikf=66.78m Xtb=1.5 Br=.7371 Nc=2 Isc=0 Ikr=0 Rc=1
+                    Cjc=3.638p Mjc=.3085 Vjc=.75 Fc=.5 Cje=4.493p Mje=.2593 Vje=.75
+                    Tr=239.5n Tf=301.2p Itf=.4 Vtf=4 Xtf=2 Rb=10)
** Analysis Requests **
.OP
.AC DEC 10 1Hz 100MegHz
** Output Requests **
.PLOT AC VdB(5)
.probe
.end
```

Figure 7.15   Spice input file for computing the frequency response of the common-collector common-emitter cascade amplifier circuit shown in Fig. 7.14.

with those in Table 7.5 obtained using Spice. For the hand calculations, we use the small-signal model parameters for Q_1 and Q_2 generated by Spice:

```
Common-Collector Common-Emitter Cascade Amplifier
```

NAME	Q1	Q2
MODEL	Q2N3904	Q2N3904
IB	6.66E-06	6.83E-06
IC	9.32E-04	9.22E-04
VBE	6.62E-01	6.63E-01

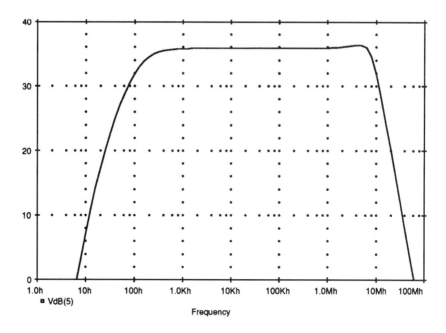

Figure 7.16 Magnitude response of the common-collector common-emitter cascade shown in Fig. 7.14 as computed by Spice.

VBC	-5.33E+00	-2.31E+00
VCE	6.00E+00	2.97E+00
BETADC	1.40E+02	1.35E+02
GM	3.55E-02	3.52E-02
RPI	4.52E+03	4.41E+03
RX	1.00E+01	1.00E+01
RO	8.52E+04	8.28E+04
CBE	1.72E-11	1.70E-11
CBC	1.91E-12	2.36E-12
CBX	0.00E+00	0.00E+00
CJS	0.00E+00	0.00E+00
BETAAC	1.61E+02	1.55E+02
FT	2.97E+08	2.88E+08

As can been seen from Table 7.5, the value of the midband gain calculated by hand is in good agreement with that obtained from Spice (-0.3% error). Both the lower and upper 3 dB frequencies (f_L and f_H) computed by the hand analysis differ from those computed by Spice with an absolute error of about 20%. We attribute most of the error in the upper 3 dB frequency to the high-frequency magnitude peaking that seems to extend the high-frequency operation of the CC-CE amplifier stage. The formula derived for the upper 3 dB frequency used in the hand analysis assumes that all circuit poles are real; magnitude peaking is an indication that some of the poles are complex.

Table 7.5 The midband gain and 3 dB frequencies of the common-collector common-emitter cascade amplifier as computed by hand and by exact analysis using Spice. The rightmost column presents the relative error (in percentage) between the value predicted by hand and by Spice.

Amplifier Type	Amplifier Parameter	Hand Analysis	Spice	% Error
Common-collector	A_M	−62.0 V/V	−61.9 V/V	−0.3
Common-emitter	f_L	143.1 Hz	122 Hz	− 17.3
Casade	f_H	7.17 kHz	9.2 MHz	+ 22.1

7.6

Spice Tips

- Spice is extremely useful for calculating the frequency response of amplifier circuits. Unlike hand analysis, Spice need not make any assumptions about the location of the poles or zeros, and thus provides exact (or as accurate as the model provides) frequency response behavior. Also, Spice is not limited by the circuit complexity.

- Spice not only provides more accurate results than hand analysis, it also allows the user an easy way of investigating component trade-offs in a design. Changing various component values in a known manner and observing the effects on the frequency response enables the designer to select the components that best suit the design requirements.

- PSpice allows the gain of a dependent source to be expressed as an arbitrary function of the complex frequency variable s.

- The gain of only a VCVS and a VCCS (designated with the letters E and G, respectively) can be represented in PSpice as a Laplace transform function. However, both these controlled sources can be used to describe the behavior of a CCVS or a CCCS using the special syntax of PSpice for describing a VCVS and a VCCS (see Fig. 7.1).

- If a circuit is excited by a 1 V AC input signal, the magnitude and phase of the voltage appearing across the output terminals computed by Spice are directly equal to the magnitude and phase of the transfer function of that circuit evaluated at the frequency of the input signal.

- An AC analysis of Spice is by definition a small-signal analysis. The level of the input AC signal will not alter this basic assumption. This is, of course, not true for a transient analysis in which the level of the input signal affects the linearity of the circuit.

- All capacitances used by Spice in the small-signal equivalent circuit of a transistor are available for the user to see. They are placed in the Spice output file under the heading *Operating Point Information* and are included among the small-signal parameters of each transistor.

7.7

Problems

7.1 If PSpice is available, use the Laplace transform function of either the voltage-controlled voltage or current source to plot the magnitude and phase response of the following transfer functions:

(a) $T(s) = \dfrac{10s}{(1 + s/10^1)(1 + s/10^4)}$

(b) $T(s) = \dfrac{10^4(1 + s/10^5)}{(1 + s/10^3)(1 + s/10^4)}$

(c) $T(s) = \dfrac{5s(1 + s/10^5)}{(1 + s/10^3)}$

(d) $T(s) = \dfrac{10^4}{(1 + s/10^2)(1 + s/10^4)(1 + s/10^6)}$

7.2 If the circuit of Fig. P7.2 has an ideal voltage amplifier of gain 100, use Spice to determine A_M, ω_L, and ω_H of this amplifier.

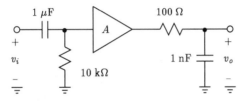

Figure P7.2

7.3 Figure P7.3 shows the high-frequency equivalent circuit of a common-source FET amplifier driven by a signal generator having a resistance of 100 kΩ. Using Spice, compute the upper 3 dB frequency f_H using the method of open-circuit time constants. Compare this result with that obtained directly from the input–output voltage transfer characteristics.

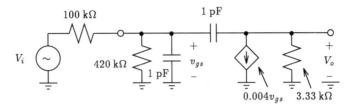

Figure P7.3

7.4 Using the circuit setup shown in Fig. P7.4, calculate the small-signal current gain i_c/i_b of the commercial 2N3904 *npn* transistor using Spice. The Spice model parameters for the 2N3904 can be found in the Spice listing of Fig. 7.11. What is the corresponding f_T of this transistor?

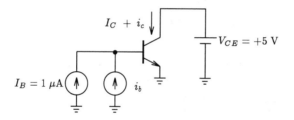

Figure P7.4

7.5 Using the circuit setup shown in Fig. P7.5, have Spice calculate the small-signal current gain i_d/i_g as a function of frequency for an n-channel MOSFET characterized by $V_t = 1$ V, $\mu_n C_{OX} = 20\mu A/V^2, L = 10\mu m, W = 30\mu m, \lambda = 0.04 V^{-1}, C_{GSO} = 1 pF,$ and $C_{GDO} = 1$ pF. What is the corresponding unity current gain frequency of this transistor?

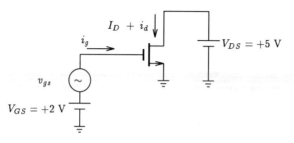

Figure P7.5

7.6 With the aid of Spice, determine the frequency of the pole and zero of the CMOS amplifier shown in Fig. P7.6 in the linear region of the amplifier. For M_1, assume the Spice parameters are $V_t = 1$ V, $\mu_n C_{OX} = 20 \, \mu A/V^2, L = 10 \, \mu m, W = 640 \, \mu m, \lambda = 0.02 \, V^{-1}$, $C_{GSO} = 1$ pF, and $C_{GDO} = 1$ pF. For M_2 and M_3, the Spice parameters are $V_t = -0.8$ V, $\mu_p C_{OX} = 10 \, \mu A/V^2, L = 10 \, \mu m, W = 1000 \, \mu m, \lambda = 0.02 \, V^{-1}, C_{GSO} = 1$ pF, and $C_{GDO} = 1$ pF.

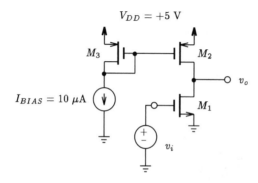

Figure P7.6

7.7 Consider the common-emitter amplifier of Fig. 7.10(a). With the aid of Spice, plot the amplifier's input impedance Z_{in} (magnitude and phase) as seen by the source excluding R_S.

7.8 For the common-emitter amplifier shown in Fig. 7.14, if C_E is increased to 30 μF, what happens to the bandwidth of this amplifier? Quantify your answer using Spice.

7.9 For a 10 kHz, 100 mV sine wave applied to the input of the CC-CE cascade amplifier shown in Fig. 7.10(b), using Spice plot the voltage waveform appearing at the output for at least one complete period of the input signal. Compare these results with those obtained for an input signal of the same frequency increased to 5 V.

7.10 For the emitter-follower shown in Fig. P7.10, with the aid of Spice, calculate the midband gain and the upper 3 dB frequency of the amplifier under the following source resistance conditions:
(a) $R_S = 1$ kΩ
(b) $R_S = 10$ kΩ
(c) $R_S = 100$ kΩ
Assume the transistor is modeled after the commercial 2N3904 *npn* transistor. In addition, for simulation purposes, replace the infinite-valued decoupling capacitor by a very large value of 10^6 pF.

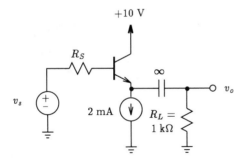

Figure P7.10

8

Feedback

In this chapter we demonstrate how Spice is used to determine the loop or transmission gain $A\beta$ of an arbitrary feedback circuit. Several examples from which the stability behavior of the network can be deduced are presented to illustrate this use of Spice. Further, we demonstrate how Spice can be used to aid the design of the frequency compensation network required to stabilize an unstable feedback network. Results in both the time and frequency domains illustrate the effectiveness of the selected frequency compensation.

8.1

The General Feedback Structure

The basic structure of a system including some form of negative feedback is shown in block diagram form in Fig. 8.1. Here the block depicted by A represents the feedforward gain stage of the closed-loop system, and the block denoted by β represents

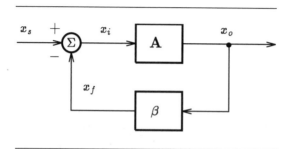

Figure 8.1   General form of a negative feedback structure.

the feedback network. The summing node is used to subtract the feedback signal x_f from the input signal x_s to create what is known as the error signal x_i. This error signal is then applied to the input of the feedforward gain stage, thus creating the output signal x_o. The closed-loop input–output signal gain A_f can be expressed in terms of A and β according to

$$A_f = \frac{A}{1 + A\beta} \tag{8.1}$$

8.2

Determining Loop Gain with Spice

In this section we demonstrate two Spice methods for obtaining the loop gain of feedback amplifiers of the type shown in Fig. 8.1. Both methods are based on breaking the loop at an appropriate place, injecting a signal into the loop, and measuring the returned signal.

To demonstrate this, consider the feedback amplifier circuit shown in Fig. 8.2. In principle, any point in the loop containing the feedforward and feedback networks can be broken. However, the circuit conditions of the loop must remain the same as when the loop was closed. This includes maintaining the same bias conditions on each transistor and terminating the feedback loop with the impedance seen by the loop under closed-loop conditions.

For the circuit shown in Fig. 8.2 a convenient location for breaking the feedback loop is in the AC coupled feedback network. By doing so the DC bias conditions

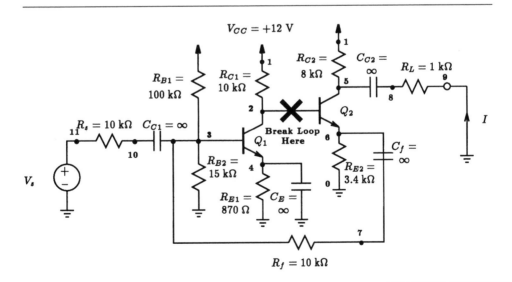

Figure 8.2 Breaking the feedback loop of an amplifier at a point in the network that affects its DC bias.

of each transistor are not disturbed. However, in many circuit situations the feedback network is directly coupled, and therefore a method that is amenable to such situations is required.

Consider breaking the feedback loop at a location that disturbs the DC biasing of the circuit. Such a location is highlighted by a big **X** in the circuit diagram of Fig. 8.2. For Spice, we insert a very large inductor of 1 GH in series with the terminals of the break. For DC, the inductor behaves as a short circuit, and the DC bias conditions of the transistors would not change. For AC signals, the large inductor presents a large impedance in series with the loop, which for all intents and purposes opens it.

To find the impedance required to terminate the loop, we calculate the impedance seen looking into the terminal where the loop was broken. Here we apply a 1 V AC voltage signal V_t to the base of Q_2 through a DC blocking capacitor as shown in Fig. 8.3(a) and calculate the current supplied by this source. The ratio V_t/I_t then gives the input impedance. Of course these results will be frequency dependent, so we limit our discussion to midband frequencies where the input impedance is purely resistive. In this example, a frequency of 1 Hz is considered to be midband. The DC blocking capacitor is necessary to prevent the input voltage source from disturbing the DC bias conditions on Q_2. Its value should be made large so as to minimize the voltage drop across it (for most purposes, a value of 1 GF should suffice).

The input file for determining the input impedance of the feedback loop is shown in Fig. 8.4. Spice produces the following results in the output file:

```
FREQ         VM(22)       VP(22)       IM(Vt)       IP(Vt)
1.000E+00    1.000E+00    0.000E+00    2.482E-06    1.800E+02
```

from which we can determine that the input resistance is 402.9 kΩ.

The output of the feedback loop can now be terminated with a resistance of 402.9 kΩ. A DC blocking capacitor is necessary once again to prevent this resistance from disturbing the amplifier's DC biasing as illustrated in the circuit diagram in Fig. 8.3(b).

To calculate the loop gain, we modify the input file given in Fig. 8.4 by attaching this resistance and capacitor combination to the output of the feedback loop by adding the following two element statements:

```
Cto 2 23 1GF
Rt 23 0 402.9k
```

and revising the output print statement as follows:

```
.PRINT AC Vm(22) Vp(22) Vm(23) Vp(23)
```

The results are then found in the output file:

```
FREQ         VM(22)       VP(22)       VM(23)       VP(23)
1.000E+00    1.000E+00    0.000E+00    5.954E+01    1.800E+02
```

Hence, one finds the loop gain $A\beta = -V_r/V_t = 59.54$.

8.2.1 An Alternative Method

An alternative method for determining loop gain, and one that better lends itself to Spice, is attributed to Rosenstark [Rosenstark, 1986]. We will not supply proof, but the

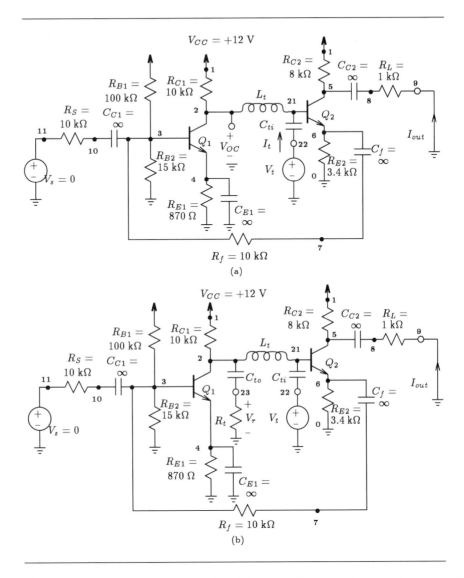

Figure 8.3 (a) Opening a feedback loop with a very large inductor L_t. The impedance seen by the feedback loop in which we must terminate the loop is found by calculating the current supplied by the AC coupled test voltage source V_t. (b) Feedback loop terminated in AC coupled load resistance R_t.

Rosenstark method determines the loop gain by two independent calculations. First, we calculate the voltage transfer function from the input of the broken loop to its output with the loop unterminated (i.e., $R_t = \infty$). This we denote as the open-circuit transfer function, T_{OC}. Second, we calculate the current transfer function from the input of the loop to its short-circuited output. This we call the short-circuit transfer function, T_{SC}.

Computing the Loop Terminating Resistance Rt

```
** Circuit Description **
* input source set to zero for output impedance calculation
Vs 11 0 DC 0 AC 0
Rs 11 10 10k
* short-circuit output port
Vout 0 9 0
* power supply
Vcc 1 0 DC +12V
* amplifier circuit
* 1st stage
Rc1 1 2 10k
Q1 2 3 4 Q2N3904
Re1 4 0 870
Ce1 4 0 1GF
Rb1 1 3 100k
Rb2 3 0 15k
* 2nd stage
Rc2 1 5 8k
Q2 5 21 6 Q2N3904
* decoupling capacitors
Cc1 10 3 1GF
Cc2 5 8 1GF
* load
Rl 8 9 1k
* feedback circuit
Re2 6 0 3.4k
Rf 3 7 10k
Cf 6 7 1GF
* inject signal into feedback loop without disturbing DC bias
Lt 2 21 1GH
Cti 21 22 1GF
Vt 22 0 AC 1V
* transistor model statement for 2N3904
.model Q2N3904  NPN (Is=6.734f Xti=3 Eg=1.11 Vaf=74.03 Bf=416.4 Ne=1.259
+                    Ise=6.734f Ikf=66.78m Xtb=1.5 Br=.7371 Nc=2 Isc=0 Ikr=0 Rc=1
+                    Cjc=3.638p Mjc=.3085 Vjc=.75 Fc=.5 Cje=4.493p Mje=.2593 Vje=.75
+                    Tr=239.5n Tf=301.2p Itf=.4 Vtf=4 Xtf=2 Rb=10)
** Analysis Requests **
.AC LIN 1 1Hz 1Hz
** Output Requests **
* print resistance seen by Vt: Rt=V(22)/I(Vt)
.PRINT AC Vm(22) Vp(22) Im(Vt) Ip(Vt)
.end
```

Figure 8.4 Spice input file for calculating the input impedance $R_t = V(22)/I(Vt)$ of the feedback loop for the amplifier circuit shown in Fig. 8.3(a).

Figure 8.5 shows these two situations for an arbitrary broken feedback loop. The two transfer functions are then combined to compute the loop transmission as follows:

$$A\beta = -T_{OC} \| T_{SC} = -\frac{1}{1/T_{OC} + 1/T_{SC}} \tag{8.2}$$

It is obvious from this formula that the smaller of the two transfer functions will dominate the loop gain.

The advantage of this approach over the previous method is that both T_{OC} and T_{SC} can be calculated over a wide range of frequencies without worrying whether the loop is terminated with the correct impedance.

To demonstrate we shall return to the opened feedback loop of the last example in Fig. 8.3(a). Here the loop is shown unterminated and driven by a voltage source V_t. The input file for this circuit is given in Fig. 8.4, which can be used to calculate the open-circuit voltage V_{OC}. Adding the following print statement to this file:

```
.PRINT AC Vm(22) Vp(22) Vm(2) Vp(2),
```

and resubmitting to Spice will result in the following open-circuit voltage in the output file:

```
FREQ          VM(22)        VP(22)        VM(2)        VP(2)
1.000E+00     1.000E+00     0.000E+00     6.086E+01    1.800E+02
```

Thus we obtain $T_{OC} = -60.86$.

To determine the short-circuit transfer function, we must remove the input voltage source V_t and replace it with an AC current source, and we must also AC short-circuit the output port by adding a zero-valued voltage source in series with a large capacitor. The zero-valued voltage source provides the means of monitoring the

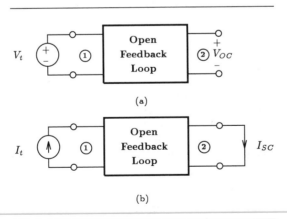

(a)

(b)

Figure 8.5 (a) Determining the open-circuit voltage transfer function $T_{OC} = V_{OC}/V_t$. (b) Determining the short-circuit current transfer function $T_{SC} = I_{SC}/I_t$.

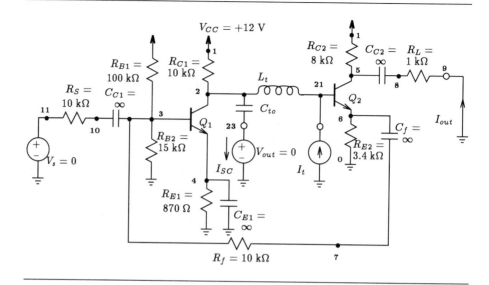

Figure 8.6 Determining the short-circuit current transfer function $T_{SC} = I_{SC}/I_t$.

short-circuit current. Figure 8.6 depicts this circuit, and Fig. 8.7 is the corresponding input file. Submitting this file to Spice results in the following short-circuit output current:

```
FREQ          IM(Vsc)        IP(Vsc)
1.000E+00     2.753E+03      1.800E+02
```

Therefore we get $T_{SC} = -2753$.

From Eq. (8.2), we calculate the loop gain to be 59.54, which is exactly the same as the loop gain calculated by the method that terminates the loop.

8.3

Stability Analysis Using Spice

From the loop gain, or more appropriately the loop transmission, we can determine the stability of a feedback amplifier. Consider the filter circuit in Fig. 8.8(a) where the feedback loop is frequency dependent and the amplifier has unity gain ($K = 1$). The most convenient point for breaking the loop is probably at the amplifier's input because the loop is not loaded by the amplifier input and therefore does not need to be terminated when it is opened. The loop transmission is then found by calculating the return voltage V_r, as indicated in Fig. 8.8(b), with a 1 V AC signal applied to the amplifier input. The sign of the input signal is made negative so that the loop transmission $A\beta$ equals V_r (instead of $-V_r$). In Fig. 8.9's Spice description of the Fig. 8.8(b) circuit, we are requesting an AC analysis of the circuit beginning at 1 Hz and extending up to 1 MHz. This frequency range should be large enough to capture the important behavior of the loop transmission.

```
Computing the Short-Circuit Current Transfer Function

** Circuit Description **
* power supply
Vcc 1 0 DC +12V
* input source set to zero for output impedance calculation
Vs 11 0 DC 0 AC 0
Rs 11 10 10k
* amplifier circuit
* 1st stage
Rc1 1 2 10k
Q1 2 3 4 Q2N3904
Re1 4 0 870
Ce1 4 0 1GF
Rb1 1 3 100k
Rb2 3 0 15k
* 2nd stage
Rc2 1 5 8k
Q2 5 21 6 Q2N3904
* decoupling capacitors
Cc1 10 3 1GF
Cc2 5 8 1GF
* load
Rl 8 9 1k
* output current
Vout 0 9 DC 0
* feedback circuit
Re2 6 0 3.4k
Rf 3 7 10k
Cf 6 7 1GF
* inject signal into feedback loop
Lt 2 21 1GH
It 0 21 AC 1A
Cto 2 23 1GF
Vsc 23 0 0
* transistor model statement for 2N3904
.model Q2N3904  NPN (Is=6.734f Xti=3 Eg=1.11 Vaf=74.03 Bf=416.4 Ne=1.259
+                    Ise=6.734f Ikf=66.78m Xtb=1.5 Br=.7371 Nc=2 Isc=0 Ikr=0 Rc=1
+                    Cjc=3.638p Mjc=.3085 Vjc=.75 Fc=.5 Cje=4.493p Mje=.2593 Vje=.75
+                    Tr=239.5n Tf=301.2p Itf=.4 Vtf=4 Xtf=2 Rb=10)
** Analysis Requests **
.AC LIN 1 1Hz 1Hz
** Output Requests **
* print short-circuit current gain
.PRINT AC Im(Vsc) Ip(Vsc)
.end
```

Figure 8.7   Spice input file for calculating the short-circuit transfer function T_{SC} = $I(V_{SC})/I_t$ in the opened feedback loop shown in Fig. 8.6.

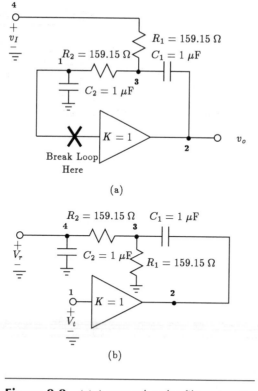

(a)

(b)

Figure 8.8 (a) A second-order filter circuit with a frequency-dependent feedback loop. (b) Opened feedback loop for calculating loop transmission.

```
Calculating Loop Transmission of Second-Order Active Filter (K=1)

** Circuit Description **
* input signal source
Vt 1 0 AC 1V 180degrees
* filter circuit
Egain 2 0 1 0 1
Inull 1 0 0 ; redundant connection
* feedback network
C1 2 3 1uF
R1 3 0 159.15
R2 3 4 159.15
C2 4 0 1uF
** Analysis Requests
.AC DEC 10 1Hz 1MegHz
** Output Requests **
.PLOT AC VdB(4) Vp(4)
.PROBE
.end
```

Figure 8.9 Spice input file for calculating the frequency-dependent loop transmission of the filter circuit shown in Fig. 8.8(b).

Spice produces plots of the magnitude and phase of the loop transmission in the output file, as drawn by the Probe facility of PSpice and displayed in Fig. 8.10(a). We see that the gain margin is $\simeq 10$ dB, which suggests that the filter circuit will be stable when the loop is closed. In Fig. 8.10(b) we use these results to draw a Nyquist plot and show that the critical point $(-1, 0)$ is not encircled, thus confirming our intuition.

We wish to consider the unstable filter circuit, so we increase the gain of the amplifier in the filter circuit of Fig. 8.8(a) to 5 (i.e., $K = 5$). Modifying the input file of Fig. 8.9 to reflect this change and rerunning this file will result in the new Bode plot in Fig. 8.11(a). Here we observe that at the $180°$ phase frequency, the loop gain will be greater than unity, indicating unstable performance. A Nyquist plot can also be drawn from the same frequency information, as shown in Fig. 8.11(b). We see that the critical point $(-1, 0)$ is encircled, which also confirms that the filter circuit will be unstable when the loop is closed.

It would be highly instructive to investigate the closed-loop transient behavior of these stable and unstable filter circuits. The input file in Fig. 8.12 provides for the appropriate analysis of the original filter circuit seen in Fig. 8.8(a) with $K = 1$. A 1 V peak amplitude sine-wave signal of 1 kHz frequency is applied to the input filter terminal using the following transient voltage source statement:

```
Vi 4 0 sin( 0V 1V 1kHz )
```

A transient analysis is to be performed over five periods of the input signal (i.e., 5 ms) with 100 points collected per period. The command appears in the input file:

```
.TRAN 10us 5ms 0ms 10us
```

Finally, a .PLOT statement is added for both the input and output voltage signals:

```
.PLOT TRAN V(2) V(4)
```

Another input file is created for the unstable filter circuit ($K = 5$) that is almost identical to Fig. 8.12 except that the amplifier gain is changed to 5. This file is added to the end of the file in Fig. 8.12. Spice produces the transient responses of both filter circuits as shown in Fig. 8.13. The top curve illustrates the sine-wave signal applied to the input of each filter circuit. The middle waveform represents the signal at the output of the stable filter circuit and, except for an amplitude and phase shift, is identical to the input signal shown above it. In contrast, the signal at the output of the unstable filter, which is shown in the bottom graph, bears no resemblance to the input sine-wave signal at all. Moreover, the output level has approached 10^{10} V in about 3 ms. Clearly, this is unstable behavior. If Spice were left running much longer than 3 ms, numbers would get so large that they would overflow the internal registers of the computer, causing Spice to terminate the job.

Novice users sometimes think that if Spice can produce a frequency response plot of a closed-loop circuit, then that circuit must be stable. This certainly is not true. To demonstrate, Fig. 8.14 shows the magnitude and phase responses of both the stable and the unstable closed-loop filter circuits. The responses were computed using a Spice

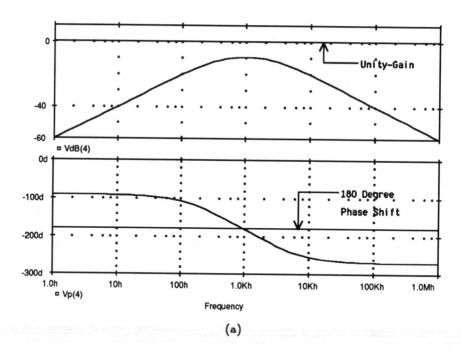

(a)

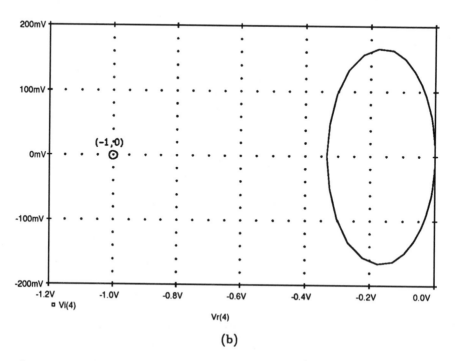

(b)

Figure 8.10   Various forms of representing the loop transmission: (a) Bode plot (b) Nyquist plot.

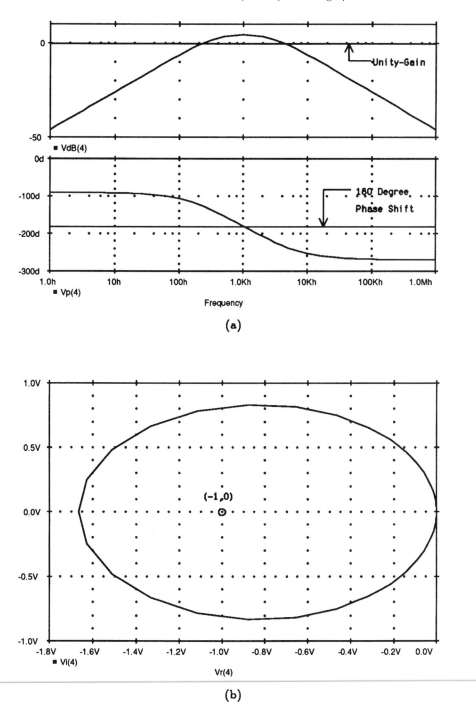

Figure 8.11   (a) Bode and (b) Nyquist plots of the loop transmission of an unstable filter ($K = 5$). Here the Nyquist plot encircles the critical point of $(-1, 0)$.

```
Closed-Loop Transient Response of Stable Filter Circuit (K=1)

** Circuit Description **

* input signal source
Vi 4 0 sin( 0V 1V 1kHz )

* filter circuit
Egain 2 0 1 0 1
* feedback stage
C1 2 3 1uF
R1 3 4 159.15
R2 3 1 159.15
C2 1 0 1uF

** Analysis Requests
.TRAN 10us 5ms 0ms 10us
** Output Requests **
.PLOT TRAN V(2) V(4)
.probe
.end
```

Figure 8.12 Spice input file for calculating the transient response of the filter circuit shown in Fig. 8.8(a) to a 1 V, 1 kHz sine-wave input.

deck in which the input signal of Fig. 8.12 was replaced by an AC signal source having the form:

```
Vi 4 0 AC 1V
```

and the transient analysis command was replaced with the AC analysis command:

```
.AC DEC 10 1Hz 1MegHz
```

The following .PLOT command was also added:

```
.PLOT AC VdB(2)
```

In Fig. 8.14 we see that both circuits, regardless of stability, have a corresponding magnitude and phase response. The magnitude response behavior of the filters is quite similar, except for a vertical shift resulting from the different amplifier gains. It is the phase that shows the greatest difference in the behavior of the two filters. The stable filter has a phase response that goes from $0°$ to $-180°$ as the input signal frequency increases from DC. In contrast, the phase response of the unstable filter goes from $0°$ to $+180°$ with increasing input signal frequency. The reason for this behavior should be obvious; in the first case the poles appear on the left side of the s-plane's $j\omega$-axis, and

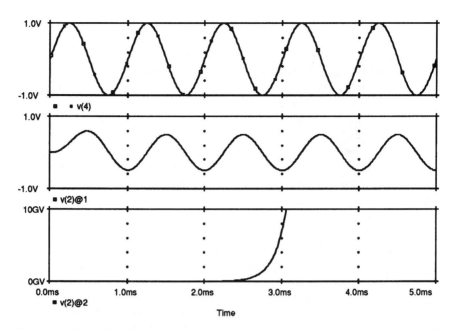

Figure 8.13 The transient response to a 1 V, 1 kHz sine-wave signal applied to the input of the filter circuit shown in Fig. 8.8(a) with $K = 1$ and $K = 5$. The top curve represents the input sine-wave signal; the middle curve represents the output signal from the stable filter circuit ($K = 1$); the bottom curve represents the output signal from the unstable filter circuit ($K = 5$).

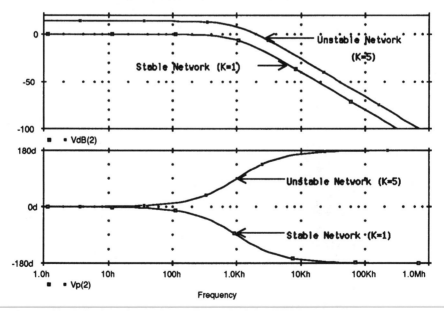

Figure 8.14 Magnitude and phase response of a stable and an unstable filter circuit. The top graph displays the magnitude responses, and the bottom graph displays the corresponding phase responses.

in the latter they appear on the right side. From the location of the poles the stability of the network can be easily deduced. Unfortunately, in more complicated circuits having many more poles and zeros the phase behavior of the amplifier would not so clearly identify poles in the right side of the s-plane. In general, without any special insight into circuit behavior the stability of a circuit cannot be determined from the frequency response of the closed-loop circuit. It is therefore recommended that the stability of a closed-loop circuit be investigated using a Spice transient analysis, preferably with a step input signal that is rich in a wide band of signal frequencies, not a single sine-wave input signal as was used in this example.

8.4

Investigating the Range of Amplifier Stability

In this section we use Bode plots generated by Spice to investigate the range of stability of an amplifier with complex frequency response behavior. We chose the same amplifier characteristics used in Section 8.10 of Sedra and Smith so that readers can compare the results given here with those calculated by hand. Furthermore, a Spice transient analysis will be used to verify circuit stability.

Consider the differential-input amplifier shown in Fig. 8.15(a) with an input–output transfer function $A(s)$. Let us assume that $A(s)$ is characterized by three real poles distributed along the negative axis of the s-plane at frequency locations of 0.1

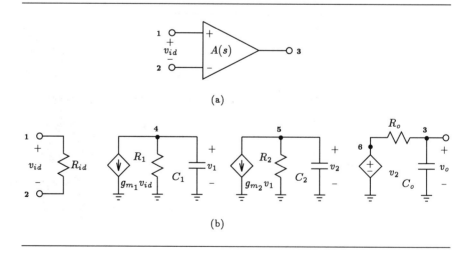

(a)

(b)

Figure 8.15   (a) Differential-input amplifier having an open-loop response $A(s)$ with three separate poles. (b) Small-signal equivalent circuit.

MHz, 1 MHz, and 10 MHz. Further, let the DC gain of the amplifier be 10^5 V/V or 100 dB. The open-loop transfer function $A(s)$ can then be described by

$$A(s) = \frac{10^5}{(s/0.2\pi \times 10^6 + 1)(s/2\pi \times 10^6 + 1)(s/20\pi \times 10^6 + 1)} \tag{8.3}$$

Figure 8.15(b) shows a small-signal equivalent circuit of a typical three-stage amplifier configuration consisting of two transconductance gain stages and an output buffer

```
Open-Loop Amplifier Characteristics

** Circuit Description **

* input signals

Vin+ 1 0 AC 1V
Vin- 2 0 AC 0V

* open-loop amplifier configuration
* connections: 1 2 3
*              | | |
*          in+ | |
*            in- |
*                out

* first stage
Rid 1 2 1MegOhm
Gm1 4 0 1 2 4.93m
R1 4 0 15.915k
C1 4 0 100pF
* second stage
Gm2 5 0 4 0 40m
R2 5 0 31.83k
C2 5 0 5pF
* output buffer stage
E3 6 0 5 0 1
Ro 6 3 100
Co 3 0 159.15pF

** Analysis Requests
.AC DEC 10 1Hz 100MegHz
** Output Requests **
.PLOT AC VdB(3) Vp(3)
.PROBE
.end
```

Figure 8.16 Spice input file for calculating the open-loop frequency response behavior of the three-pole amplifier circuit shown in Fig. 8.15(b).

stage. With the following component values, the transfer function of this small-signal equivalent circuit is described by that given in Eq. 8.3:

First stage:

$$R_{id} = 1 \text{ M}\Omega, g_{m_1} = 4.93 \text{ mA/V}, R_1 = 15.915 \text{ k}\Omega, C_1 = 100 \text{ pF}$$

Second stage:

$$g_{m_2} = 40 \text{ mA/V}, R_2 = 31.83 \text{ k}\Omega, C_2 = 5 \text{ pF}$$

Third stage:

$$R_o = 100 \text{ }\Omega, C_o = 159.15 \text{ pF}$$

To observe the amplifier's frequency response behavior, we use the input file in Fig. 8.16. The amplifier input is excited with a 1V AC signal applied to the positive input terminal. The negative input terminal is set to 0 V. An AC analysis is then requested between 1 Hz and 100 MHz. Both the magnitude and phase of the signal appearing at the amplifier output are to be plotted. Because the input signal is set at 1 V, the magnitude and phase of this output signal also represent the magnitude and phase of the input–output transfer function $A(s)$.

The Spice results are plotted in Fig. 8.17. We see that the gain of the amplifier is 100 dB at low frequencies and has a 3 dB frequency located quite close to 100 kHz.

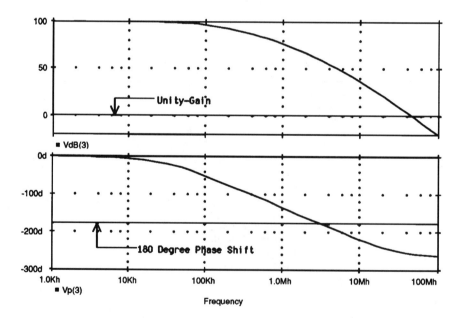

Figure 8.17 Magnitude and phase response of the differential-input amplifier shown in Fig. 8.15(b) with $A(s) = 10^5/[(s/0.2\pi \times 10^6 + 1)(s/2\pi \times 10^6 + 1)(s/20\pi \times 10^6 + 1)]$.

Near the unity-gain frequency of 46 MHz, the slope of the magnitude response is about -60 dB per decade. The phase response is provided in the lower plot. The majority of the phase shift occurs between 10 kHz and 100 MHz, one decade below the first pole and one decade above the highest-frequency pole. According to the cursor facility of Probe, the frequency at which the phase shift equals 180° is 3.31 MHz.

For a unity feedback factor independent of frequency (i.e., $\beta = 1$) the loop gain $A\beta$ would have a gain margin of $GM = -58.4$ dB or, alternatively, a phase margin (PM) of $-76.5°$. Clearly, connecting this amplifier in a unity-gain configuration would result in unstable operation. In fact, it is not until the loop gain drops below 41.6 dB (e.g., $A_o - GM = 100$ dB $- 58.4$ dB) that the closed-loop amplifier configuration would result in stable operation. This corresponds to the point at which the gain margin reduces to zero for a feedback factor $20\log(1/\beta) = 58.4$ dB, or $\beta = 1.202 \times 10^{-3}$. Thus, for this amplifier frequency-independent feedback factors β of less than 1.202×10^{-3} will maintain stable circuit operation in a closed-loop configuration. Correspondingly, the closed-loop gain, under large loop gain conditions, is approximately $1/\beta$, so that this amplifier is limited to applications requiring a closed-loop gain greater than 831.82 V/V or 58.4 dB.

To verify this result, let us investigate the step response of the noninverting amplifier configuration shown in Fig. 8.18 with closed-loop gains of $+2$ V/V and $+1001$ V/V. The differential amplifier will be modeled after the small-signal equivalent circuit shown in Fig. 8.15(b). In the first case we select R_A to be 100 kΩ, and in the second case R_A will be 100 Ω. For both cases R_B will be 100 kΩ. Figure 8.19 shows the input deck for the amplifier with the $+2$ V/V gain. (The other amplifier's input deck is identical except for the value of R_A.) A 1 mV step input is applied to the input of the closed-loop amplifier. We expect a 2 mV output signal if the amplifier is stable.

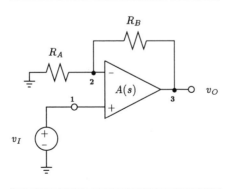

Figure 8.18 Noninverting amplifier configuration with closed-loop gain $A_f = (1 + R_B/R_A)$. The differential-input amplifier has transfer function $A(s)$ defined by Eq. (8.3) and $\beta = R_A/(R_A + R_B)$.

```
Step Response of a Noninverting Amplifier with a Gain of +2 V/V

** Subcircuits **

.subckt diffopamp 1 2 3
* open-loop amplifier configuration
* connections: 1 2 3
*              | | |
*          in+ | |
*            in- |
*                out
* first stage
Rid 1 2 1MegOhm
Gm1 4 0 1 2 4.93m
R1 4 0 15.915k
C1 4 0 100pF
* second stage
Gm2 5 0 4 0 40m
R2 5 0 31.83k
C2 5 0 5pF
* output buffer stage
E3 6 0 5 0 1
Ro 6 3 100
Co 3 0 159.15pF
.ends diffopamp

** Main Circuit **

* 1-V step input signal
Vstep 1 0 PWL (0 0V 1ns 0V 2ns 1mV 1s 1mV)
* noninverting amplifier
Xdiffamp 1 2 3 diffopamp
RA 2 0 100k
RB 2 3 100k

** Analysis Requests
.TRAN 100ns 12us 0s 100ns
** Output Requests **
.PLOT TRAN V(3) V(1)
.PROBE
.end
```

Figure 8.19 Spice input file for calculating the step response of a noninverting amplifier having a closed-loop gain of +2 V/V. The differential-input amplifier is modeled by the small-signal equivalent circuit shown in Fig. 8.15(b).

The step input signal is approximated by a series of piecewise linear segments described to Spice using the PWL source statement. It appears in the input file as

```
Vstep 1 0 PWL ( 0 0V 1ns 0V 2ns 1mV 1s 1mV)
```

The input voltage signal is held low for 1 ns, then made to rise to 1 mV with a rise time of 1 ns, and then held at 1 mV for one complete second. A transient analysis is to be performed over a 12 μs interval with a point collected every 100 ns. This should provide sufficient time resolution to see the important transient behavior of the output signal.

The step response of the noninverting amplifier with a gain of +2 V/V is plotted in Fig. 8.20. The top curve displays the 1 mV step input, and the graph below it illustrates the corresponding output signal. Clearly, the output signal bears no resemblance to the input signal. Moreover, the output signal is seen to oscillate between +15 V and −15 V. This therefore confirms that the closed-loop amplifier configuration is unstable as was predicted by our earlier Bode analysis.

If we repeat this step analysis on the noninverting amplifier with a gain of +1001 V/V, according to the Bode analysis our amplifier should be stable. This implies that with a 1 mV step input signal, the output signal should settle to a voltage level of 1 V after its transient response. The results of the Spice analysis shown in Fig. 8.21 confirm this, showing that this closed-loop amplifier configuration is indeed stable. The

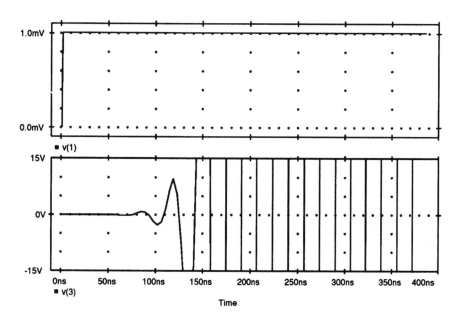

Figure 8.20   The 1 mV step response of a noninverting amplifier having a closed-loop gain of +2 V/V. The top graph displays the 1 mV step input. The bottom graph displays the corresponding output signal from the amplifier. Clearly, this output signal bears no resemblance to the input signal and indicates unstable behavior.

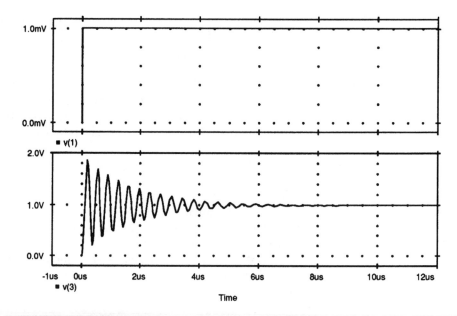

Figure 8.21   The 1 mV step response of a noninverting amplifier having a closed-loop gain of +1001 V/V. The top graph displays the 1 mV step input. The bottom graph displays the corresponding amplifier output signal. Here the output signal settles to a 1 V output level as expected for stable amplifier operation.

excessive ringing seen in the transient response is indicative of a closed-loop amplifier with a low phase margin. For this amplifier ($A_f = +1001$), the phase margin is about 4.3°. Designers of feedback amplifiers usually desire phase margins greater than 45° in order to obtain reasonable settling times and avoid excessive ringing.

8.5

The Effect of Phase Margin on Transient Response

In this section we investigate the step response of the noninverting amplifier previously shown in Fig. 8.18 for a wide range of phase margins. The open-loop frequency response behavior of the differential-input amplifier will be identical to that given previously in Eq. (8.3). The phase margin can be altered by changing its feedback factor β. For instance, in Table 8.1 we list the values of the two feedback resistors R_A and R_B that correspond to phase margins varying from 10° to 100° in steps of 10°. The resistances were obtained indirectly from the frequency response behavior graph shown in Fig. 8.17. First, the value of $20 \log(1/\beta)$ (the middle column of Table 8.1) was found for the desired phase margin, then from $\beta = R_A/(R_A + R_B)$ the values for R_A and R_B were found.

Using the values for R_A and R_B in Table 8.1 with the input deck shown in Fig. 8.19 we can change the values of RA and RB in the Spice deck and compute the step response of the closed-loop amplifier for various phase margins. To compare the

Table 8.1 Feedback resistors R_A and R_B of the noninverting amplifier shown in Fig. 8.18 for different amplifier phase margins.

Phase Margins	20 log(1/β) (dB)	Feedback Resistors	
		R_A (Ω)	R_B (kΩ)
10°	61.8	81.5	100
20°	66.3	48.5	100
30°	70.5	29.8	100
40°	74.2	19.4	100
50°	77.4	13.4	100
60°	80.6	9.4	100
70°	83.4	6.8	100
80°	86.1	4.9	100
90°	88.7	3.7	100
100°	91.1	2.8	100

different step responses we shall concatenate all 10 files corresponding to the different phase margins and submit that single file to Spice for analysis.

Figure 8.22 shows the step responses of the closed-loop amplifier for different phase margins. Clearly, the larger the phase margin, the lower the amount of ringing in the output signal.

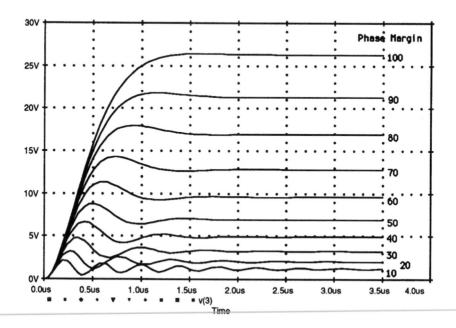

Figure 8.22 Transient response of closed-loop amplifier having a wide range of phase margins. The larger the phase margin, the less the ringing in the output signal. A voltage step signal of 1 mV is applied to the input of each amplifier.

8.6

Frequency Compensation

In this section we use Spice to demonstrate how an amplifier whose closed-loop behavior is unstable can be made stable by modifying its open-loop transfer function. This technique is referred to as frequency compensation. There are many ways to perform frequency compensation. Here we shall address only the dominant-pole method, but in two different forms: the shunt-capacitor and the pole-splitting techniques.

Consider the differential-input amplifier used in the two previous sections (Fig. 8.15). It was found there that this amplifier was stable only when connected in closed-loop configurations using resistive feedback with gains greater than 58.4 dB or, conversely, with a feedback factor β less than 1.202×10^{-3}. Let us consider modifying the open-loop frequency response behavior of this amplifier so that it has a positive gain or phase margin. The amplifier could then be used in closed-loop configurations with frequency-independent feedback factors as large as unity. The resulting differential-input amplifier is considered to be *unity-gain stable*.

First we consider the shunt-capacitor method: we place a 1 μF capacitor C_C across the output terminal of the first internal stage of the differential-input amplifier, as seen in Fig. 8.23(a). Adding this shunt capacitor will alter the pole originally formed by R_1

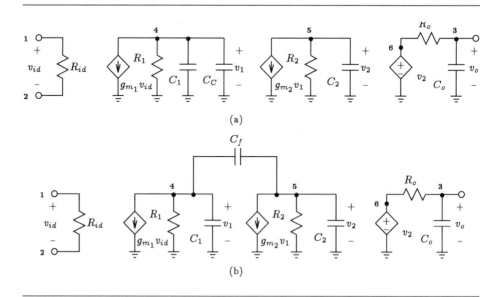

Figure 8.23   Two different versions of a dominant-pole compensation: (a) adding a shunt capacitor C_C across the output of the first stage and (b) introducing a pole-splitting compensation capacitor C_f between the first and second stages.

and C_1 located at 0.1 MHz and will move it down in frequency to 10 Hz (i.e., $f_D' = 1/2\pi(C_1 + C_C)R_1$). The remaining poles formed by the other resistors and capacitors will remain unchanged.

The pole-splitting technique requires that a bridging capacitor C_f be connected between the output of the first and second stages of the amplifier as shown in Fig. 8.23(b). In Example 8.6 of Sedra and Smith a bridging capacitor of 78.5 pF was found to move the first pole to a frequency of 100 Hz (i.e., $f_{P1}' \simeq 1/2\pi g_{m_2} R_2 C_f R_1$) and the second pole to about 57 MHz $[f_{P2}' \simeq g_{m_2}/2\pi(C_1 + C_2)]$.

To see the effect of adding these two compensation capacitors we simulate the behavior of these two circuits by adding the appropriate compensation capacitor to the input decks seen previously in Fig. 8.16. In the case of the circuit in Fig. 8.23(a) we add the element statement:

```
* compensation capacitor
Cc 4 0 1uF
```

and in the case of the circuit shown in Fig. 8.23(b) we add the statement:

```
* compensation capacitor
Cf 4 5 78.5pF
```

The Spice decks of these amplifiers were combined into one file, together with the Spice deck for the uncompensated amplifier, and then were submitted to Spice. The magnitude and phase results are plotted in Fig. 8.24.

In Fig. 8.24 we see that the two compensated amplifiers have frequency response behavior very different from the original uncompensated amplifier. Each of the two compensated amplifiers is dominated by a low-frequency pole; furthermore, the second pole of each of these amplifiers occurs very near the unity-gain frequency. In the case of the shunt capacitor–compensated amplifier, the first break frequency is at 10 Hz and the unity-gain frequency is at about 800 kHz. The phase margin associated with this amplifier is 47.7°, and the gain margin is 20.8 dB.

The first break frequency of the pole splitting–compensated amplifier occurs at 100 Hz and its unity-gain frequency at about 8 MHz. The phase margin for this amplifier is 38.1°, and its gain margin is 11.7 dB. Since the gain and phase margins of each amplifier are positive, they are both unity-gain stable.

To further convince ourselves that the amplifiers are indeed unity-gain stable, let us compute the step response of each compensated amplifier connected in the unity-gain configuration shown in Fig. 8.25. The input file for the shunt capacitor–compensated amplifier in this circuit setup is shown in Fig. 8.26. The input deck for the pole splitting–compensated amplifier connected in a unity-gain configuration is almost identical to that given in Fig. 8.26. The element statement for the compensation capacitor C_f should appear as previously indicated, and the transient analysis request statement should read:

```
.TRAN 10ns 500ns 0s 10ns
```

The results of this transient analysis are shown in Figs. 8.27 and 8.28, which plot both the input and the output signals. Both responses clearly indicate stable behavior. We see that the settling time of the pole splitting–compensated amplifier is much

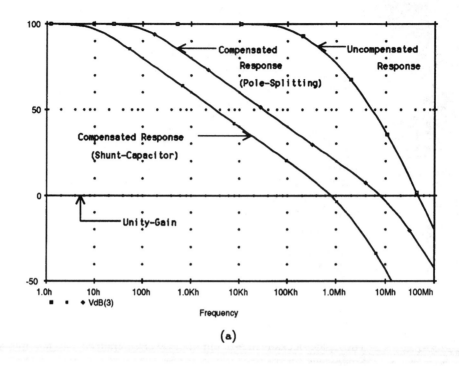

(a)

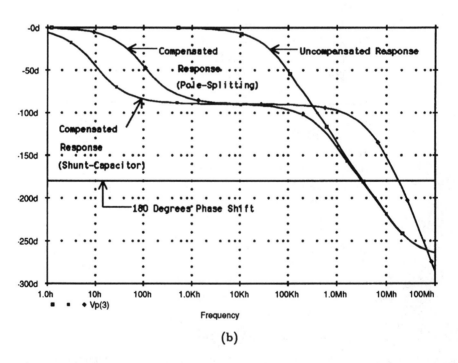

(b)

Figure 8.24 Comparing the open-loop frequency response behaviors of an amplifier compensated using a shunt-capacitor and of another compensated by the pole-splitting technique. The open-loop frequency response behavior of the uncompensated amplifier is also shown. (a) Magnitude response. (b) Phase response.

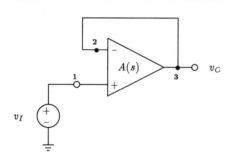

Figure 8.25 Unity-gain configuration. A zero-valued voltage source is used to create the short-circuit connection between the negative input terminal of the amplifier and its output.

shorter than that of the shunt capacitor–compensated amplifier. (Note the time scale on the *x*-axis of each graph.) This result is not too surprising given that the 3 dB bandwidth of the pole splitting–compensated amplifier is about 10 times larger than the other one.

8.7

Spice Tips

- When a feedback loop is opened for investigative purposes (loop gain, stability, etc.), the DC conditions of the loop must remain unchanged. This can be accomplished using a large-valued inductor in series with the feedback loop and a large-valued capacitor in series with the signal source that injects the signal into the loop.
- Loop transmission $A\beta(s)$ in a single-loop feedback amplifier can be determined from the open-circuit and short-circuit transfer functions ($T_{OC}(s)$ and $T_{SC}(s)$, respectively) according to the equation:

$$A\beta = -T_{OC} \parallel T_{SC} = -\frac{1}{1/T_{OC} + 1/T_{SC}}$$

- The stability of a closed-loop linear circuit can be determined from the magnitude and phase behavior of the loop transmission. Spice can compute and display both these quantities in the form of a Bode plot. The Probe facility of PSpice can produce either Bode or Nyquist plots of this information.
- The stability of a closed-loop circuit cannot be determined from its frequency response behavior. Instead, the closed-loop stability can be observed from the output behavior of the closed-loop circuit when subjected to a fast-rising or fast-falling step input using the transient analysis capability of Spice.

```
Step Response of a Unity-Gain Amplifier Compensated with a Shunt Capacitor

** Subcircuits **

.subckt diffopamp 1 2 3
* open-loop amplifier configuration
* connections: 1 2 3
*              | | |
*            in+ | |
*              in- |
*                 out

* first stage
Rid 1 2 1MegOhm
Gm1 4 0 1 2 4.93m
R1 4 0 15.915k
C1 4 0 100pF
* second stage
Gm2 5 0 4 0 40m
R2 5 0 31.83k
C2 5 0 5pF
* output buffer stage
E3 6 0 5 0 1
Ro 6 3 100
Co 3 0 159.15pF
* compensation capacitor
Cc 4 0 1uF
.ends diffopamp

** Main Circuit **

* 1 V step input signal
Vstep 1 0 PWL (0 0V 1ns 0V 2ns 1V 1s 1V)
* noninverting amplifier
Xdiffamp 1 2 3 diffopamp
Vshort 2 3 0

** Analysis Requests
.TRAN 100ns 5us 0s 100ns
** Output Requests **
.PLOT TRAN V(3) V(1)
.end
```

Figure 8.26 Spice input file for calculating the step response of the shunt capacitor–compensated amplifier shown in Fig. 8.23(a) connected in a unity-gain configuration.

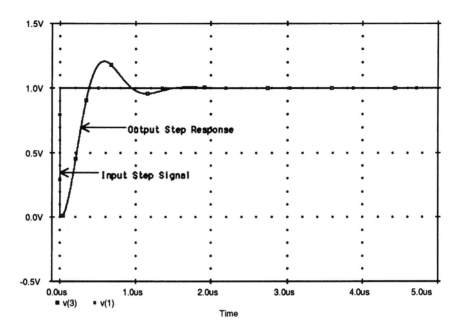

Figure 8.27 A 1 V step response of a shunt capacitor–compensated amplifier connected in a unity-gain configuration. Both the input and output signals are shown. Notice that the time scale ranges between 0 and 5 μs.

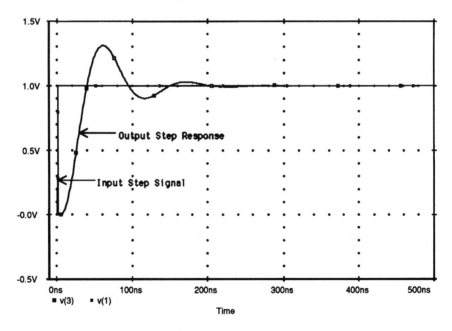

Figure 8.28 A 1 V step response of a pole splitting–compensated amplifier connected in a unity-gain configuration. Both the input and output signals are shown. Notice that the time scale ranges between 0 and 500 ns.

8.8

Bibliography

S. Rosenstark, *Feedback Amplifier Principles,* New York: Macmillan Publishing Company, 1986.

8.9

Problems

8.1 Determine the loop gain of the amplifier in Fig. P8.1 by breaking the loop at the gate of M_2 and finding the return voltage across the 100 kΩ resistor (while setting V_s to zero). The devices have $|V_t| = 1$ V, $\mu_n C_{OX} = 1.0$ mA/V², and $h_{fe} = 100$. The Early voltage magnitude for all devices (including those that implement the current sources) is 100 V. Determine the output resistance R'_{of}.

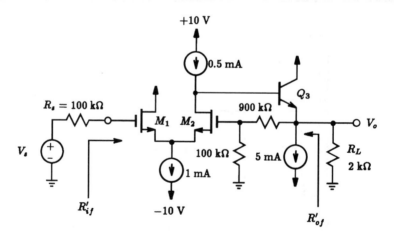

Figure P8.1

8.2 Repeat Problem 8.1, but this time break the loop at the base of Q_3 and terminate the loop by the resistance seen looking into the base of Q_3. How does the loop gain compare with that found in Problem 8.1?

8.3 Determine the loop gain of the amplifier circuit shown in Fig. P8.3 using (a) the method of breaking the loop at the gate of M_2 and terminating the loop with the proper impedance and (b) the method of open- and short-circuit transfer functions. How do the results compare?

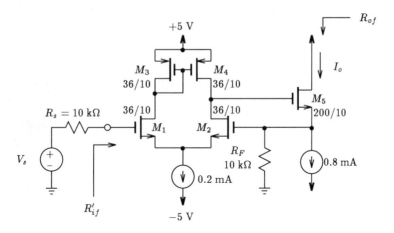

Figure P8.3

8.4 A single-pole DC amplifier has an equivalent circuit representation shown in Fig. P8.4. If this amplifier is operated in a loop whose frequency-independent feedback factor is 0.1, with the aid of Spice determine the low-frequency gain, the 3 dB frequency, and the unity-gain frequency of the closed-loop amplifier. Compare these results with those of the open-loop amplifier. By what factor does the single pole shift?

$R_1 = 5\,k\Omega$

V_{id} R_{id} V_{id} C_1 10 nF $+ V_1 -$ $10^3 v_1$ $+ V_o -$

Figure P8.4

8.5 For the feedback amplifier described in Problem 8.4, plot both the magnitude and phase of the loop gain. Is the amplifier stable? Confirm your conclusions by performing a transient analysis of the closed-loop amplifier subjected to a 1 V step input signal.

8.6 An amplifier has a DC gain of 10^5 and poles at 10^5 Hz, 3.16×10^5 Hz, and 10^6 Hz. An equivalent circuit representation of this amplifier is shown in Fig. P8.6. Using Spice, plot the magnitude and phase of the voltage gain of this amplifier. What is the value of β and the corresponding closed-loop gain for which a phase margin of $45°$ is obtained?

$R_1 = 159.15\,\Omega$ $R_2 = 50.37\,\Omega$ $R_3 = 15.92\,\Omega$

V_{id} $R_{id} = 1\,M\Omega$ V_{id} C_1 10 nF $+ V_1 -$ V_1 C_2 10 nF $+ V_2 -$ V_2 C_3 10 nF $+ V_3 -$ $10^5 v_3$ $+ V_o -$

Figure P8.6

8.7 Synthesize an amplifier whose open-loop gain $A(s)$ is given by

$$A(s) = \frac{1000}{(1 + s/10^4)(1 + s/10^5)^2}$$

Plot its magnitude and phase response using PSpice. If this amplifier is to be used in a feedback configuration where the feedback factor β is independent of frequency, find the frequency at which the phase shift is 180°, and find the critical value of β at which oscillation will commence. Confirm your conclusion using a Spice transient analysis.

8.8 The op amp in the circuit of Fig. P8.8 has an open-loop gain of 10^5 and a single-pole roll-off with $\omega_{3dB} = 10$ rad/s.
(a) Plot the Bode plot for the loop gain using Spice.
(b) Find the frequency at which $|A\beta| = 1$, and find the corresponding phase margin.

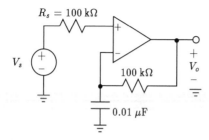

Figure P8.8

8.9 An op amp designed to have a low-frequency gain of 10^5 and a high-frequency response dominated by a single pole at 100 rad/s acquires, through a manufacturing error, a pair of additional poles at 10,000 rad/s. Using Spice, plot the magnitude and phase response of the open-loop gain. At what frequency does the total phase shift reach 180°? At this frequency for what value of β, assumed to be frequency independent, does the loop gain reach a value of unity? What is the corresponding value of closed-loop gain at low frequencies?

8.10 For the situation described in Problem 8.9, with the aid of Spice obtain Nyquist plots of the loop gain for $\beta = 0.1$ and 10^{-3} in the frequency range of 1 Hz to 100 MHz.

9

Output Stages and Power Amplifiers

Output stages are used to supply large amounts of power with very little signal distortion and with maximum efficiency. Power amplifiers are simply amplifiers with a high power-output stage. In this chapter we demonstrate how to use Spice to compute various performance measures of an output stage, including power efficiency and harmonic distortion.

9.1

Emitter–Follower Output Stage

Figure 9.1 displays an emitter–follower output stage biased with a constant current source of $I_{bias} \simeq (15-0.7)/5000 = 2.86$ mA realized by transistors Q_2 and Q_3 and resistor R. The output terminal is connected to a load resistance of 1 kΩ. Each transistor is modeled after the widely available commercial 2N2222A *npn* transistor. The input file needed for Spice to determine the transfer characteristic of the emitter–follower is shown in Fig. 9.2. A DC sweep of the input voltage source is to be performed between -20 V and $+20$ V. Although this particular voltage sweep extends beyond the power supply limits of the output stage, it is intended to highlight the saturation limits of the emitter–follower.

Spice produces the emitter–follower transfer characteristic shown in Fig. 9.3. The output voltage saturates at -3.25 V for inputs below -2.7 V and saturates at $+14.9$ V for input levels beyond $+15.6$ V. This output stage also exhibits an input offset voltage of $+0.67$ V. The slope of the transfer characteristic within the linear region of operation is very nearly unity.

It is interesting to note that the lower saturation limit of the output voltage is somewhat greater than that predicted by the simple expression $I_{bias}R_L$, which says the lower saturation limit should be -2.86 mA $\times$ 1 kΩ $= -2.86$ V instead of the simulation result of -3.25 V. This deviation occurs because the output current of the current source Q_2 varies with output voltage.

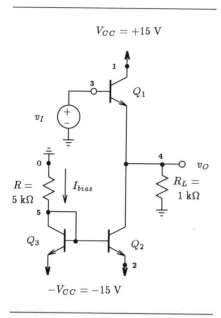

Figure 9.1 An emitter–follower output stage with current-source bias.

To better understand the behavior of the voltage-follower output stage, at the circuit input we apply several 1 kHz sine-wave signals with various amplitudes. In this way we can visualize the emitter–follower circuit's expected output signals as we would see them in the laboratory using a function generator and an oscilloscope. To be more specific, let us apply three different sine waves with amplitudes of 1 V, 2.7 V, and 10 V to the input of the amplifier. According to the transfer characteristic in Fig. 9.3, the 1 V and the 2.7 V sine-wave signals should (just barely) pass undistorted. The 10 V signal should become clipped on the negative portion of the output waveform. To see this using Spice, for each input level we use a modified version of the input deck in Fig. 9.2. The input voltage source statement is replaced by one that describes a sine-wave input. In the case of the 1 V input signal the Spice statement would appear:

```
Vi 3 0 sin ( 0V 1V 1kHz )
```

Statements for the other two input levels would be similar.

The following two statements are included in each input deck to compute the transient response of the circuit over three periods of the input signal:

```
.TRAN 10us 3ms 0ms 10us
.PLOT TRAN V(4)
```

One hundred points per period of the output waveform will be collected and plotted. The three input decks are concatenated into one file for easy comparison of the results.

```
Class A Output Stage

** Circuit Description **
* power supplies
Vcc+ 1 0 DC +15V
Vcc- 2 0 DC -15V
* input signal source
Vi 3 0 DC 0V
* output transistor
Q1 1 3 4 Q2N2222A
* load resistance
Rl 4 0 1k
* bias circuit
R 0 5 5k
Q2 4 5 2 Q2N2222A
Q3 5 5 2 Q2N2222A
* transistor model statement for the 2N2222A
.model Q2N2222A  NPN (Is=14.34f Xti=3 Eg=1.11 Vaf=74.03 Bf=255.9 Ne=1.307
+                    Ise=14.34f Ikf=.2847 Xtb=1.5 Br=6.092 Nc=2 Isc=0 Ikr=0 Rc=1
+                    Cjc=7.306p Mjc=.3416 Vjc=.75 Fc=.5 Cje=22.01p Mje=.377 Vje=.75
+                    Tr=46.91n Tf=411.1p Itf=.6 Vtf=1.7 Xtf=3 Rb=10)
** Analysis Requests **
.DC Vi -20V +20V 100mV
** Output Requests **
.PLOT DC V(4)
.Probe
.end
```

Figure 9.2   Spice input file for computing the transfer characteristic of the emitter–follower output stage shown in Fig. 9.1.

Figure 9.4 shows the plots of the output of the emitter–follower and confirms that for sine wave inputs less than 2.7 V the output waveform shows no obvious distortion. For an input sine wave of 10 V peak the negative portion of the output sine wave is clipped at -3.25 V.

Other waveforms associated with the emitter–follower as calculated by Spice are shown in Fig. 9.5. The input to this circuit is the largest that can be applied before distortion sets in—a 2.7 V sine-wave signal of 1 kHz frequency. The input deck for this analysis is identical to that just described, except that the plot command is changed to

```
.PLOT TRAN V(4) V(1,4) I(Vcc+)
```

The top graph shows the voltage waveform that appears across the load resistance. The output signal is sinusoidal with a peak value of 2.62 V riding on a DC level of -0.614 V. The second and third graphs from the top display the collector–emitter voltage and the collector current of transistor Q_1, respectively. Both of these waveforms appear sinusoidal with amplitudes of 2.62 V and 2.703 mA, respectively.

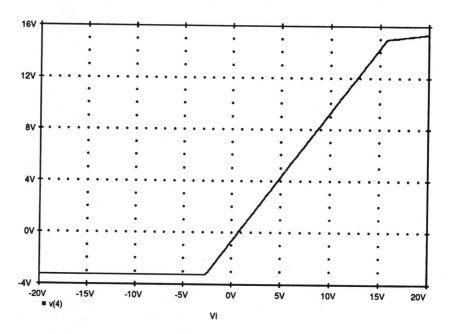

Figure 9.3 Transfer characteristic of the emitter–follower circuit shown in Fig. 9.1. As evident, the output voltage can swing between −3.25 V and 14.9 V without experiencing significant distortion.

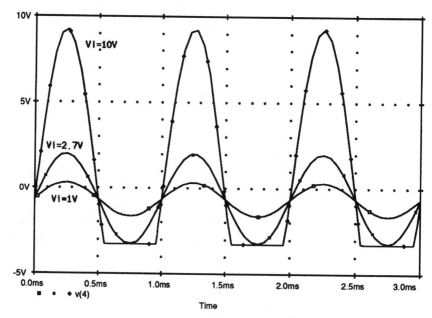

Figure 9.4 Output voltage waveform of the voltage-follower circuit of Fig. 9.1 for three different input levels: 1 V, 2.7 V, and 10 V. Excessive voltage clipping occurs on the negative portion of the output voltage waveform corresponding to an input amplitude of 10 V.

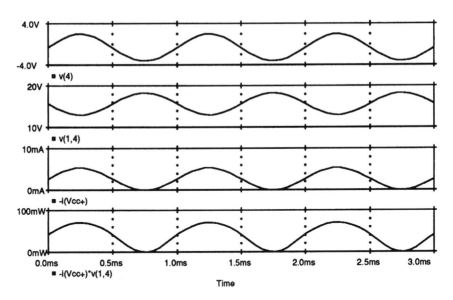

Figure 9.5 Several waveforms associated with the emitter–follower circuit of Fig. 9.1 when excited by a 2.7 V, 1 kHz sinusoidal signal. The top graph displays the voltage across the load resistance, the next graph displays the collector–emitter voltage of Q_1, the third graph displays the collector current of Q_1, and the bottom graph displays the instantaneous power dissipated by Q_1.

The average value of the voltage waveform appearing between the collector and emitter terminals is 15.61 V, and the average value of the collector current is 2.73 mA. The bottom graph displays the instantaneous power dissipated by Q_1. This waveform was obtained by using Probe to multiply the collector current and collector-emitter voltage of Q_1; its peak value is 70.58 mW, and its average is about 40 mW.

9.2

Class B Output Stage

Figure 9.6 shows a class B output stage loaded by an 8 Ω resistor and driven by a voltage source V_i. It consists of a pair of complementary transistors assumed to have similar characteristics. This output stage is noted for its relatively high power-conversion efficiency (at most, 78.5%) and its large crossover distortion. In the following discussion we use Spice to investigate these two attributes. We will assume transistors modeled after National Semiconductor's complementary NA51 and NA52 transistors.

9.2.1 Power Conversion Efficiency

With a 17.9 V, 1 kHz sinusoidal signal applied to the input of the output stage, we would like to determine the average power supplied by the two 23 V power supplies

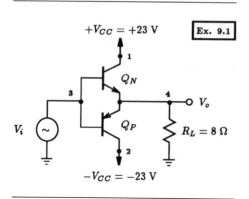

Figure 9.6   Class B output stage.

(P_S), the average power provided to the load resistance (P_L), and the power-conversion efficiency (i.e., $\eta = P_L/P_S$). In addition, we would also like to determine the power that each transistor must dissipate. This problem, in a somewhat different form, is in Example 9.1 of Sedra and Smith.

Figure 9.7 shows the input file describing the class B output stage of Fig. 9.6 and includes the Spice models of the two complementary transistors. A 1 kHz input signal of 17.9 V amplitude is applied to the input, and a transient analysis is to be performed from time 0 ms to 3 ms later. This will simulate the behavior of the output stage for three complete periods of the input signal. For output we have requested plots of various node voltages and branch currents. Note that we are accessing the current through the load resistance R_L directly using i(Rl) rather than in the usual way of inserting a zero-valued voltage source in series with R_L to sense the current flowing through it. This approach is possible with PSpice and is more convenient to use than the ammeter approach.

The results of this analysis then can be combined to provide the instantaneous power supplied by the DC sources, as well as the power dissipated by the load and by the transistors. Unfortunately, this computation cannot be performed directly by Spice. We could add additional circuitry to perform these computations, but this is not the most direct approach. Instead, we use the built-in computational abilities of Probe to obtain the desired quantities.

In Fig. 9.8 we display the different waveforms. The upper and middle graphs display the voltage and current, respectively, that appear across the load. Peak voltage amplitude is 17.05 V, and peak current amplitude is 2.13 A. Notice that both the voltage and current waveforms exhibit a small crossover distortion. We investigate this further in the next subsection. The bottom graph displays the instantaneous and the average power dissipated by the load resistor as computed by Probe using $V(4) \times i(Rl)$ for instantaneous power and taking a running average for the average power (P_L). The rather transient behavior of the average load power, which eventually settles into

```
Class B Output Stage

** Circuit Description **
* power supplies
Vcc+ 1 0 DC +23V
Vcc- 2 0 DC -23V
* input signal source
Vi 3 0 sin ( 0V 17.9V 1kHz )
* output buffer
Qn 1 3 4 NA51
Qp 2 3 4 NA52
* load resistance
Rl 4 0 8
* transistor model statement for National Semiconductor's
* complementary transistors NA51 and NA52
.model NA51 NPN (Is=10f Xti=3 Eg=1.11 Vaf=100 Bf=100 Ise=0 Ne=1.5 Ikf=0
+               Nk=.5 Xtb=1.5 Br=1 Isc=0 Nc=2 Ikr=0 Rc=0 Cjc=76.97p Mjc=.2072
+               Vjc=.75 Fc=.5 Cje=5p Mje=.3333 Vje=.75 Tr=10n Tf=1n Itf=1 Xtf=0
+               Vtf=10)
.model NA52 PNP (Is=10f Xti=3 Eg=1.11 Vaf=100 Bf=100 Ise=0 Ne=1.5 Ikf=0
+               Nk=.5 Xtb=1.5 Br=1 Isc=0 Nc=2 Ikr=0 Rc=0 Cjc=112.6p Mjc=.1875
+               Vjc=.75 Fc=.5 Cje=5p Mje=.3333 Vje=.75 Tr=10n Tf=1n Itf=1 Xtf=0
+               Vtf=10)
** Analysis Requests **
.Tran 10us 3ms 0ms 10us
** Output Requests **
.Plot Tran V(1) i(Vcc+)
.Plot Tran V(2) i(Vcc-)
.Plot Tran V(4) i(Rl)
.Plot Tran V(1,4) i(Vcc+)
.Plot Tran V(2,4) i(Vcc-)
.probe
.end
```

Figure 9.7 Spice input file for computing the transient response of the class B output stage shown in Fig. 9.6. A 17.9 V, 1 kHz sinusoid is applied to the input.

a quasi-constant steady state of about 17.7 W, is an artifact of the PSpice algorithm used to compute the running average of a waveform.

The upper two graphs of Fig. 9.9 show the voltage and current waveforms, respectively, supplied by the positive voltage supply $+V_{CC}$. The bottom graph shows the instantaneous and average power supplied by $+V_{CC}$. (Similar waveforms are found for the negative power supply $-V_{CC}$ except that current is supplied in complementary time slots.) The average power supplied by either $+V_{CC}$ or $-V_{CC}$ is found to be approximately 15 W for a total supply power (P_S) of 30 W. The power-conversion efficiency is then computed to be $\eta = P_L/P_S = 17.7/30 \times 100\% = 59\%$.

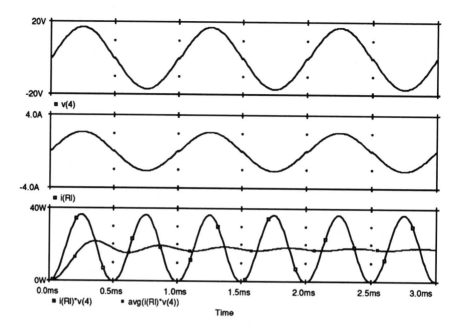

Figure 9.8 Several waveforms associated with the class B output stage shown in Fig. 9.6 when excited by a 17.9 V, 1 kHz sinusoidal signal. The upper graph displays the voltage across the load resistance, the middle graph displays the load current, and the lower graph displays the instantaneous and average power dissipated by the load.

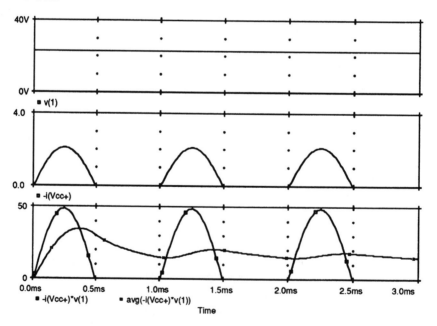

Figure 9.9 The voltage (upper graph), current (middle graph), and instantaneous and average power (bottom graph) supplied by the positive voltage supply ($+V_{CC}$).

Another important quantity in the design of an output stage is the amount of power dissipated by the two complementary transistors Q_N and Q_P. Figure 9.10 shows plots of the voltage, current, and power waveforms associated with transistor Q_P. Similar waveforms are associated with Q_N. Both the voltage and current waveforms appear as expected. That is, the voltage waveform is a sinusoid, and the current waveform is a half-wave sinusoid. However, the instantaneous power waveform shown in the bottom graph of Fig. 9.10 has a rather strange shape. It appears periodic but is non-sinusoidal, almost pulselike (less a few higher-order harmonic components). This is because the two transistors are being driven quite hard, and although it is not apparent in the corresponding voltage and current waveforms, some distortion is present and is manifesting itself very clearly in the instantaneous power waveform. If the input voltage level of 17.9 V is reduced slightly, say to 17 V, the ripple in the power waveform will disappear. The average power dissipated by transistor Q_P is seen to be approximately 6 W, likewise for transistor Q_N.

It is reassuring to confirm that the average power supplied by the two power supplies (30 W) roughly equals the sum of the power dissipated by the output stage (12 W) and that dissipated by the load (17.7 W).

Table 9.1 compares the results found through a Spice analysis with those calculated by a hand analysis. The hand-calculated results agree quite well, with a maximum 8.3% error.

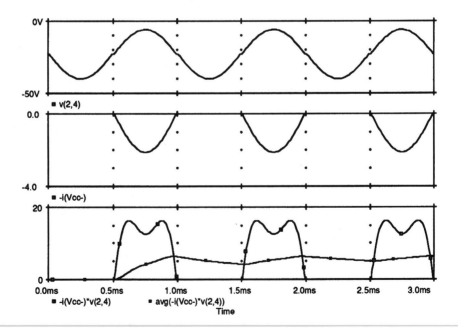

Figure 9.10   Waveforms of the voltage across, the current through, and the power dissipated in the *pnp* transistor Q_P of the output stage shown in Fig. 9.6.

Table 9.1 Various power terms associated with the class B output stage shown in Fig. 9.6 as computed by hand and by Spice analysis. The rightmost column presents the relative percentage error between the values predicted by hand and by Spice.

Power/Efficiency	Equation	Hand Analysis	Spice	Error %
P_S	$\dfrac{2}{\pi}\dfrac{\hat{V}_o}{R_L}V_{CC}$	31.2 W	30.0 W	4
P_D	$\dfrac{2}{\pi}\dfrac{\hat{V}_o}{R_L}V_{CC}-\dfrac{1}{2}\dfrac{\hat{V}_o^2}{R_L}$	13.0 W	12.0 W	8.3
P_L	$\dfrac{1}{2}\dfrac{\hat{V}_0^2}{R_L}$	18.2 W	17.7 W	2.8
η	$\dfrac{P_L}{P_S}\times100\%$	58.3%	59.0%	-1.2

9.2.2 Transfer Characteristics and a Measure of Linearity

The output signal of a class B output stage experiences crossover distortion, as barely can be seen in the load voltage and current waveforms shown in Fig. 9.8. Another way of investigating this crossover distortion is to plot the output voltage as a function of the input voltage level. To use Spice for this we replace the transient analysis command in the input file of Fig. 9.7 with the following DC sweep command:

```
.DC Vi -10V +10V 50mV
```

We limit the swing of the input DC sweep level to ± 10 V to highlight the deadband region of this amplifier. In addition, the following command should replace the transient plot command in order to obtain a plot of the output stage transfer characteristic:

```
.Plot DC V(4)
```

Spice produces the transfer characteristic shown in Fig. 9.11. The slope of the line of input versus output voltage is very near unity, and the deadband region extends between -0.72 V and $+0.72$ V.

Although the deadband effect is very visible in the amplifier transfer characteristic, engineers prefer to quantify its effect, and possibly other sources of distortion, in terms of harmonic distortion. Spice provides this analysis through the **Fourier analysis** command, which decomposes a waveform generated through a transient analysis into its Fourier series components. These components include the fundamental, a DC component, and the next eight harmonics. Spice also will compute the total harmonic distortion (THD) of the waveform.

The syntax of the .FOUR command is given in Table 9.2. The key word .FOUR specifies that a Fourier series decomposition is to be performed on each variable indicated in the variable list at the end of the statement, after the field indicating the specified frequency of the fundamental (i.e., *fundamental frequency*). It is

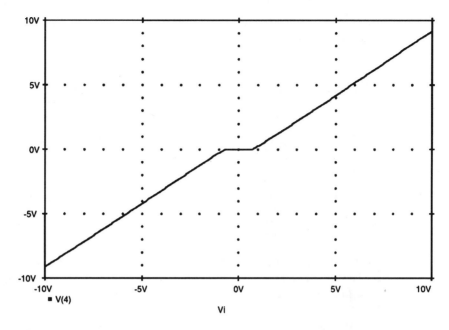

Figure 9.11 Transfer characteristic of a class B output stage constructed with National Semiconductor's NA51 and NA52 complementary transistors.

important to note that the Fourier analysis is performed by Spice on the waveform defined only between its end and one period of the fundamental back from this end (i.e., *one/fundamental_frequency*). It is not done on the entire waveform. This implies that the transient analysis used in conjunction with a Fourier analysis must be at least *one/fundamental_frequency* long. The .FOUR command requires neither a .PRINT nor a .PLOT command; the results of the analysis are automatically printed into the Spice output file in table form.

We can add a Fourier analysis command to the input file for the output stage listed in Fig. 9.7 and compute the total harmonic distortion (THD) contained in the output voltage waveform:

```
.FOUR 1kHz V(4)
```

where the fundamental frequency of 1 kHz is set equal to the frequency of the input sinusoidal signal.

Table 9.2 General syntax of the Fourier analysis command in Spice.

Analysis Request	Spice Command
Fourier analysis	.FOUR *fundamental_frequency output_variable_list*

The results of the Fourier analysis as computed by Spice are shown here:

```
FOURIER COMPONENTS OF TRANSIENT RESPONSE V(4)

DC COMPONENT =   9.648686E-05
```

HARMONIC NO	FREQUENCY (HZ)	FOURIER COMPONENT	NORMALIZED COMPONENT	PHASE (DEG)	NORMALIZED PHASE (DEG)
1	1.000E+03	1.683E+01	1.000E+00	-8.163E-05	0.000E+00
2	2.000E+03	1.373E-04	8.162E-06	-9.067E+01	-9.067E+01
3	3.000E+03	3.387E-01	2.013E-02	-1.800E+02	-1.800E+02
4	4.000E+03	6.927E-05	4.117E-06	-8.964E+01	-8.964E+01
5	5.000E+03	1.977E-01	1.175E-02	-1.800E+02	-1.800E+02
6	6.000E+03	5.794E-05	3.443E-06	-8.846E+01	-8.846E+01
7	7.000E+03	1.378E-01	8.188E-03	-1.800E+02	-1.800E+02
8	8.000E+03	5.281E-05	3.139E-06	-8.740E+01	-8.740E+01
9	9.000E+03	1.045E-01	6.213E-03	-1.800E+02	-1.800E+02

```
TOTAL HARMONIC DISTORTION =   2.547618E+00 PERCENT
```

The output voltage from the class B output stage is rich in odd-order harmonics, resulting in a rather high THD measure of 2.55%.

Generally, the distortion behavior of a class B output stage is much poorer than that which can be achieved by an emitter–follower circuit with the same input voltage. Let us return to the emitter–follower example of Section 9.1 and compute its harmonic content when excited by a 17.7 V amplitude, 1 kHz input signal. Recall from Section 9.1 that the emitter–follower circuit given there could only handle input signals with peak values less than 2.7 V. Therefore, let us increase the load resistor from 1 kΩ to 5 kΩ and the level of the two power supplies to ±23 V. This will increase the range of input signals that can be applied to this amplifier before the output becomes distorted. (Computing the transfer characteristic of this revised voltage-follower circuit using Spice indicates that the linear region of this amplifier is between −21.4 V and +23 V.)

The results of the Fourier analysis as computed by Spice for the voltage-follower circuit are as follows:

```
FOURIER COMPONENTS OF TRANSIENT RESPONSE V(4)

DC COMPONENT =  -6.772019E-01
```

HARMONIC NO	FREQUENCY (HZ)	FOURIER COMPONENT	NORMALIZED COMPONENT	PHASE (DEG)	NORMALIZED PHASE (DEG)
1	1.000E+03	1.752E+01	1.000E+00	2.912E-03	0.000E+00
2	2.000E+03	5.782E-03	3.300E-04	-8.921E+01	-8.921E+01
3	3.000E+03	2.596E-03	1.482E-04	1.971E+01	1.971E+01
4	4.000E+03	1.738E-03	9.922E-05	8.523E+01	8.523E+01
5	5.000E+03	8.973E-04	5.121E-05	9.995E+01	9.995E+01
6	6.000E+03	7.882E-04	4.499E-05	7.459E+01	7.459E+01
7	7.000E+03	9.442E-04	5.389E-05	7.217E+01	7.217E+01
8	8.000E+03	9.679E-04	5.525E-05	7.415E+01	7.415E+01
9	9.000E+03	9.591E-04	5.474E-05	7.316E+01	7.316E+01

```
TOTAL HARMONIC DISTORTION =   3.928177E-02 PERCENT
```

Here we see that the THD for the emitter–follower circuit is 0.039%, obviously much less than that seen for the class B stage at 2.55%.

9.3

Spice Tips

- The current through a resistor is directly accessible in PSpice. No zero-valued voltage source needs to be placed in series with the resistor. This simplifies the creation of the input Spice deck.

- The Fourier analysis command of Spice decomposes a time-domain waveform into its Fourier series components. These include the fundamental, a DC component, and the next eight harmonics. A total harmonic distortion (THD) measure is also provided.

- Fourier analysis is performed on the last cycle of a time-varying waveform computed by Spice. It is important that the waveform has reached steady state by the final cycle if the Fourier analysis results are to be interpreted correctly.

- Amplifier power efficiency can be computed using the results obtained through a transient analysis, that is, voltages and currents. The postprocessing capabilities of Probe are helpful in generating graphic displays of the various power waveforms by multiplying different voltage and current signals together and by computing a running average.

9.4

Problems

9.1 A class A emitter–follower, biased as in Fig. 9.1, uses $V_{CC} = 5$ V and $R = R_L = 1$ kΩ, with all transistors (including Q_3) identical. Assume β is very large. For linear operation, what are the upper and lower limits of the output voltage and the corresponding inputs? How do these values change if the emitter–base junction area of Q_3 is made twice as big as that of Q_2? Half as big? Repeat for $\beta = 50$.

9.2 A source–follower circuit using enhancement NMOS transistors is constructed following the pattern shown in Fig. 9.1. All three transistors used are identical, with $V_t = 1$ V and $\mu_n C_{OX} = 20$ mA/V^2. In addition, $V_{CC} = 5$ V and $R = R_L = 1$ kΩ. For linear operation, what are the upper and lower limits of the output voltage and the corresponding inputs?

9.3 The BiCMOS follower shown in Fig. P9.3 uses devices for which $I_S = 100$ fA, $\beta = 100$, $\mu_n C_{OX} = 20$ mA/V^2, and $V_t = -2$ V. For linear operation, what is the range of output voltages obtained with $R_L = \infty$ as calculated by Spice? With $R_L = 100$ Ω? What is the smallest load resistance allowed for which a 1 V peak sine-wave output is available? What is the corresponding power-conversion efficiency?

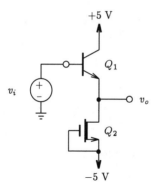

Figure P9.3

9.4 Consider the feedback configuration with class B output shown in Fig. P9.4. Let the amplifier gain $A_o = 100$ V/V and the bipolar transistors be modeled after the NA51 and NA52 commercial transistors. Spice parameters for these devices can be obtain from Fig. 9.7. Compute the input–output voltage transfer characteristics using Spice. Compare these results with those generated by Spice for a corresponding class B output stage without feedback.

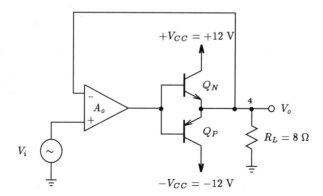

Figure P9.4

9.5 For the feedback amplifier shown in Fig. P9.4 with pertinent parameters described in Problem 9.4, compute the output voltage transient waveform for an input 5 kHz sine-wave voltage signal of 6 V peak using Spice. Also, have Spice compute the Fourier series coefficients of this output signal. What is the resulting total harmonic distortion (THD) of this amplifier? Repeat the analysis on a corresponding class B output stage without feedback present. How do the THD results compare?

9.6 Consider the class B output stage using enhancement MOSFETs shown in Fig. P9.6. Let the devices have $|V_t| = 1$ V and $\mu_n C_{OX} = 2\mu_p C_{OX} = 200$ mA/V^2. With a 10 kHz sine-wave input of 5 V peak and a high value of load resistance, what peak output results? What fraction of the sine-wave period does the crossover interval represent? For what value of load resistance is the peak output voltage reduced to half the input?

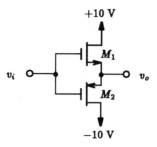

Figure P9.6

9.7 Consider the complementary BJT class B output stage constructed from 2N2222A commercial transistors. (See Section 9.1 for Spice model parameters.) For ± 10 V power supplies and a 100 Ω load resistance, what is the maximum sine-wave output power available? What is the power conversion efficiency? For output signals of half this amplitude, find the output power, the supply power, and the power-conversion efficiency.

9.8 The circuit shown in Fig. P9.8 uses four matched transistors for which $I_S = 10$ fA and $\beta \geq 50$. What quiescent current flows in the output transistors? What bias current flows in the bases of the input transistors? Where does it flow? What is the net input current (the offset current) for a β mismatch of 10%? For a load resistance $R_L = 100\ \Omega$, what is the input resistance? What is the small-signal voltage gain?

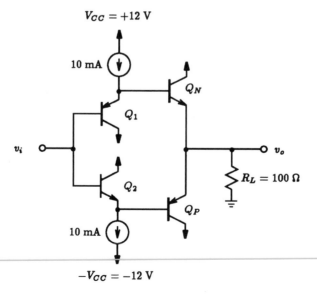

Figure P9.8

9.9 Characterize a Darlington compound transistor formed from two *npn* BJTs modeled after the 2N3904 commercial transistor using Spice. For operation at 10 mA, what is the equivalent β_{eq}, V_{BEeq}, $r_{\pi eq}$, and g_{meq}?

9.10 For the current conveyor circuit shown in Fig. P9.10, assuming all transistors have large β and $A_o = 10^6$, use Spice to compute the output current i_o when the output is shorted directly to ground. Compare this to the situation where $\beta = 100$. By what percentage has i_o changed? The diodes are meant to be realized using a diode-connected transistor.

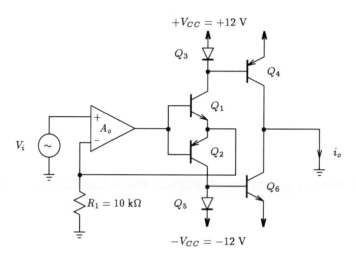

Figure P9.10

10

Analog Integrated Circuits

So far we have looked at various circuit building blocks that were combined in a rather simplistic way to form operational amplifiers whose large- and small-signal static behavior we analyzed using Spice. We can use this same approach for more sophisticated analog ICs, such as the very popular bipolar 741 op amp circuit that appears in Chapter 2. In this chapter we shall perform a detailed Spice analysis of the 741 op amp at the transistor level, including an investigation into its static and dynamic circuit behaviors. We also will take a brief look at its expected thermal noise behavior using the noise analysis capabilities of Spice. Then we shall investigate a CMOS op amp design that is finding important application in VLSI systems.

10.1

A Detailed Analysis of the 741 Op Amp Circuit

A detailed circuit schematic of the 741 op amp is shown in Fig. 10.1. It consists of five main parts: the bias circuit, the input gain stage, the second gain stage, the output buffer, and the short-circuit protection circuit. Each of these stages, in one form or another, was analyzed with Spice in previous chapters. For a detailed explanation of each section of the 741 op amp, see Chapter 10 of Sedra and Smith.

We shall demonstrate how to use Spice to analyze the behavior of a complex analog circuit such as this 741 op amp circuit. In this more complex setting we will review many procedures and setups similar to those in our earlier examples. New information, however, is provided on the Spice method used to compute the thermal noise behavior of the 741 op amp.

Assuming the typical set of device parameters for both the *npn* and *pnp* transistors shown in Table 10.1, we shall compute the amplifier's DC operating point gain and frequency response, slew rate, and noise behavior. Reduced versions of these

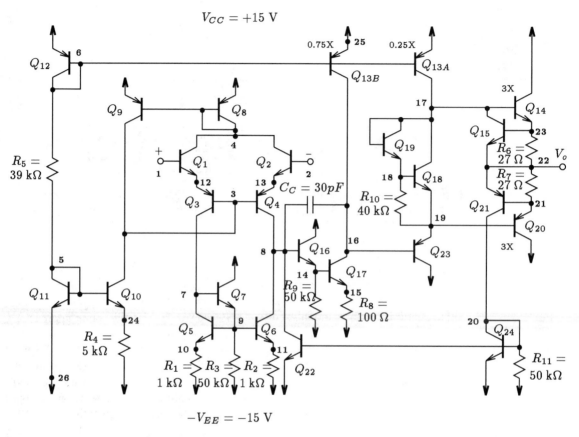

Figure 10.1   The 741 op amp circuit.

parameters were used by Sedra and Smith in their Chapter 10 hand analysis of the 741 op amp circuit. We can compare their results with those obtained with Spice.

The Spice input file describing the circuit of the 741 op amp is listed in Fig. 10.2. The differential input is driven by the familiar multiple-input voltage-source arrangement of Section 6.1, shown again in Fig. 10.3. This arrangement provides maximum flexibility when performing different types of analysis on a differential amplifier (e.g., differential-mode and common-mode transfer characteristics).

10.1.1   DC Analysis of the 741 Op Amp

A DC operating-point analysis will provide us with some insight into the DC behavior of all the transistors in the op amp circuit for grounded inputs, in case such insight is needed. We also shall request a DC sweep of the input differential voltage to indicate the boundaries of the high-gain region of the amplifier. Thus, other analyses can be performed on the amplifier while it is biased around a known operating point, prefer-

Table 10.1 Typical Spice parameters for integrated *npn* and *pnp* transistors [Gray and Meyer, 1984].

Parameter	*npn* Transistor	*pnp* Transistor
β_F	200	50
β_R	2	4
V_{AF}	125 V	50 V
I_S	10 fA	10 fA
τ_F	0.35 ns	30 ns
r_b	200 Ω	300 Ω
r_c	200 Ω	100 Ω
r_e	2 Ω	10 Ω
C_{je}	1.0 pF	0.3 pF
V_{je}	0.7 V	0.55 V
M_{je}	0.33	0.5
C_{jc}	0.3 pF	1.0 pF
V_{jc}	0.55 V	0.55 V
M_{jc}	0.5	0.5
C_{js}	3.0 pF	3.0 pF
V_{js}	0.52 V	0.52 V
M_{js}	0.5	0.5

ably a point within the linear region of the amplifier. Initially we set up the DC sweep so that the input differential voltage is varied between the limits of the two power supplies. However, after the first submission to Spice we refine the sweep range to vary between -400 μV and -200 μV in 10 μV increments so that we obtain better resolution around the high-gain region of the amplifier.[†] During this analysis the input common-mode level is held at ground potential.

Spice produces the large-signal differential-mode transfer characteristic and the DC operating-point information of the 741 op amp. The former is shown in Fig. 10.4. From this graph we observe several characteristics of the op amp. First, the linear region of the amplifier, as seen from the input terminals of the op amp, is bounded between -360 μV and -268 μV. Conversely, this corresponds to a maximum output voltage swing bounded in the negative direction by -13.2 V and in the positive direction by $+13.2$ V. This suggests that the small-signal DC gain of this amplifier is approximately $[13.2 - (-13.2)]/[-268\mu - (-360\mu)] = +287$ kV/V. We also notice that the transfer characteristics of this amplifier cross the 0 V output axis somewhere between -320 μV and -310 μV. A careful look using Probe indicates that the crossover point occurs at -314.1 μV. Thus, this particular amplifier has an

---

[†]To quickly converge on the location of the high-gain region of the op amp, one can connect the op amp in a unity-gain configuration with the positive terminal of the op amp connected to ground and compute the input-referred offset voltage of the op amp V_{OS}. Knowing this voltage, we can infer that the differential transfer characteristics of the op amp will cross the horizontal voltage axis ($V_o = 0$ V) at $-V_{OS}$.

```
The 741 Op-Amp

** Circuit Description **

* power supplies                    * short-circuit protection circuitry
Vcc 25 0 DC +15V                    Q15 17 23 22 npn
Vee 26 0 DC -15V                    Q21 20 21 22 pnp
                                    Q22 8 20 26 npn
* differential-mode                 Q24 20 20 26 npn
  signal level                      R11 20 26 50k
Vd 101 0 DC 0V
Rd 101 0 1                          * biasing stage
EV+ 1 100 101 0 +0.5                Q10 3 5 24 npn
EV- 2 100 101 0 -0.5                Q11 5 5 26 npn
* common-mode signal level          Q12 6 6 25 pnp
Vcm 100 0 DC 0V                     R4 24 26 5k
                                    R5 6 5 39k

* first or input stage              * transistor model statements
Q1 4 1 12 npn                       .model npn NPN ( Bf=200 Br=2.0 VAf=125V Is=10fA Tf=0.35ns
Q2 4 2 13 npn                       + Rb=200 Rc=200 Re=2 Cje=1.0pF Vje=0.70V Mje=0.33 Cjc=0.3pF
Q3 7 3 12 pnp                       + Vjc=0.55V Mjc=0.5 Cjs=3.0pF Vjs=0.52V Mjs=0.5)
Q4 8 3 13 pnp                       .model pnp PNP ( Bf=50 Br=4.0 VAf=50V Is=10fA Tf=30ns
Q5 7 9 10 npn                       + Rb=300 Rc=100 Re=10 Cje=0.3pF Vje=0.55V Mje=0.5 Cjc=1.0pF
Q6 8 9 11 npn                       + Vjc=0.55V Mjc=0.5 Cjs=3.0pF Vjs=0.52V Mjs=0.5)
Q7 25 7 9 npn
Q8 4 4 25 pnp                       ** Analysis Requests **
Q9 3 4 25 pnp                       .OP
R1 10 26 1k                         .DC Vd -400uV -200uV 10uV
R2 11 26 1k
R3 9 26 50k                         ** Output Requests **
                                    .PLOT DC V(22)
* second stage                      .probe
Q13B 16 6 25 pnp 0.75               .end
Q16 25 8 14 npn
Q17 16 14 15 npn
R8 15 26 100
R9 14 26 50k
Cc 8 16 30p

* output or buffer stage
Q13A 17 6 25 pnp 0.25
Q14 25 17 23 npn 3
Q18 17 18 19 npn
Q19 17 17 18 npn
Q20 26 19 21 pnp 3
Q23 26 16 19 pnp
R6 22 23 27
R7 21 22 27
R10 18 19 40k
```

Figure 10.2 Spice input deck for analyzing the DC circuit behavior of the 741 op amp. Because the number of transistors exceeds the limit imposed by the student version of PSpice, this particular Spice input deck is to be processed by the professional version of PSpice, which has no transistor count limit.

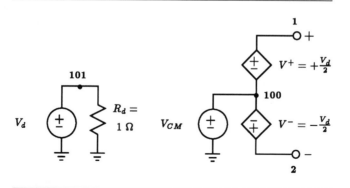

Figure 10.3 Multiple-input voltage source used to drive the differential input of the 741 op amp.

input-referred DC offset voltage of $+314.1\ \mu\text{V}$. This is the systematic offset of the op amp design and does not include offsets that result from component mismatches.

The DC operating-point information (which is not shown here) reveals that for grounded amplifier inputs, transistors Q_{13A} and Q_{13B} are saturated, causing transistors Q_{14}, Q_{18} to Q_{20}, and Q_{23} to cut off. Except for these transistors and those of the

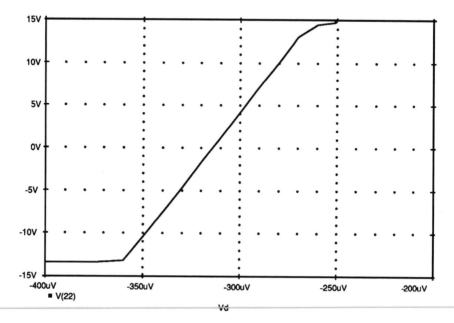

Figure 10.4 Large-signal differential-input transfer characteristic for the 741 op amp circuit. The input common-mode voltage level is set to 0 V.

protection circuit, the remaining transistors are operating in their active regions. Other DC information, such as input bias currents, static power dissipation, and so on, was calculated through the .OP command; however, these results are obtained when the op amp is biased outside of its normal operating region (i.e., its linear region). Since the op amp is generally operated in its linear region, it is more appropriate to repeat the DC operating-point command with the op amp biased there. This is easily accomplished by applying a differential-input offset voltage of $-314.1 \mu V$ across the two inputs to the op amp. We can then repeat the DC analysis with the following element statement replacing the one listed in Fig. 10.2:

```
Vd 101 0 DC -314.1uV
```

On completion of the Spice job, we obtain the following small-signal bias solution:

```
****      SMALL SIGNAL BIAS SOLUTION        TEMPERATURE =   27.000 DEG C
****************************************************************************

 NODE    VOLTAGE      NODE    VOLTAGE      NODE    VOLTAGE      NODE    VOLTAGE

(   1)-157.1E-06  (    2) 157.1E-06  (    3)   -1.0496  (    4)   14.4540
(   5)  -14.3510  (    6)   14.3420  (    7)  -13.9310  (    8)  -13.7450
(   9)  -14.4640  (   10)  -14.9920  (   11)  -14.9920  (   12)    -.5265
(  13)    -.5263  (   14)  -14.2900  (   15)  -14.9340  (   16)   -1.1909
(  17)     .5834  (   18)     .0348  (   19)    -.5789  (   20)  -15.0000
(  21)    -.0040  (   22) 592.6E-06  (   23)     .0052  (   24)  -14.9020
(  25)   15.0000  (   26)  -15.0000  (  100)    0.0000  (  101) -314.1E-06


   VOLTAGE SOURCE CURRENTS
   NAME            CURRENT

   Vcc            -1.841E-03
   Vee             1.841E-03
   Vd              3.141E-04
   Vcm            -6.897E-08


   TOTAL POWER DISSIPATION    5.52E-02  WATTS

**** VOLTAGE-CONTROLLED VOLTAGE SOURCES

 NAME            EV+          EV-
 V-SOURCE       -1.571E-04    1.571E-04
 I-SOURCE       -3.440E-08   -3.456E-08
```

From this output we see that the 741 op amp dissipates a static power of 55.2 mW. We can also calculate the input bias and offset currents of the op amp from the currents passing through the two VCVSs, EV+ and EV−. The input bias current I_B is $[(-34.40 \text{ n}) + (-34.56 \text{ n})]/2 = 34.48 \text{ nA}$, and the offset current I_{OS} is $|(-34.40 \text{ n}) - (-34.56 \text{ n})| = 0.16 \text{ nA}$. This particular offset current is due solely to the systematic offset of the op amp design itself, as opposed to any random offset effect that can occur during manufacture.

Also included in the Spice output file is a detailed listing of the operating point of each transistor. Table 10.2 shows the collector current (in microamperes) of each transistor in the 741 op amp circuit, compared with the current levels computed by hand in Sedra and Smith. Agreement is reasonable.

To determine the input common-mode range (CMR) of the 741 op amp, we shall sweep the input common-mode level V_{CM} between the two power supply limits, maintaining the input differential offset voltage at $-314.1\ \mu V$ to ensure that the op amp is biased in the linear region. To do this, we simply alter the .DC sweep command listed in the Spice deck of Fig. 10.2 as follows:

```
.DC Vcm -15V +15V 0.1V
```

No other alterations are necessary.

With the revised input file Spice produces the large-signal common-mode DC transfer characteristic displayed in Fig. 10.5. We see that linear operation is maintained over an input CMR between -10.0 V and $+8.05$ V, corresponding to an output voltage swing varying between -13.26 V and 14.27 V. The lower limit of the CMR is caused by saturation of Q_{13A} of the second stage. Similarly, the upper limit of the CMR is caused by saturation of transistor Q_{17}, also of the second stage. We also can estimate the common-mode voltage gain in the linear region of the op amp to be $[(-13.26) - 14.27]/[8.05 - (-10.0)] = -1.53$ V/V.

These results are quite different from those normally encountered in an actual 741 op amp. Typically, the input CMR is determined by the transistors of the front-end stage leaving their linear regions rather than by transistors of the second stage doing so, as we have just seen. We attribute this difference to the fact that the 741 op amp consists of six different types of transistor (i.e., small *npn*, large *npn*, lateral *pnp*, substrate *pnp*, dual-emitter *pnp*, and a dual-collector *pnp*), each having different terminal characteristics. The preceding Spice simulation of the 741 op amp uses only

Table 10.2 DC collector currents of the 741 circuit in microamperes as computed by hand analysis and by Spice.

Transistor	Hand Analysis	Spice	Transistor	Hand Analysis	Spice
Q_1	9.5	7.68	Q_{13B}	550	658
Q_2	9.5	7.71	Q_{14}	154	170
Q_3	9.5	7.59	Q_{15}	0	~ 0
Q_4	9.5	7.63	Q_{16}	16.2	17.1
Q_5	9.5	7.55	Q_{17}	550	644
Q_6	9.5	7.56	Q_{18}	165	198
Q_7	10.5	10.8	Q_{19}	15.8	16.2
Q_8	19	14.8	Q_{20}	154	168
Q_9	19	19.4	Q_{21}	0	~ 0
Q_{10}	19	19.6	Q_{22}	0	~ 0
Q_{11}	730	732	Q_{23}	180	213
Q_{12}	730	708	Q_{24}	0	~ 0
Q_{13A}	180	215			

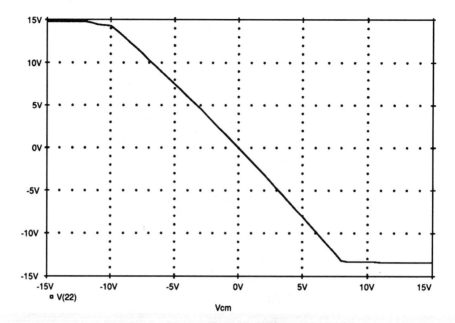

Figure 10.5 Large-signal common-mode DC transfer characteristic of the 741 op amp circuit shown in Fig. 10.1. An input differential offset voltage of $-314.1\ \mu V$ is applied to the input of the amplifier input to prevent any premature saturation.

two model types, and some behavior of the actual 741 op amp is not properly captured in our simulations.

The foregoing examples are all of a large-signal analysis. Much of Sedra and Smith's discussion is based on small-signal analysis. We use the .TF command of Spice to obtain the small-signal parameters of the op amp. For instance, the small-signal differential voltage gain and small-signal output resistance of the op amp can be computed by adding the following command to the Spice deck in Fig. 10.2:

```
.TF V(22) Vd
```

It is important to maintain the input differential offset voltage at $-314.1\ \mu V$ in order to keep the amplifier in its linear region. The input common-mode level is also held at ground potential. The results are as follows:

```
****      SMALL-SIGNAL CHARACTERISTICS

          V(22)/Vd =  2.946E+05

          INPUT RESISTANCE AT Vd =  1.000E+00

          OUTPUT RESISTANCE AT V(22) =  1.064E+02
```

Here we see that the small-signal differential voltage gain of +294.6 kV/V is quite close to the +287 kV/V estimated from the slope of the op amp large-signal transfer characteristic in Fig. 10.4. The input resistance stated here is the $1\,\Omega$ resistance connected directly to V_d rather than the input differential resistance of the op amp. To obtain an estimate of the op amp's input resistance, we use the results of the AC analysis that follows. The output resistance of the op amp is seen to be quite low at $106.4\,\Omega$.

We can repeat the preceding .TF analysis with respect to the input common-mode level around the operating point defined by $V_{CM} = 0$ V and $V_d = -314.1\,\mu$V. We use the command:

```
.TF V(22) Vcm
```

The results are then found in the output file:

```
****      SMALL-SIGNAL CHARACTERISTICS

    V(22)/Vcm = -1.555E+00

    INPUT RESISTANCE AT Vcm =  6.064E+08

    OUTPUT RESISTANCE AT V(22) =  1.064E+02
```

The common-mode voltage gain is -1.555 V/V, the input common-mode resistance is a very large 606.4 MΩ, and the output resistance is, as expected, the same as the previous calculation of $106.4\,\Omega$.

When we divide the differential voltage gain of 2.946×10^5 V/V by the magnitude of common-mode voltage gain of 1.555 V/V, we obtain 105.6 dB as the common-mode rejection ratio (CMRR) for the 741 op amp.

10.1.2 Gain and Frequency Response of the 741 Op Amp

With several alterations made to the Spice input file in Fig. 10.2 we can carry the small-signal analysis over to the frequency domain and have Spice compute the differential magnitude and phase response of the op amp. First, the input differential excitation should be changed to include an AC voltage component. The amplitude of this AC input is not important because AC analysis is performed using a linearized model of the transistor circuit. A 1 V amplitude is commonly chosen for the input signal because the output voltage then directly represents the transfer function of the circuit. The input DC offset voltage remains at $-314.1\,\mu$V in order to keep the op amp in its linear region. Thus the input excitation statement should be changed to read:

```
Vd 101 0 DC -314.1uV AC 1V
```

We also request that Spice compute the frequency response of the amplifier over a frequency interval beginning at 0.1 Hz and ending at 100 MHz using the following statement:

```
.AC DEC 10 0.1Hz 100MegHz
```

A .PLOT statement can be included to request graphical display of the frequency response analysis's output voltage:

```
.PLOT AC VdB(22) Vp(22)
```

```
The 741 Op-Amp

** Circuit Description **

* power supplies                    * short-circuit protection circuitry
Vcc 25 0 DC +15V                    Q15 17 23 22 npn
Vee 26 0 DC -15V                    Q21 20 21 22 pnp
                                    Q22 8 20 26 npn
* DC & AC differential-mode         Q24 20 20 26 npn
  signal level                      R11 20 26 50k
Vd 101 0 DC -314.1uV AC 1V
Rd 101 0 1                          * biasing stage
EV+ 1 100 101 0 +0.5               Q10 3 5 24 npn
EV- 2 100 101 0 -0.5               Q11 5 5 26 npn
* common-mode signal level         Q12 6 6 25 pnp
Vcm 100 0 DC 0V                     R4 24 26 5k
                                    R5 6 5 39k

* first or input stage
Q1 4 1 12 npn                       * transistor model statements
Q2 4 2 13 npn                       .model npn NPN ( Bf=200 Br=2.0 VAf=125V Is=10fA Tf=0.35ns
Q3 7 3 12 pnp                       + Rb=200 Rc=200 Re=2 Cje=1.0pF Vje=0.70V Mje=0.33 Cjc=0.3pF
Q4 8 3 13 pnp                       + Vjc=0.55V Mjc=0.5 Cjs=3.0pF Vjs=0.52V Mjs=0.5)
Q5 7 9 10 npn                       .model pnp PNP ( Bf=50 Br=4.0 VAf=50V Is=10fA Tf=30ns
Q6 8 9 11 npn                       + Rb=300 Rc=100 Re=10 Cje=0.3pF Vje=0.55V Mje=0.5 Cjc=1.0pF
Q7 25 7 9 npn                       + Vjc=0.55V Mjc=0.5 Cjs=3.0pF Vjs=0.52V Mjs=0.5)
Q8 4 4 25 pnp
Q9 3 4 25 pnp                       ** Analysis Requests **
R1 10 26 1k                         .AC DEC 10 0.1Hz 100MegHz
R2 11 26 1k
R3 9 26 50k                         ** Output Requests **
                                    .PLOT  AC VdB(22) Vp(22)
* second stage                      .PRINT AC Im(EV+) Im(EV-)
Q13B 16 6 25 pnp 0.75              .probe
Q16 25 8 14 npn                     .end
Q17 16 14 15 npn
R8 15 26 100
R9 14 26 50k
Cc 8 16 30p

* output or buffer stage
Q13A 17 6 25 pnp 0.25
Q14 25 17 23 npn 3
Q18 17 18 19 npn
Q19 17 17 18 npn
Q20 26 19 21 pnp 3
Q23 26 16 19 pnp
R6 22 23 27
R7 21 22 27
R10 18 19 40k
```

Figure 10.6   Spice input deck for computing the frequency response of the 741 op amp.

We also include a .PRINT statement for the magnitude of the small-signal current flowing through the two VCVSs in series with the op amp input terminals. This will allow us to calculate the input differential resistance of the op amp with the small-signal input differential voltage applied to the amplifier known to be 1 V. The command line for this is

```
.PRINT AC Im(EV+) Im(EV-)
```

The revised input file for calculating the small-signal frequency response of the 741 op amp is shown in Fig. 10.6.

The results of the frequency-domain analysis are shown in Fig. 10.7. The low-frequency behavior is dominated by a single-pole response having a low-frequency gain of 109.4 dB and a 3 dB bandwidth of approximately 2.6 Hz. The unity-gain frequency f_t of 0.652 MHz is read directly from the graph using Probe. We also find that this amplifier has a phase margin of 66.7°. A second pole is also evident, located at approximately 1.82 MHz.

The low-frequency current drawn by the op amp input terminals was found to be

```
FREQ          IM(EV+)       IM(EV-)
1.000E-01    2.766E-07    2.732E-07
```

We see here a small difference between the two currents drawn by the op amp input terminals, a difference due to op amp systematic offset. To determine the input

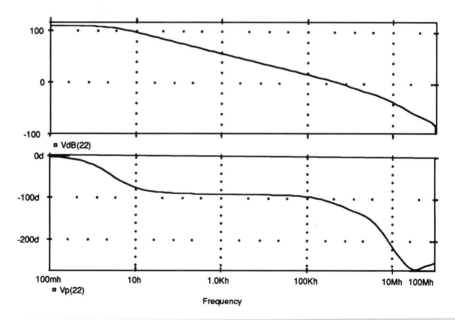

Figure 10.7   Differential magnitude and phase response of the 741 op amp. The low-frequency response behavior of the 741 op amp is dominated by a single-pole roll-off beginning at 2.6 Hz. A second pole is also evident at 1.82 MHz.

differential resistance to the amplifier under such asymmetric conditions, we average the two base currents so that we can eliminate the presence of the offset current in the resistance calculation. We find the average input base current to be 274.9 nA and obtain the input differential resistance at 3.64 MΩ.

10.1.3 Slew-Rate Limiting of the 741 Op Amp

Slew-rate limiting is an important attribute of op amp behavior and usually limits the high-frequency operation of op amp circuits. Figure 10.8 demonstrates how we can connect the op amp in a unity-gain configuration and apply a large voltage pulse to its input so as to reveal both its positive-going and negative-going slew rates. The input file for this arrangement is in Fig. 10.9. Here a zero-valued voltage source described by

```
Vshort 2 22 0
```

is used to form a direct connection between the op amp output and its negative input terminal. This avoids having to renumber the negative input terminal of the op amp with the same number as the op amp output. Another voltage source is used to create the input pulse:

```
Vd 1 0 PWL ( 0,-5V 1ns,+5V 30000ns,+5V 30001ns,-5V 1s,-5V )
```

Notice that this input pulse signal is described in a piecewise linear fashion, beginning at a low level of −5 V, quickly rising to +5 V 1 ns after this, staying there for 30 μs, and then returning to the −5 V level 1 ns later. The input signal remains at this low level of −5 V for the remaining duration of the pulse. A transient analysis is requested to compute the response of the op amp circuit over a 100 μs interval using a 0.1 ns sampling interval.

Spice produces the plot in Fig. 10.10, which shows both the input pulse to the amplifier and its output response. Here we see that the positive-going portion of the output signal has a very different shape from the negative-going portion. In the positive-going

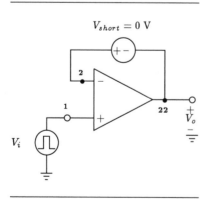

Figure 10.8 Circuit arrangement for computing op amp positive- and negative-going slew-rate limits.

```
The 741 Op-Amp

** Circuit Description **

* power supplies                    * short-circuit protection circuitry
Vcc 25 0 DC +15V                    Q15 17 23 22 npn
Vee 26 0 DC -15V                    Q21 20 21 22 pnp
                                    Q22 8 20 26 npn
* slew-rate limiting setup          Q24 20 20 26 npn
Vd 1 0 PWL ( 0,-5V 1ns,+5V          R11 20 26 50k
          30000ns,+5V
          +30001ns,-5V              * biasing stage
          1s,-5V )                  Q10 3 5 24 npn
Vshort 2 22 0                       Q11 5 5 26 npn
                                    Q12 6 6 25 pnp
* first or input stage              R4 24 26 5k
Q1 4 1 12 npn                       R5 6 5 39k
Q2 4 2 13 npn
Q3 7 3 12 pnp
Q4 8 3 13 pnp                       * transistor model statements
Q5 7 9 10 npn                       .model npn NPN ( Bf=200 Br=2.0 VAf=125V Is=10fA Tf=0.35ns
Q6 8 9 11 npn                       + Rb=200 Rc=200 Re=2 Cje=1.0pF Vje=0.70V Mje=0.33 Cjc=0.3pF
Q7 25 7 9 npn                       + Vjc=0.55V Mjc=0.5 Cjs=3.0pF Vjs=0.52V Mjs=0.5)
Q8 4 4 25 pnp                       .model pnp PNP ( Bf=50 Br=4.0 VAf=50V Is=10fA Tf=30ns
Q9 3 4 25 pnp                       + Rb=300 Rc=100 Re=10 Cje=0.3pF Vje=0.55V Mje=0.5 Cjc=1.0pF
R1 10 26 1k                         + Vjc=0.55V Mjc=0.5 Cjs=3.0pF Vjs=0.52V Mjs=0.5)
R2 11 26 1k
R3 9 26 50k                         ** Analysis Requests **
                                    .TRAN 0.1ns 100us
* second stage
Q13B 16 6 25 pnp 0.75               ** Output Requests **
Q16 25 8 14 npn                     .PLOT TRAN V(1) V(22)
Q17 16 14 15 npn                    .probe
R8 15 26 100                        .end
R9 14 26 50k
Cc 8 16 30p

* output or buffer stage
Q13A 17 6 25 pnp 0.25
Q14 25 17 23 npn 3
Q18 17 18 19 npn
Q19 17 17 18 npn
Q20 26 19 21 pnp 3
Q23 26 16 19 pnp
R6 22 23 27
R7 21 22 27
R10 18 19 40k
```

Figure 10.9 Spice input deck for computing the positive- and negative-going slew-rate limits of the 741 op amp.

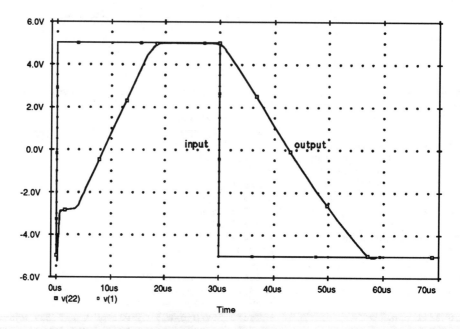

Figure 10.10 Input and output transient voltage waveforms of the 741 op amp circuit connected in a unity-gain configuration. Both the positive-going and negative-going slew-rate limits of the op amp are evident from these results.

signal, the output voltage initially experiences a small jump of 2.2 V, is held constant at −2.8 V for 3.6 μs, then rises linearly to +5 V in 18.2 μs. The positive-going slew rate is estimated at +0.55 V/μs. The negative-going response behaves in a more expected way, showing a steady decline from +5 V to −5 V. The negative-going slew rate is found to be −0.39 V/μs.

10.1.4   Noise Analysis of the 741 Op Amp

As our final analysis, we use the noise analysis capability of Spice to estimate both the output-referred and input-referred thermal-noise power spectral density $S_\eta(j\omega)$ of the 741 op amp circuit shown in Fig. 10.1. Analyzing the noise behavior of op amp circuits is a rather advanced topic and is not covered by Sedra and Smith. It is included here to demonstrate another important Spice analysis capability.

Resistors and semiconductor devices generate various types of noise. The minimization of thermal noise is an important consideration in the design of op amp circuits. Thermal noise is caused by the random movement of charge carriers, which in turn causes instantaneous voltage excursions to appear across device terminals whose average voltage is zero. These voltage variations also affect the circuit node voltages. The total effect on any one particular node is simply the root mean of the sum of the squares of each individual effect. The noise analysis capability of Spice performs this computation in exactly this manner. In general, these noise fluctuations are functions of frequency and therefore are computed in conjunction with the .AC analysis, and the

Table 10.3 General syntax of the noise analysis command in Spice.

Analysis Request	Spice Command
Noise analysis	.NOISE *output_voltage source_name*

noise level is reported in the output file in units of either $V/\sqrt{Hz}$ or $A/\sqrt{Hz}$, depending on its context.

Table 10.3 gives the syntax of the noise analysis command. The key word .NOISE specifies that a noise analysis is to be performed on a given circuit over the frequency interval specified on the .AC command line. The next field, *output_voltage,* denotes a node voltage. The node(s) across which this voltage appears indicates the output port of the circuit. Spice will compute the effective noise voltage spectral density that appears at this port because of internal noise sources. Spice will not compute the effective output noise current spectral density associated with a short-circuit output. The last field, *source_name,* specifies the name of an independent voltage or current source. The output noise voltage spectral density will be referred back to the port defined by this independent source as either an input-referred noise voltage or current spectral density, depending on the input source type.

To access the noise information computed by Spice we use either a .PRINT or .PLOT command with reference to the information stored in the Spice variable *ONOISE* (output noise) or *INOISE* (equivalent input-referred noise).

We return to the Spice input deck in Fig. 10.6, which we used to compute the frequency response of the op amp circuit. We add the following .NOISE command statement and plot request to the file:

```
.NOISE V(22) Vd
.PLOT NOISE ONOISE INOISE
```

Spice produces the output-referred and input-referred noise voltage spectral densities shown in Fig. 10.11 for the 741 op amp. As a point of reference, we see that the output-referred noise voltage density at 1 kHz is $14.7 \, \mu V/\sqrt{Hz}$. Conversely, the output noise power spectral density can be referred to the input of the op amp at this same frequency to find its value of $19.0 \, nV/\sqrt{Hz}$.

10.1.5 A Summary of the 741 Op Amp's Characteristics

Table 10.4 summarizes the results that we have obtained from Spice for the 741 op amp and compares some of these results with those computed by hand in Sections 10.1 to 10.6 of Sedra and Smith. For the most part, the results obtained through hand analysis agree quite well with those computed with Spice, especially with the parameters obtained through the small-signal analysis. The same cannot be said for many of the op amp's large-signal DC characteristics. We attribute this to Sedra and Smith's neglecting the effect of transistor Early voltage in their large-signal DC analysis, unlike their small-signal analysis. The accuracy of the hand analysis could be improved by including the Early voltage effect in the large-signal calculations, but the accuracy gained probably is not worth the additional complexity, especially when Spice could be used.

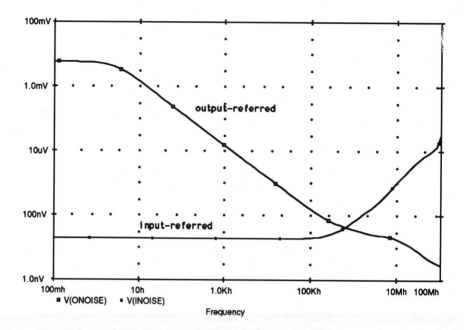

Figure 10.11 Output- and input-referred noise voltage spectral density of the 741 op amp circuit. The vertical axis is in units of $V/\sqrt{Hz}$.

Table 10.4 Comparison of the hand analysis (Sedra and Smith) and Spice results for the 741 op amp circuit shown in Fig. 10.1.

Parameter	Units	Hand Calculation	Spice
Input-referred offset voltage	μV	0	+314.1
Input bias currents	nA	47.5	34.48
Input offset current	nA	0	0.16
Quiescent power dissipation	mW	—	55.2
Differential-mode voltage gain	kV/V	243.1	294.6
Common-mode voltage gain	V/V	—	−1.55
Common-mode rejection ratio	dB	—	105.6
Input differential resistance	$M\Omega$	2.1	3.64
Input common-mode resistance	$M\Omega$	—	606.4
Output resistance	Ω	~75	106.4
Input common-mode range	V	−12.6 to +14.4	−10.0 to +8.05
Output voltage swing	V	−13.5 to +14.0	−13.2 to +13.2
3 dB bandwidth	Hz	4.1	2.6
Unity-gain bandwidth	MHz	1	0.652
Phase margin	degrees	—	66.7
Positive slew rate	$V/\mu s$	+0.63	+0.55
Negative slew rate	$V/\mu s$	−0.63	−0.39
Input-referred noise voltage @ 1 kHz	$nV/\sqrt{Hz}$	—	19.0
Output-referred noise voltage @ 1 kHz	$\mu V/\sqrt{Hz}$	—	14.7

10.2

A CMOS Op Amp

In Fig. 10.12 we show a two-stage CMOS op amp circuit with the transistor geometries seen in Table 10.5. In Example 10.2 of Sedra and Smith, this amplifier stage was analyzed by hand with rather simple transistor models, where it was found to have a DC differential gain of +3125 V/V, an input common-mode range varying between −4.5 V and +3.0 V, and an output voltage swing ranging between −4.5 V and +4.4 V.

We recalculate these parameters of the CMOS amplifier with Spice, using more realistic transistor models. In particular, we consider that each MOSFET is modeled after the transistors found in the 5 μm CMOS process at Bell Northern Research (BNR). We then compare these results with those predicted by the simple formulas presented in Sedra and Smith in their Example 10.2, using the DC operating-point information provided by Spice.

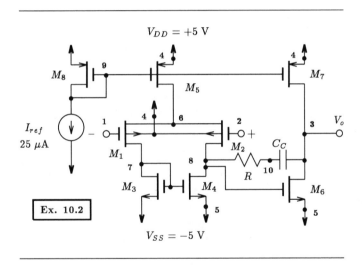

Figure 10.12   Two-stage CMOS op amp circuit with p-channel input transistors.

Table 10.5   Transistor geometries for the CMOS op amp circuit shown in Fig. 10.12.

Transistor	M_1	M_2	M_3	M_4	M_5	M_6	M_7	M_8
W/L	120/8	120/8	50/10	50/10	150/10	100/10	150/10	150/10

```
A CMOS Operational Amplifier (5um CMOS Models)

** Circuit Description **
* power supplies
Vdd 4 0 DC +5V
Vss 5 0 DC -5V
* differential-mode signal level
Vd 101 0 DC 0V
Rd 101 0 1
EV+ 2 100 101 0 +0.5
EV- 1 100 101 0 -0.5
* common-mode signal level
Vcm 100 0 DC 0V
* front-end stage
M1 7 1 6 4 pmos_transistor L=8u W=120u
M2 8 2 6 4 pmos_transistor L=8u W=120u
M3 7 7 5 5 nmos_transistor L=10u W=50u
M4 8 7 5 5 nmos_transistor L=10u W=50u
M5 6 9 4 4 pmos_transistor L=10u W=150u
* second gain stage
M6 3 8 5 5 nmos_transistor L=10u W=100u
M7 3 9 4 4 pmos_transistor L=10u W=150u
* current source biasing stage
M8 9 9 4 4 pmos_transistor L=10u W=150u
Iref 9 5 25uA
* compensation network
Cc 8 10 10pF
R 10 3 10k
* 5um BNR CMOS transistor model statements
.MODEL nmos_transistor nmos ( level=2 vto=1 nsub=1e16 tox=8.5e-8 uo=750
+ cgso=4e-10 cgdo=4e-10 cgbo=2e-10 uexp=0.14 ucrit=5e4 utra=0 vmax=5e4 rsh=15
+ cj=4e-4 mj=2 pb=0.7 cjsw=8e-10 mjsw=2 js=1e-6 xj=1u ld=0.7u )
.MODEL pmos_transistor pmos ( level=2 vto=-1 nsub=2e15 tox=8.5e-8 uo=250
+ cgso=4e-10 cgdo=4e-10 cgbo=2e-10 uexp=0.03 ucrit=1e4 utra=0 vmax=3e4 rsh=75
+ cj=1.8e-4 mj=2 pb=0.7 cjsw=6e-10 mjsw=2 js=1e-6 xj=0.9u ld=0.6u )
** Analysis Requests **
.DC Vd -4mV +4mV 100uV
** Output Requests **
.PLOT DC V(3)
.probe
.end
```

Figure 10.13 Spice input file describing the CMOS amplifier shown in Fig. 10.12. A level 2 MOSFET model of each type of transistor is given.

The Spice description of the CMOS amplifier is shown in Fig. 10.13. The differential input is driven by the multiple-source arrangement shown in Fig. 10.3. A level 2 MOSFET model of each transistor type with a lengthy list of device parameters is provided. These parameters were obtained through extensive measurements on transistors fabricated through the 5 μm CMOS process at BNR.

The first analysis requested is a DC sweep of the input differential voltage V_d between the two supply limits with the input common-mode level V_{CM} set to 0 V; it results in the plot in Fig. 10.14. We see here that the linear region of the amplifier extends between the input voltage levels of -0.9 mV and $+1.2$ mV. This gives rise to a corresponding maximum output voltage swing between -4.40 V and $+4.54$ V. Thus, the small-signal differential gain can be estimated to be about $+4.257$ kV/V. We also see that the input-referred offset voltage is about -220.0 μV.

To perform a DC sweep of the input common-mode level with the differential input to the amplifier offset by $+220.0$ μV, we replace the DC sweep command in Fig. 10.13 by

```
.DC Vcm -5V +5V 0.1V
```

and modify the input differential voltage statement V_d as follows:

```
Vd 101 0 DC +220.0uV
```

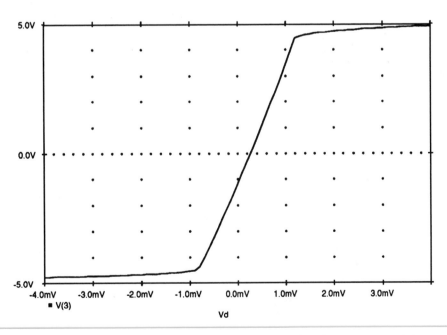

Figure 10.14 Large-signal differential-input transfer characteristic of the CMOS op amp circuit shown in Fig. 10.12. The input common-mode voltage level is set to 0 V.

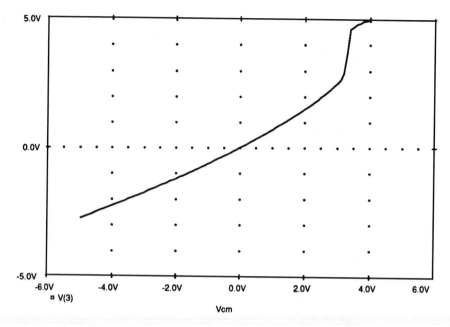

Figure 10.15 Large-signal common-mode DC transfer characteristic of the CMOS op amp circuit shown in Fig. 10.12. An input differential offset voltage of +220.0 μV is applied to prevent premature saturation.

The results of this analysis are displayed in Fig. 10.15. The input common-mode range extends from the lower limit of the power supply V_{SS} to +3.1 V. We can also add commands .TF and .OP to the input deck and obtain the small-signal DC gain of the circuit and any relevant DC operating-point information:

```
.OP
.TF V(3) Vd
```

We then find the following information in the output file:

```
****      SMALL SIGNAL BIAS SOLUTION          TEMPERATURE =   27.000 DEG C
****************************************************************************

NODE     VOLTAGE       NODE     VOLTAGE       NODE     VOLTAGE       NODE     VOLTAGE

(     1)-110.0E-06   (     2) 110.0E-06   (     3)      .0125   (     4)     5.0000
(     5)   -5.0000   (     6)    1.8828   (     7)    -3.5404   (     8)    -3.5516
(     9)    3.4242   (    10)      .0125   (   100)     0.0000   (   101) 220.0E-06

         VOLTAGE SOURCE CURRENTS
         NAME              CURRENT

         Vdd            -8.061E-05
         Vss             8.061E-05
         Vd             -2.200E-04
         Vcm             0.000E+00
```

TOTAL POWER DISSIPATION   8.06E-04 WATTS

NAME	M1	M2	M3	M4
MODEL	pmos_transistor	pmos_transistor	nmos_transistor	nmos_transistor
ID	-1.34E-05	-1.34E-05	1.34E-05	1.34E-05
VGS	-1.88E+00	-1.88E+00	1.46E+00	1.46E+00
VDS	-5.42E+00	-5.43E+00	1.46E+00	1.45E+00
VBS	3.12E+00	3.12E+00	0.00E+00	0.00E+00
VTH	-1.51E+00	-1.51E+00	9.52E-01	9.52E-01
VDSAT	-3.18E-01	-3.18E-01	2.83E-01	2.83E-01
GM	7.18E-05	7.18E-05	5.35E-05	5.34E-05
GDS	7.85E-07	7.84E-07	6.05E-07	6.07E-07
GMB	8.53E-06	8.53E-06	3.88E-05	3.88E-05

NAME	M5	M6	M7	M8
MODEL	pmos_transistor	nmos_transistor	pmos_transistor	pmos_transistor
ID	-2.69E-05	2.87E-05	-2.87E-05	-2.50E-05
VGS	-1.58E+00	1.45E+00	-1.58E+00	-1.58E+00
VDS	-3.12E+00	5.01E+00	-4.99E+00	-1.58E+00
VBS	0.00E+00	0.00E+00	0.00E+00	0.00E+00
VTH	-9.59E-01	9.38E-01	-9.54E-01	-9.64E-01
VDSAT	-4.56E-01	2.87E-01	-4.61E-01	-4.51E-01
GM	8.64E-05	1.14E-04	9.19E-05	8.08E-05
GDS	1.09E-06	6.85E-07	9.36E-07	1.42E-06
GMB	2.63E-05	8.17E-05	2.77E-05	2.48E-05

****   SMALL-SIGNAL CHARACTERISTICS

V(3)/Vd =   3.603E+03

INPUT RESISTANCE AT Vd = 1.000E+00

OUTPUT RESISTANCE AT V(3) = 6.210E+05

Here we see that the small-signal DC gain of +3.603 kV/V has the same order of magnitude as our earlier estimate. We can check this value against the gain predicted by the formula derived in Sedra and Smith: $A_o = g_{m_1}(r_{o2} \parallel r_{o4})g_{m_6}(r_{o6} \parallel r_{o7})$. Substituting the appropriate values from the preceding Spice-generated small-signal data, we see that $A_O = +3.630$ kV/V, which agrees quite well.

We can also make use of the preceding DC and small-signal information to estimate the input CMR and the output voltage swing of this amplifier. The lower limit of the input CMR is determined by M_1 leaving the saturation region, which occurs when the voltage at the gate of M_1 drops below the voltage at its drain ($V(7) = -3.5404$ V) by a single threshold level. From the Spice-generated data, we see that the threshold voltage of M_1 is -1.51 V. We can then compute the lower limit of the input CMR to be -5.1 V. This limit seems to extend beyond the V_{SS} of -5 V, unlike that calculated by hand where the lower CMR limit was found to be -4.5 V. The difference occurs because a threshold voltage of -1 V was assumed for M_1 in the hand calculation, neglecting its dependence on the source–substrate back-bias voltage (the body effect). If the hand calculation used the actual threshold voltage of M_1 of -1.51 V, we would obtain exactly the same lower limit to the CMR found through Spice. The upper limit of the input CMR is determined by M_5 leaving the saturation region. This occurs when the drain of M_5 rises one threshold voltage above its gate voltage of $+3.4242$ V. The threshold voltage of M_5 is -0.959 V. Hence, M_1 leaves saturation when its drain voltage exceeds $+4.38$ V. To relate this voltage to the input of the amplifier, we simply subtract the gate–source voltage of either M_1 or M_2. The result is the upper CMR limit of 4.38 V $- 1.88$ V $= +2.5$ V. This is somewhat lower than that of $+3$ V predicted by hand analysis. It is also lower than that seen in the plot shown in Fig. 10.15 of the output voltage of the amplifier as a function of the input common-mode voltage. There are two reasons for the discrepancy. First, the transistor body effect again has been neglected. In addition, the currents flowing through both M_1 and M_2 are somewhat reduced because of the dependence of the drain current of M_5 on its drain voltage. This in turn decreases the gate–source voltage of M_1 and M_2. Thus, a larger input common-mode voltage can appear at the input terminals of the amplifier before any internal saturation occurs.

Following similar reasoning, the maximum range of the output voltage is one threshold voltage above the gate voltage of M_7. Thus, it is simply 3.42 V $+ 0.954$ V $= +4.37$ V. Likewise, the minimum output voltage is one threshold voltage below the gate voltage of M_6 and is therefore -4.49 V. These results seem to be reasonably close to the results obtained from Fig. 10.14 using Probe ($+4.54$ V and -4.40 V) and to those computed by hand ($+4.4$ V and -4.5 V).

10.3

Spice Tips

- Inputs to differential amplifiers should consist of both a differential and a common-mode component. Arrangements of several voltage sources were given in this chapter, illustrating how the differential and common-mode components can be independently adjusted.

- Performing a proper small-signal analysis on a high-gain amplifier circuit requires that the linear region of the amplifier be located before any transfer function or frequency analysis is performed.

- The linear region of an op amp is located by first sweeping the input differential voltage V_d between the limits of the power supplies with the common-mode voltage V_{CM} equal to 0 V. This analysis is repeated with a reduced sweep range centered more closely around the high-gain linear region of the amplifier, until a smooth transition through the high-gain linear region is achieved. After this, one must check to see whether the input common-mode voltage of 0 V keeps the amplifier in its linear region.

- To quickly converge on the location of the high-gain region of an op amp, the op amp can be connected in a unity-gain configuration, with the positive terminal of the op amp connected to ground, and the input-referred offset voltage of the op amp V_{OS} computed. Knowing this voltage, we can infer that the differential transfer characteristics of the op amp will cross the horizontal voltage axis

$(V_o = 0 \text{ V})$ at $-V_{OS}$. This usually is a valid point within the high-gain region of the op amp.

- Once the linear region of the amplifier is located, the correct DC conditions can be applied to the inputs of the amplifier to establish the proper bias point about which the small-signal analysis of Spice can be performed.

- The small-signal input resistance of an op amp is computed using Spice by applying a known AC voltage across the input terminals of the differential amplifier and computing the AC currents that flow into the amplifier terminals. In many practical amplifier situations these currents will not be equal, so their average is used in the input resistance calculation.

- The output thermal noise behavior of electronic circuits can be estimated using the noise analysis capability of Spice.

10.4

Bibliography

P. R. Gray and R. G. Meyer, *Analysis and Design of Analog Integrated Circuits* (2nd ed.), New York: Wiley, 1984.

10.5

Problems

10.1 With the aid of Spice determine the differential-mode and common-mode voltage gain of the 741 op amp when the power supplies are reduced to ± 5 V.

10.2 Assume that the IC process used to fabricate the 741 op amp circuit undergoes some variation that causes the Early voltage of each device to decrease by 30%, but all other parameters remain invariant. Make the appropriate changes to the Spice parameters listed in Table 10.1, recalculate the large- and small-signal parameters of the 741 op amp listed in Table 10.4, and compare your results to those listed in this table.

10.3 One way to eliminate the input-referred offset voltage of an op amp such as the 741 is to connect what is known as a nulling resistor between the emitters of Q_5 and Q_6 of the first stage of the op amp seen in Fig. 10.1. With the aid of Spice determine the value of this nulling resistor that will cause the input offset voltage to be reduced to a value less than 10 μV. Hint: Connect the op amp in a unity-gain configuration and adjust the value of the nulling resistor to eliminate the offset voltage.

10.4 Through a processing imperfection the β of Q_4 in Fig. 10.1 is reduced to 25 while the β of Q_3 remains at its regular value of 50. Using Spice, find the input offset voltage that this mismatch introduces.

10.5 For a 741 op amp employing ± 7.5 V supplies, determine the maximum output voltage swing possible.

10.6 Determine the phase and gain margin of the 741 op amp when the compensation capacitor C_C is reduced from 30 pF to 20 pF. Will the amplifier be stable when it is connected in a unity-gain configuration? Use the device parameters provided in Table 10.1.

10.7 This problem involves investigating the short-circuit protection capability of the 741 op amp. Consider the op amp connected in a unity-gain configuration with the output terminal connected directly to ground. With the input to the voltage follower connected to +5 V, compare the power dissipated by the output stage with and without the internal short-circuit protection circuitry in place.

10.8 In a particular design of the CMOS op amp of Fig. 10.12 the designer wishes to investigate the effects of increasing the W/L ratio of both M_1 and M_2 by a factor of 4. Assume that all other parameters are kept unchanged.

(a) Find the resulting change in $(|V_{GS}| - |V_t|)$ and in g_m of M_1 and M_2.

(b) What change results in the voltage gain of the input stage? Also, how does this affect the overall voltage gain?

(c) What is the effect on the input offset voltage?

(d) If f_t is to remain the same as before the change to M_1 and M_2, what is the new value of C_C?

10.9 Consider a CMOS amplifier that is complementary to that in Fig. 10.12 and in which each device is replaced by its complement of the same physical size with the supplies reversed. Use the overall conditions as specified in Section 10.2. For all devices have Spice compute I_D and the small-signal parameters g_m and r_o. Compute the gain of the first and second stages, the overall amplifier voltage gain, the input common-mode range, and the output voltage range.

10.10 Using Spice, determine the frequency response behavior of the CMOS op amp shown in Fig. 10.12 for a load capacitance of 1 pF, 5 pF, 10 pF, 20 pF, and 100 pF. How does the amplifier's phase margin vary?

11

Filters and Tuned Amplifiers

In this chapter we investigate the frequency response of various active-*RC* filter circuits and an *LC* tuned amplifier. We also demonstrate how Spice or PSpice can be used in computer-aided design.

11.1

The Butterworth and Chebyshev Transfer Functions

The Butterworth and Chebyshev transfer functions are filter functions commonly used to approximate low-pass transmission characteristics. In Chapter 7 we demonstrated how PSpice can be used to compute the frequency response of an arbitrary transfer function by specifying the gain of a dependent source as a Laplace transform function. Adopting a similar approach, here we verify that the two filter functions calculated in Examples 11.1 and 11.2 of Sedra and Smith do indeed meet the required specifications.

The following ninth-order Butterworth transfer function was calculated by Sedra and Smith in their Example 11.1:

$$T(s) = \frac{6.773 \times 10^4}{(s + 6.773 \times 10^4)} \frac{(6.773 \times 10^4)^2}{(s^2 + s * 1.8794 * 6.773 \times 10^4 + (6.773 \times 10^4)^2)}$$
$$\times \frac{(6.773 \times 10^4)^2}{(s^2 + s * 1.5321 * 6.773 \times 10^4 + (6.773 \times 10^4)^2)}$$
$$\times \frac{(6.773 \times 10^4)^2}{(s^2 + s * 1.0 * 6.773 \times 10^4 + (6.773 \times 10^4)^2)}$$
$$\times \frac{(6.773 \times 10^4)^2}{(s^2 + s * 0.3472 * 6.773 \times 10^4 + (6.773 \times 10^4)^2)} \quad (11.1)$$

It was calculated so that its magnitude response would satisfy the following filter specifications: $f_p = 10$ kHz, $A_{max} = 1$ dB, $f_s = 15$ kHz, $A_{min} = 25$ dB, and DC gain $= 1$. This transfer function poses a problem, however. PSpice limits an expression in the input file to a single line of 131 characters when specifying the gain of a dependent source as a Laplace transform expression (see Section 7.1). The transfer function specified in Eq. (11.1) is very long and cannot be specified on a single line.

To get around this problem, we break up the transfer function into several first- and second-order transfer functions and assign each one as the gain of a separate VCVS. The resulting set of VCVSs are then connected in cascade as shown in Fig. 11.1. Figure 11.2 shows the input file for this circuit arrangement. We are requesting that the transfer function be evaluated linearly over the frequency interval from 1 Hz to 20 kHz with 100 data points collected.

Spice produces the magnitude response of the ninth-order Butterworth transfer function shown in Fig. 11.3. The top graph displays an expanded view of the passband region, and the bottom graph displays a view of both the passband and stopband regions. We see that at $f = f_p = 10$ kHz the gain is 1 dB below the DC value (and is monotonically decreasing). At $f = f_s = 15$ kHz the gain is about 25 dB below the DC value. (Using Probe we find that at this frequency the gain is -25.8 dB.) We can conclude that the transfer function of Eq. (11.1) meets the required specifications.

We would like to repeat the foregoing analysis for the following fifth-order Chebyshev transfer function given in Example 11.2 of Sedra and Smith:

$$T(s) = \frac{6.2832 \times 10^4}{8.1408(s + 0.2895 * 6.2832 \times 10^4)}$$

$$\times \frac{(6.2832 \times 10^4)^2}{(s^2 + s * 0.4684 * 6.2832 \times 10^4 + 0.4293 * (6.2832 \times 10^4)^2)}$$

$$\times \frac{(6.2832 \times 10^4)^2}{(s^2 + s * 0.1789 * 6.2832 \times 10^4 + 0.9883 * (6.2832 \times 10^4)^2)} \quad (11.2)$$

We would also like to verify that the magnitude of this transfer function meets the same specifications as did the Butterworth transfer function.

Adopting the same approach, we describe the Chebyshev transfer function to PSpice with the input file shown in Fig. 11.4. PSpice produces the plot of the magnitude response in Fig. 11.5, which shows both an expanded view of the passband and a broader view of both the passband and stopband regions. We see that the gain in the

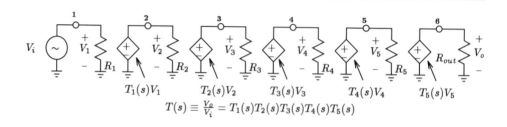

Figure 11.1   Cascade of several VCVSs whose gain is described by either a first-order or a second-order transfer function, forming an overall transfer function $T(s)$.

```
Nineth-Order Butterworth Filter Function

** Circuit Description **
* input signal
Vin 1 0 AC 1V
* a cascade of biquads forming a nineth-order Butterworth filter function
* first biquad
R1 1 0 100Meg
E1 2 0 Laplace V(1) = (6.773e+4)/(s+6.773e+4)
* second biquad
R2 2 0 100Meg
E2 3 0 Laplace V(2) = (6.773e+4*6.773e+4)/(s*s+1.8794*6.773e4*s+6.773e4*6.773e4)
* third biquad
R3 3 0 100Meg
E3 4 0 Laplace V(3) = (6.773e+4*6.773e+4)/(s*s+1.5321*6.773e4*s+6.773e4*6.773e4)
* fourth biquad
R4 4 0 100Meg
E4 5 0 Laplace V(4) = (6.773e+4*6.773e+4)/(s*s+1.0*6.773e4*s+6.773e4*6.773e4)
* fifth biquad
R5 5 0 100Meg
E5 6 0 Laplace V(5) = (6.773e+4*6.773e+4)/(s*s+0.3472*6.773e4*s+6.773e4*6.773e4)
* output
Rout 6 0 1k
** Analysis Requests **
.AC LIN 100 1Hz 20kHz
** Output Requests **
.PLOT AC VdB(6) Vp(6)
.probe
.end
```

Figure 11.2 PSpice input deck for computing the magnitude response of the ninth-order Butterworth transfer function given in Eq. (11.1).

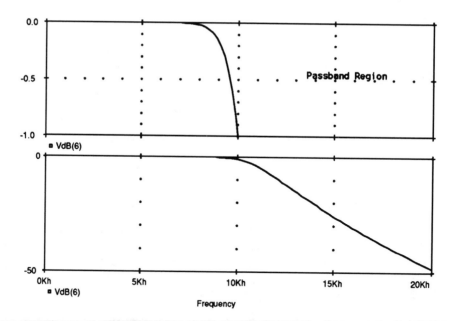

Figure 11.3   Magnitude response of the ninth-order Butterworth transfer function given in Eq. (11.1) for $f_p = 10$ kHz, $A_{max} = 1$ dB, $f_s = 15$ kHz, and $A_{min} = 25$ dB. The top graph displays an expanded view of the passband region, and the bottom graph displays a view of both the passband and stopband regions.

passband region (0 Hz to 10 kHz) oscillates between 0 and -1 dB. For frequencies above 10 kHz the gain rolls off monotonically with increasing frequency. At the stopband edge of $f = f_s = 15$ kHz, we see from the lower graph in Fig. 11.5 that the gain is -30 dB. Therefore, we can conclude that the lower-order Chebyshev filter function given by Eq. (11.2) satisfies the same specifications as the higher-order Butterworth filter function described by Eq. (11.1).

11.2

Second-Order Active Filters Based on Inductor Replacement

In Fig. 11.6 we display a circuit consisting of a cascade of two second-order simulated-*LCR* resonator circuits and a single first-order op amp *RC* circuit. The components of this circuit were selected so that it realizes the fifth-order Chebyshev function in Eq. (11.2). Using Spice we would like to verify this by comparing the magnitude response of this circuit with that computed directly from its transfer function, shown in Fig. 11.5. We assume that the op amps are quasi-ideal and are modeled after a VCVS with a gain of 10^6 V/V as shown in Fig. 11.6(b). This process normally would be the first step after the circuit has been designed, and we want to verify the synthesis procedure.

```
Fifth-Order Chebyshev Filter Function

** Circuit Description **
* input signal
Vin 1 0 AC 1V
* a cascade of biquads forming a fifth-order Chebyshev filter function
* first biquad
R1 1 0 100Meg
E1 2 0 Laplace V(1) =  (6.2832e+4/8.1408)/(s+0.2895*6.2832e+4)
* second biquad
R2 2 0 100Meg
E2 3 0 Laplace V(2) =  (6.2832e+4*6.2832e+4)/(s*s+0.4684*6.2832e+4*s+0.4293*6.2832e+4*6.2832e+4)
* third biquad
R3 3 0 100Meg
E3 4 0 Laplace V(3) =  (6.2832e+4*6.2832e+4)/(s*s+0.1789*6.2832e+4*s+0.9883*6.2832e+4*6.2832e+4)
* output
Rout 4 0 1k
** Analysis Requests **
.AC LIN 100 1Hz 20kHz
** Output Requests **
.PLOT AC VdB(4) Vp(4)
.probe
.end
```

Figure 11.4 PSpice input deck for computing the magnitude response of the fifth-order Chebyshev transfer function given in Eq. (11.2).

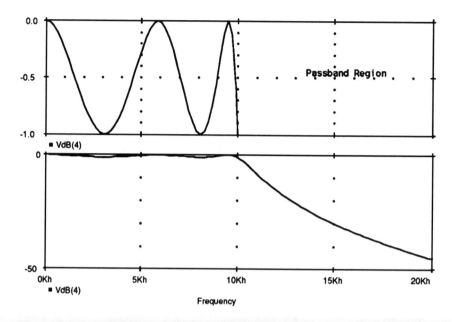

Figure 11.5   Magnitude response of the fifth-order Chebyshev transfer function given in Eq. (11.2) for $f_p = 10$ kHz, $A_{max} = 1$ dB, $f_s = 15$ kHz, and $A_{min} = 25$ dB. The top graph displays an expanded view of the passband region, and the bottom graph displays a view of both the passband and stopband regions.

The Spice input file describing this circuit is shown in Fig. 11.7. Because the same type of quasi-ideal op amp is repeated many times in this circuit, we have chosen to represent it with a single subcircuit named *ideal_opamp* (see Chapter 2 for subcircuits). An AC analysis is requested over the linear frequency interval from 1 Hz to 20 kHz using 100 data points. The input to the filter is driven by a 1 V AC signal.

Spice produces the magnitude response plotted in Fig. 11.8. Both an expanded view of the passband and a view of the passband and stopband regions are shown. When we compare these results with those in Fig. 11.5 computed directly from Eq. (11.2), we see that they are identical for all practical purposes. We can conclude that we carried out the synthesis procedure correctly. The next step is to investigate how the filter magnitude response is affected by the frequency characteristics of a real op amp; this is the subject of the next section on the Tow-Thomas biquad circuit.

11.3

Second-Order Active Filters Based on the Two-Integrator-Loop Topology

Another important class of biquadratic active-*RC* filter circuits are those formed by cascading two integrators in an overall feedback loop. In Fig. 11.9 we show the Tow-Thomas biquad based on this idea. The components were selected so that this circuit realized a second-order bandpass filter with $f_o = 10$ kHz, $Q = 20$, and unity

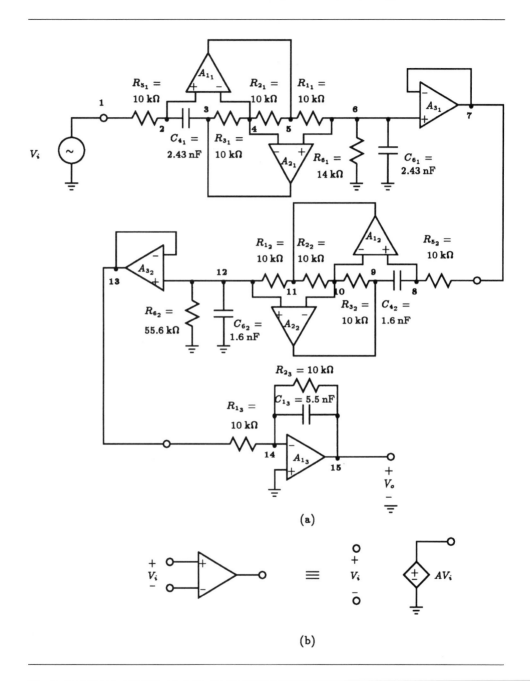

(a)

(b)

Figure 11.6 (a) Fifth-order Chebyshev filter circuit implemented as a cascade of two second-order simulated-*LCR* resonator circuits and a single first-order op amp *RC* circuit. (b) VCVS representation of an ideal op amp (*A* = 10⁶).

```
Fifth-Order Chebyshev Filter Circuit

** Circuit Description **

* op amp subcircuit
.subckt ideal_opamp 1 2 3
* connections:       | | |
*              output | |
*             +ve input |
*              -ve input
Eopamp 1 0 2 3 1e6
.ends ideal_opamp

** Main Circuit **
* input signal source
Vi 1 0 DC 0V AC 1V
* first biquad stage (Wo=41.17k rad/s   Q=1.4)
X_A1_1 5 2 4 ideal_opamp
X_A2_1 3 6 4 ideal_opamp
R1_1 5 6 10k
R2_1 4 5 10k
R3_1 3 4 10k
C4_1 2 3 2.43nF
R5_1 1 2 10k
C6_1 6 0 2.43nF
R6_1 6 0 14k
X_A3_1 7 6 7 ideal_opamp
* second biquad stage (Wo=62.46k rad/s   Q=5.56)
X_A1_2 11 8 10 ideal_opamp
X_A2_2 9 12 10 ideal_opamp
R1_2 11 12 10k
R2_2 10 11 10k
R3_2  9 10 10k
C4_2  8  9 1.6nF
R5_2  7  8 10k
C6_2 12  0 1.6nF
R6_2 12  0 55.6k
X_A3_2 13 12 13 ideal_opamp
* first-order stage
X_A1_3 15 0 14 ideal_opamp
R1_3 13 14 10k
R2_3 14 15 10k
C1_3 14 15 5.5nF
** Analysis Requests **
.AC LIN 100 1Hz 20kHz
** Output Requests **
.PLOT AC VdB(15) Vp(15)
.probe
.end
```

Figure 11.7   Spice input deck for calculating the frequency response of the low-pass filter circuit shown in Fig. 11.6.

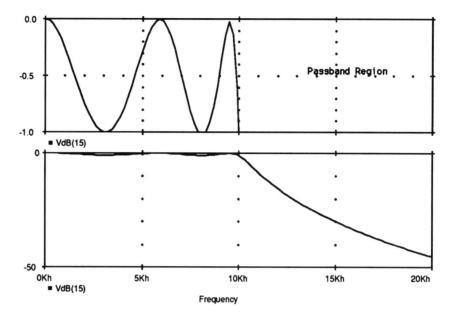

Figure 11.8 Magnitude response of the fifth-order lowpass filter circuit shown in Fig. 11.6. The top graph displays an expanded view of the passband region, and the bottom graph displays a view of both the passband and stopband regions.

center frequency gain. Op amps are assumed to be of the 741 type. Using Spice we would compare the frequency response of this filter circuit to its ideal frequency response (i.e., one with ideal op amps).

One possible approach for evaluating this nonideal frequency response behavior is to describe the entire circuit in Fig. 11.9, including 741 op amps, at the transistor level. This would certainly lead to the most accurate representation of the bandpass filter circuit, but it would demand a lot of computer time and storage to compute the frequency response. A more reasonable approach, and the one we will undertake here, is to use a **macromodel** of the 741 op amp.

We model the terminal behavior of the 741 op amp with the single–time constant linear network shown in Fig. 11.10. Other time constants also can be included in this model by extension of the middle stage. We may be tempted to add nonlinearities to this model, but Spice linearizes a circuit about its operating point prior to the start of the AC analysis, so the inclusion of any op amp nonlinearity in AC analysis is ignored by Spice and serves no purpose. Nonlinearities can, of course, be included in the model when performing transient analysis.

It is a simple matter to derive the open-circuit transfer function of the equivalent op amp circuit shown in Fig. 11.10 and to determine that it has a one-pole response given by

$$\frac{V_o}{V_{id}}(s) = \frac{G_m R_1}{1 + sR_1 C_1} \tag{11.3}$$

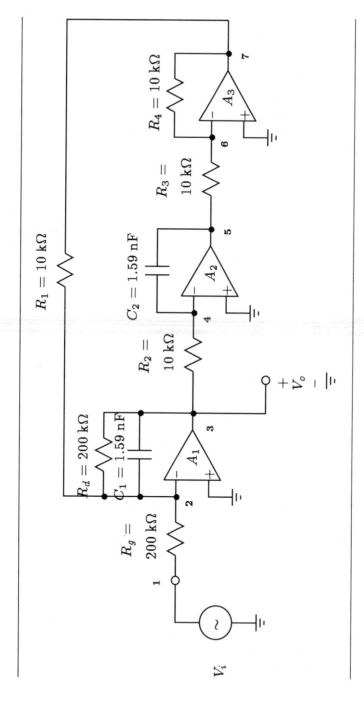

Figure 11.9 Second-order bandpass filter implemented with a Tow-Thomas biquad circuit with $f_o = 10$ kHz, $Q = 20$, and unity center frequency gain. The op amps are assumed to be of the 741 type.

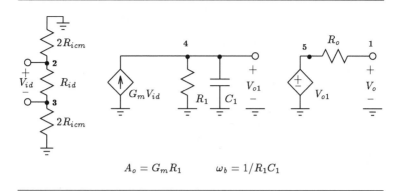

$$A_o = G_m R_1 \qquad \omega_b = 1/R_1 C_1$$

Figure 11.10 One-pole equivalent circuit representation of an op amp operated within its linear region.

The 741 op amp is an internally compensated op amp that has a frequency response characterized by a one-pole frequency roll-off:

$$A(s) = \frac{A_o}{1 + s/\omega_b} \tag{11.4}$$

where A_o denotes the DC gain and ω_b is the 3 dB frequency. The 741 op amp has a nominal DC gain of 2.52×10^5 V/V and a 3 dB frequency of 4 Hz. From Eqs. (11.3) and (11.4) we can write two equations in three unknowns: $A_o = G_m R_1$ and $\omega_b = 1/R_1 C_1$. Assigning $C_1 = 30$ pF, we can solve for $G_m = 0.19$ mA/V and $R_1 = 1.323 \times 10^9$. We can also add some input and output resistances to account for the effects of loading. Here we shall choose $R_{id} = 2$ MΩ, $R_{icm} = 500$ MΩ, and $R_o = 75$ Ω.

Figure 11.11 shows the input file describing the Tow-Thomas biquad circuit of Fig. 11.9. This listing also includes the op amp macromodel shown in Fig. 11.10 representing the small-signal nominal frequency response behavior of the 741 op amp described by the subcircuit *nonideal_opamp*. A 1 V AC signal is applied to the input of the filter so that the output voltage of the filter will also represent the filter transfer function. The frequency response of the filter is computed between 8 kHz and 12 kHz using 100 points linearly separated from one another. We concatenate another Spice deck to this one to generate the "ideal" frequency response. The nonideal subcircuit is replaced with one containing a model of an ideal op amp, such as that used in the previous example in Fig. 11.7.

The results of the two circuit simulations are shown in Fig. 11.12. We see that there are significant differences in their magnitude responses. In fact, we see that the center frequency of the circuit using the 741 op amp has shifted to the left by about 100 Hz, and its 3 dB bandwidth has decreased from 500 Hz to about 110 Hz. This, in effect, is an increase in the intended filter Q of 20 up to 90. In addition, the center frequency gain has also increased from 0 dB to 14 dB.

```
Second-Order Bandpass Filter Circuit (Nonideal Op-Amp)

** Circuit Description **

* op-amp subcircuit
.subckt nonideal_opamp 1 2 3
* connections:          | | |
*                  output | |
*                 +ve input |
*                  -ve input
Ricm+ 2 0 500Meg
Ricm- 3 0 500Meg
Rid 2 3 2Meg
Gm 0 4 2 3 0.19m
R1 4 0 1.323G
C1 4 0 30pF
Eoutput 5 0 4 0 1
Ro 5 1 75
.ends nonideal_opamp


** Main Circuit **
* input signal source
Vi 1 0 DC 0V AC 1V
* Tow-Thomas biquad
X_A1 3 0 2 nonideal_opamp
X_A2 5 0 4 nonideal_opamp
X_A3 7 0 6 nonideal_opamp
Rg 1 2 200k
R1 2 7 10k
R2 3 4 10k
R3 5 6 10k
R4 6 7 10k
Rd 2 3 200k
C1 2 3 1.59nF
C2 4 5 1.59nF
** Analysis Requests **
.AC LIN 100 8kHz 12kHz
** Output Requests **
.PLOT AC VdB(3) Vp(3)
.probe
.end
```

Figure 11.11 Spice input deck for calculating the frequency response of the second-order bandpass filter circuit shown in Fig. 11.9. The op amp is assumed to have a DC gain of 252 kV/V and a 3 dB frequency of 4 Hz—much like the small-signal frequency response characteristics of a 741 op amp circuit.

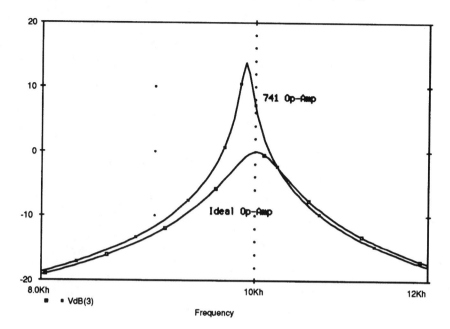

Figure 11.12   Comparing the magnitude response of the Tow-Thomas biquad circuit shown in Fig. 11.9 constructed with 741-type op amps with its ideal magnitude response. These results illustrate the effect of the finite DC gain and bandwidth of the 741 op amp on the frequency response of the Tow-Thomas biquad circuit.

These results illustrate the adverse effects that actual op amps such as the 741 type have on the frequency response of a Tow-Thomas filter circuit. One possible approach for minimizing these effects is to add what is known as a compensation capacitor to the circuit. There are several ways to add such a capacitor; here we shall consider adding it across resistor R_2 in Fig. 11.9. The actual value of the capacitor is not known yet, so we must search for the capacitance that improves the overall filter response and makes the magnitude response of the filter most closely approach the desired or ideal response. We begin with a capacitance of 20 pF and simulate the magnitude response of the filter. We compare this result with the ideal magnitude response and determine whether the added capacitance improves the filter response. If it does, we shall continue increasing the capacitance until we no longer improve the filter response. Here we are using Spice for computer-aided *design* and not just analysis.

To demonstrate this design process, Fig. 11.13 shows plots of the magnitude responses that result from varying the value of the compensation capacitor from 0 pF to 80 pF in 20 pF increments. We see from these results that, as the compensation capacitance increases from 0 pF, both the filter Q and the resonant peak of the filter response tend more closely toward the desired response. Once the value of the compensation

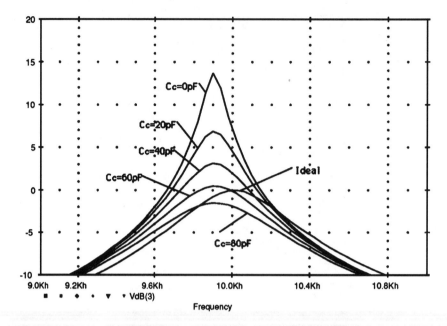

Figure 11.13   Magnitude response of the Tow-Thomas biquad circuit with different values of compensation capacitance. For comparison the ideal response of the Tow-Thomas biquad circuit is also shown.

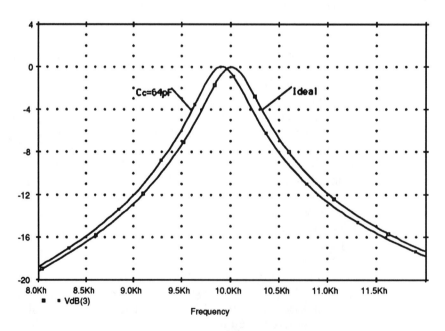

Figure 11.14   Comparing the magnitude response of the Tow-Thomas biquad circuit using a 64 pF compensation capacitor against the ideal response.

capacitor exceeds 60 pF, the response begins to deviate from the desired one. This suggests that the best result lies between 60 pF and 80 pF. More refinement of the same approach between these two values reveals that the best results occur when the compensation capacitor is set at 64 pF.

In Fig. 11.14, we compare the filter magnitude response using a compensation capacitor of 64 pF against the ideal response. The actual filter magnitude response has about the same filter Q and center frequency gain as the ideal response, but they have slightly different center frequencies.

We can conclude from this computer-aided design process that much of the non-ideal effect caused by the limited bandwidth of the 741 op amp has been eliminated by the addition of this single compensation capacitor. Finally, we should note that a closed-form expression [Sedra and Brackett, 1978] exists for determining the required value of the compensation capacitor. This expression, however, relies on knowledge of the value of the op amp f_t. In practical situations the exact value of f_t would not be known, and the appropriate value of the compensation capacitance would be found using a variable capacitor, in much the same way as using Spice.

11.4

Tuned Amplifiers

In Fig. 11.15 we show a special kind of frequency selective network called the LC-tuned amplifier. Loading the drain of a typical resistor-biased transistor circuit with an LC tank circuit results in an amplifier with a highly selective amplitude response. This class of circuit finds extensive application in high-frequency communication systems such as radio.

The tuned circuit portion of the amplifier in Fig. 11.15 was designed in Example 11.4 of Sedra and Smith with a center frequency of 1 MHz, a 3 dB frequency of 10 kHz, and a center frequency gain of -10 V/V. The resistor biasing network was added to bias transistor M_1 at $g_m = 5$ mA/V and $r_o = 10$ kΩ, much like the biasing network assumed by Sedra and Smith in their example. MOSFET M_1 is assumed to be 10 μm long and 1250 μm wide. It also has the following device parameters: $\mu_n C_{OX} = 100$ μA/V^2, $V_T = +2$ V, and $\lambda = 0.1$ V^{-1}. We would like to use Spice to calculate the magnitude response of this circuit and to compare it with our expectations.

Figure 11.16 is the input file for this circuit. The input signal is a 1 V AC signal, and two types of analysis are requested. The first is an operating-point calculation (.OP), which we use to compute the bias point of the MOSFET and to determine whether it has been biased correctly. Second, an AC command computes the frequency response of this amplifier in a frequency range from 0.99 MHz to 1.01 MHz with 100 data points computed in a linear fashion.

From Spice we obtained the magnitude response of the tuned amplifier shown in Fig. 11.17. The center frequency of the magnitude response is almost exactly at the intended design value of 1 MHz, but neither the gain of the amplifier at its center frequency of 15.86 V/V nor its 3 dB bandwidth of 9.01 kHz agrees with their design

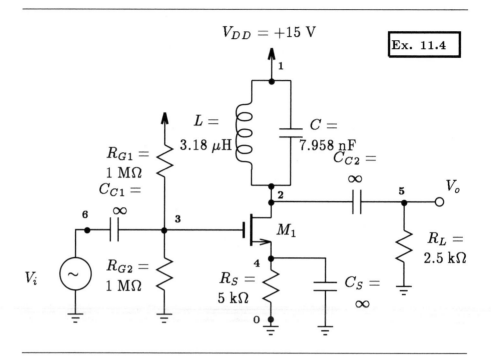

Figure 11.15　Single-MOSFET tuned amplifier with bias circuit.

values. We can account for these differences by reviewing the small-signal model for the MOSFET calculated through the .OP command:

```
****     OPERATING POINT INFORMATION     TEMPERATURE =   27.000 DEG C
*****************************************************************************
**** MOSFETS

NAME          M1
MODEL         NMOS
ID            1.04E-03
VGS           2.29E+00
VDS           9.79E+00
VBS           0.00E+00
VTH           2.00E+00
VDSAT         2.90E-01
GM            7.18E-03
GDS           5.27E-05
GMB           0.00E+00
```

Here we see that both the g_m and r_o (or 1/GDS) of the transistor have deviated from our initial design of 1 mA/V and 10 kΩ, respectively. Since both the center

```
A Bandpass Tuned Amplifier

** Circuit Description **

* power supplies
Vdd 1 0 DC +15V
* input signal source
Vi 6 0 DC 0V AC 1V
* amplifier
Cc1 6 3 1uF
RG1 1 3 1Meg
RG2 3 0 1Meg
M1 2 3 4 4 NMOS L=10u W=1250u
Rs 4 0 5k
Cs 4 0 1uF
Cc2 2 5 1uF
* tuned circuit
L 1 2 3.18uH
C 1 2 7.958nF
* output load
Rl 5 0 2.5k
* MOSFET model statement
.model NMOS nmos (kp=100u Vto=+2V lambda=0.1)
** Analysis Requests **
.OP
.AC LIN 100 0.98MegHz 1.02MegHz
** Output Requests **
.PLOT AC Vm(5)
.probe
.end
```

Figure 11.16 Spice input deck for calculating the magnitude response of the tuned amplifier shown in Fig. 11.15.

frequency gain and 3 dB bandwidth of the tuned amplifier shown in Fig. 11.15 are functions of these two parameters (see Sedra and Smith) we would expect them also to deviate from the intended design values. To obtain better estimates of the center frequency gain and the 3 dB bandwidth, we use the actual bias data generated by Spice. For instance, the equivalent resistance seen at the drain of M_1 is $R = r_o \parallel R_L = (1/5.27 \times 10^{-5}) \parallel 2.5 \times 10^3 = 2.208$ kΩ. Thus, the magnitude of the center frequency gain is $g_m R = 7.18 \times 10^{-3} \times 2.208 \times 10^3 = 15.85$ V/V. Likewise, the 3 dB bandwidth is $B = 1/RC = 1/(2.208 \times 10^3 \times 7.958 \times 10^{-9})$ or 9.057 kHz. Both these values are now much closer to those we obtained through simulation.

For another computer-aided design, let us adjust the value of the load resistor R_L in the tuned amplifier in Fig. 11.15 so that the required 10 kHz bandwidth is achieved. To

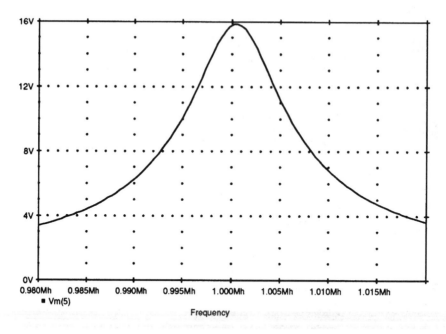

Figure 11.17   Frequency response of the tuned bandpass amplifier shown in Fig. 11.15.

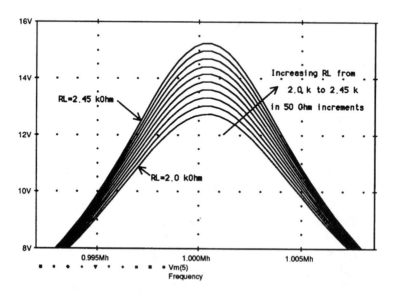

Figure 11.18   Expanded view of the magnitude response of the the tuned amplifier for various load resistances.

make matters more realistic, we shall account for the MOSFET parasitic capacitances by including the following model statement for the device:

```
.model NMOS nmos ( kp=100u Vto=+2V lambda=0.1 tox=100e-10 tt=100n
+                  cgso=100p cgdo=100p cgbo=50p cj=4e-4 cjsw=8e-10 )
```

The DC parameters of this MOSFET are identical to the previous ones. The 1 kΩ source resistance will be included in order to properly account for the effect of these capacitances on the overall circuit behavior. The input deck for this example is quite similar to Fig. 11.16 and is not shown here.

We shall first determine the 3 dB bandwidth of this amplifier with the MOSFET parasitic capacitances present and with the load resistance set at its nominal value of 2.5 kΩ. We resubmit the revised Spice input for analysis, plot the magnitude response, and find that the 3 dB bandwidth is 9.06 kHz. To increase the bandwidth to 10 kHz, we must decrease the load resistance. The amount by which we should decrease R_L is not known yet, so we perform a search using Spice. Beginning with $R_L = 2.45$ kΩ, we repeatedly compute the 3 dB bandwidth while incrementally decreasing the value of R_L by an arbitrary 50 Ω until we reach 2.0 kΩ. If the amplifier 3 dB bandwidth has not increased to 10 kHz or more within this range of load resistance, we must continue the search with R_L less than 2 kΩ.

In Fig. 11.18 we present an expanded view of the magnitude response of the tuned amplifier for the various load resistances. Although it is not directly evident, with the Probe cursor we were able to determine the 3 dB bandwidth in each case. We found that for a load resistance of 2.20 kΩ the 3 dB bandwidth is 10.1 kHz. For a load resistance of 2.25 kΩ, we found that the 3 dB bandwidth is 9.966 kHz. We can therefore conclude that a 10 kHz bandwidth for this amplifier will occur with a load resistance somewhere between 2.20 kΩ and 2.25 kΩ. If we repeat the process for load resistance between 2.20 kΩ and 2.25 kΩ with a 10 Ω increment, we find that the tuned amplifier will have a 10 kHz bandwidth when the load resistance is 2.24 kΩ. The resistor cannot be specified more accurately, so more refinement is pointless.

11.5

Spice Tips

- The element statement of a dependent source with a gain expressed in terms of a Laplace transform function cannot be longer than 131 characters.

- The small-signal frequency response behavior of the 741 op amp can be modeled in Spice with a macromodel that consists of a single–time constant linear network representing the dominant pole behavior of the op amp.

- Spice is useful not only for analyzing a design, but also for fine-tuning a design.

11.6

Bibliography

A. S. Sedra and P. O. Brackett, *Filter Theory and Design: Active and Passive,* Portland, OR: Matrix, 1978.

11.7

Problems

11.1 With the aid of PSpice prepare a plot of both the magnitude and phase of the following filter transfer functions that captures the most important frequency information:

(a) $T(s) = \dfrac{1}{(s^2 + \sqrt{2}s + 1)}$

(b) $T(s) = \dfrac{5\,725\,600}{(s + 125.3)(s^2 + 125.3s + 45\,698)}$

(c) $T(s) = \dfrac{0.2816(s^2 + 3.2236)}{(s + 0.7732)(s^2 + 0.4916s + 1.1742)}$

(d) $T(s) = \dfrac{0.083\,947(s^2 + 17.485\,28)}{(s^2 + 1.357\,15s + 1.555\,32)}$

(e) $T(s) = \dfrac{s^3}{(s^2 + 1000s + 10^6)(s + 1000)}$

11.2 For the first-order op amp circuit shown in Fig. P11.2, select the values of its components so that a transmission zero is formed at a frequency of 1 kHz, a pole occurs at a frequency

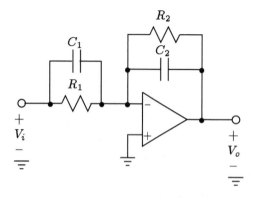

Figure P11.2

of 100 kHz, and the circuit has a DC gain magnitude of unity. The low-frequency input resistance is to be 1 kΩ. Use Spice to plot the magnitude of the transfer function of the circuit that results.

11.3 The circuit of Fig. P11.3 has been designed to realize an allpass transfer function that provides a phase shift of 180° at 1 kHz and to have $Q = 1$. Using Spice verify that this is indeed the case.

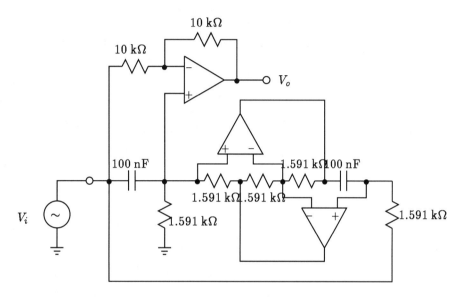

Figure P11.3

11.4 The KHN circuit shown in Fig. P11.4 is used to realize a pair of complex poles located at $\omega = 10^4$ rad/s and $Q = 2$. With the aid of Spice analyze the circuit to determine the transfer functions V_1/V_i, V_2/V_i, and V_3/V_i. Classify each as lowpass, bandpass, bandstop, and so on.

11.5 For the KHN circuit shown in Fig. P11.4 compare the magnitude response $|V_2/V_i(j\omega)|$ between 1 Hz and 10 MHz for op amp DC gain A_o of 10^6, 10^4, and 10^2.

11.6 The circuit of Fig. P11.6 has been designed to realize a lowpass notch filter with $\omega_o = 10^3$ rad/s, $Q = 10$, DC gain = 1, and $\omega_n = 1.2 \times 10^4$. Verify that this is indeed the case with Spice, assuming that the op amps are almost ideal with a frequency-independent voltage gain of 10^6 V/V.

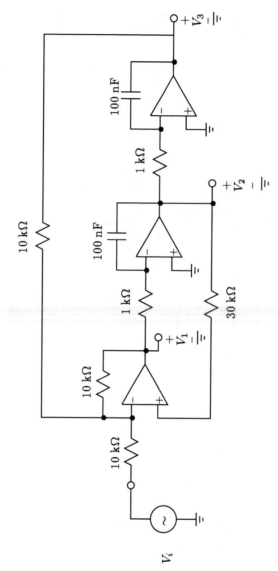

Figure P11.4

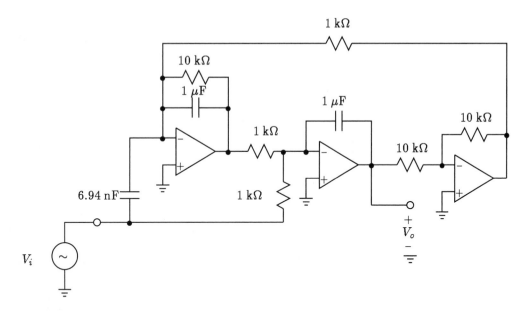

Figure P11.6

11.7 Repeat Problem 11.6, but this time model each op amp after the small-signal frequency characteristics of the 741 op amp. Compare your results with those found previously in Problem 11.6.

11.8 Consider the application of an input signal consisting of two 1 V peak sine waves of 100 and 1909.86 Hz to the input of the notch filter shown in Fig. P11.6. Compare the voltage waveform appearing at the input of the filter with the voltage waveform appearing at its output.

11.9 Using Spice plot the magnitude response $|V_2/V_i|$ of the KHN filter circuit shown in Fig. P11.4, assuming that each op amp has a unity-gain bandwidth of 10 MHz and a DC gain of 10^6 V/V. Compare these results with the ideal response (i.e., ideal op amp case).

11.10 A bandpass amplifier is shown in Fig. P11.10 whose output is coupled to the load resistor of 10 kΩ through a transformer. The transformer consists of a primary and a secondary inductor, $L_p = 100$ μH and $L_s = 10$ μH, respectively, which have a coupling coefficient of 0.8. To specify this coupling, the following Spice statement is included in the Spice input file together with the two inductor statements: K1 Lp Ls 0.8. Using Spice determine the frequency response of this amplifier. What is the center frequency and bandwidth of this amplifier? Assume that the BJTs have model parameters $I_S = 10^{-16}$ A, $\beta_F = 100$, and $V_A = 75$ V.

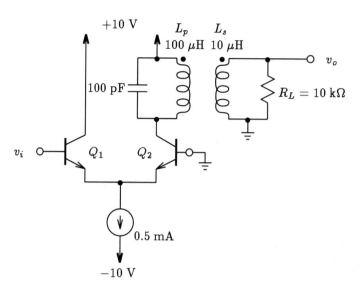

Figure P11.10

Signal Generators and Waveform-Shaping Circuits

In this chapter we investigate the behavior of various types of signal generators and waveform-shaping circuits for both sinusoidal and square waves. First, we investigate the sinusoidal behavior of a Wien bridge oscillator and an active filter–tuned sinusoidal oscillator constructed with the 741 op amp. The spectral purity of the signals generated will be calculated using Spice's Fourier series analysis. Then we investigate several multivibrator circuits that produce square waves, including analysis of bistable, astable, and monostable multivibrator circuits constructed with op amps.

The chapter concludes with the analysis of several precision rectifier op amp circuits commonly employed in instrumentation systems, including a precision rectifier, a peak detector, and a clamping circuit.

12.1

Op Amp–*RC* Sinusoidal Oscillators

Here we investigate the behavior of Wien bridge and active filter–tuned oscillator circuits assumed constructed with 741 op amps. For the examples in the previous chapter, it was sufficient to represent the terminal behavior of the op amp with a linear network (see Section 11.3), but the examples of this chapter rely on the nonlinear operation of the op amp (e.g., output voltage saturation), and thus their terminal behavior must be modeled more completely. Fortunately, many IC component manufacturers are making available macromodels of their op amps in the form of Spice subcircuit descriptions. These macromodels capture much of the op amps' large-signal behavior in a circuit representation that is much simpler than detailed description at the transistor level [Texas Instruments, 1990]. This makes the task of circuit simulation more efficient without sacrificing accuracy. As an example a Spice subcircuit description of the 741 op amp is listed in Fig. 12.1. We shall use this subcircuit description for the 741 op amp in all of the circuit simulations of this chapter.[†]

12.1.1   The Wien Bridge Oscillator

Figure 12.2 presents a Wien bridge oscillator circuit with a diode limiter in its negative feedback path. The diode limiter maintains the loop gain at unity and stabilizes the amplitude of the oscillations. The components were selected so that oscillations occur at 1 kHz. The potentiometer P is to be adjusted so that the oscillations just begin to grow.

We would like to investigate the behavior of this oscillator circuit for different potentiometer settings, which is equivalent to selecting different values for R_{1a} and R_{1b} such that $R_{1a} + R_{1b} = 50$ kΩ. Because oscillations begin when $(R_2 + R_{1b})/R_{1a} = 2$, or when $R_{1a} = 20$ kΩ and $R_{1b} = 30$ kΩ, we consider three possible settings: (a) $R_{1a} = 15$ kΩ, $R_{1b} = 35$ kΩ; (b) $R_{1a} = 18$ kΩ, $R_{1b} = 32$ kΩ; and (c) $R_{1a} = 25$ kΩ, $R_{1b} = 25$ kΩ. The first setting establishes a loop gain of 1.33 at the frequency of oscillation and should be more than enough for the oscillations to begin. The second setting reduces the loop gain to 1.1, just above the level that is necessary for oscillations to begin. Finally, the last setting creates a loop gain of 0.8 and should prevent the circuit from beginning or sustaining oscillations.

Figure 12.3 shows the Spice input file describing the Wien bridge oscillator. The potentiometer is set using the settings previously given in (a) for R_{1a} and R_{1b}. Notice that the voltage across each capacitor is initialized at 0 V. This is to demonstrate that the offset voltage of the op amp will begin oscillations in this circuit on its own without the need of any start-up circuit. To force Spice to use the supplied initial conditions, the .TRAN statement includes the key word UIC (use initial conditions). Settings (b) and (c) for the potentiometer are in separate Spice input file descriptions but are attached to the file in Fig. 12.3 before submission to Spice. This will enable us to compare the output behavior for the three different circuit setups.

Figure 12.4 shows the simulation results. In the upper graph, the output oscillation for the Wien bridge has a loop gain of 1.33. The oscillation frequency is found to be 926.78 Hz, slightly less than the intended value of 1 kHz. The peaks of the output signal are clipped, indicating that the loop gain is too high. The large distortion means

---

[†]Those readers who have the PSpice student version will find that this op amp macromodel is contained in a file called NOM.LIB, along with other models of electronic devices.

```
.subckt uA741          1 2 3 4 5
* connections:         | | | | |
*                      | | | | |
*      noninverting input | | | |
*            inverting input | | |
*      positive power supply | |
*        negative power supply |
*                         output
*
*
  c1    11 12 8.661E-12
  c2     6  7 30.00E-12
  dc     5 53 dx
  de    54  5 dx
  dlp   90 91 dx
  dln   92 90 dx
  dp     4  3 dx
  egnd  99  0 poly(2) (3,0) (4,0) 0 .5 .5
  fb     7 99 poly(5) vb vc ve vlp vln 0 10.61E6 -10E6 10E6 10E6 -10E6
  ga     6  0 11 12 188.5E-6
  gcm    0  6 10 99 5.961E-9
  iee   10  4 dc 15.16E-6
  hlim  90  0 vlim 1K
  q1    11  2 13 qx
  q2    12  1 14 qx
  r2     6  9 100.0E3
  rc1    3 11 5.305E3
  rc2    3 12 5.305E3
  re1   13 10 1.836E3
  re2   14 10 1.836E3
  ree   10 99 13.19E6
  ro1    8  5 50
  ro2    7 99 100
  rp     3  4 18.16E3
  vb     9  0 dc 0
  vc     3 53 dc 1
  ve    54  4 dc 1
  vlim   7  8 dc 0
  vlp   91  0 dc 40
  vln    0 92 dc 40
.model dx D(Is=800.0E-18 Rs=1)
.model qx NPN(Is=800.0E-18 Bf=93.75)
.ends uA741
```

Figure 12.1 Spice subcircuit description of a nonlinear macromodel of the 741 op amp. We shall rely on this macromodel in all op amp examples in this chapter.

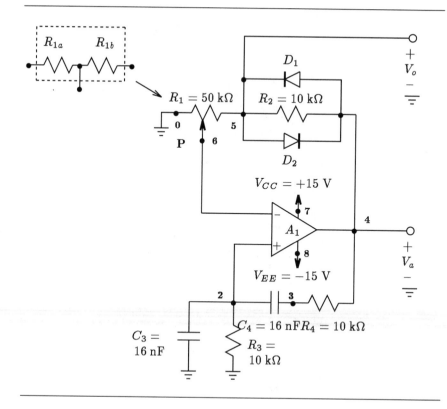

Figure 12.2 Wien bridge oscillator with a limiter used for amplitude control.

that this is obviously not very useful as a sinusoidal generator. The middle graph illustrates the voltage of the oscillator with a reduced loop gain of 1.1 and demonstrates an undistorted sinusoidal waveform of 979.43 Hz and a 8.165 V amplitude offset by -22.5 mV. The bottom graph confirms that oscillations will not begin when the loop gain is less than unity.

It is interesting to compare the spectral purity of the waveform appearing at the output of the oscillator (at V_o) with the waveform appearing at the op amp output (labeled V_a) in Fig. 12.2. We use the Fourier series analysis capabilities of Spice. First, we compute the first nine harmonics of the voltage waveform appearing across the output V_o. From these we can compute the total harmonic distortion (THD). To do this we add the following .FOUR command to the Spice file shown in Fig. 12.3 for a loop gain of 1.1:

```
.FOUR 979.43Hz V(5)
```

It is essential to use a close estimate of the oscillation frequency for the Fourier series analysis in Spice. A fractional error, say $\pm 0.1\%$, produces a noticeable change in the

```
A Wien Bridge Oscillator with Amplitude Stabilization
* loop gain is 1.1 (R1a=15k R1b=35k)

** Circuit Description **

* op-amp subcircuit

***** place uA741 op-amp subcircuit here (see Fig. 12.1) *****

** Main Circuit **
* power supplies
Vcc 7 0 DC +15V
Vee 8 0 DC -15V
* Wien bridge oscillator
XAmp 2 6 7 8 4 uA741
R1a 6 0 15k
R1b 6 5 35k
R2 5 4 10k
R3 2 0 10k
C3 2 0 16nF IC=0V
R4 3 4 10k
C4 2 3 16nF IC=0V
* diode limiter circuit
D1 4 5 D1N4148
D2 5 4 D1N4148
* model statements
.model D1N4148 D (Is=0.1p Rs=16 CJO=2p Tt=12n Bv=100 Ibv=0.1p)
** Analysis Requests **
.OPTIONS itl5=0
.OP
.TRAN 200us 20ms 0ms 200us UIC
** Output Requests **
.PLOT TRAN V(4) V(5)
.probe v(4) v(5)
.end
```

Figure 12.3   Spice input deck for computing the transient behavior of the Wien bridge oscillator shown Fig. 12.2.

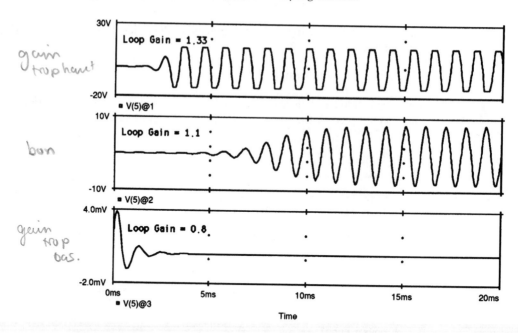

Figure 12.4 Start-up transient behavior of the Wien bridge oscillator shown in Fig. 12.2.

harmonic content in the output voltage waveform. The results of this analysis are then found in the output file:

```
FOURIER COMPONENTS OF TRANSIENT RESPONSE V(5)

DC COMPONENT =  -1.159314E-02

HARMONIC    FREQUENCY    FOURIER     NORMALIZED    PHASE        NORMALIZED
   NO         (HZ)      COMPONENT    COMPONENT    (DEG)       PHASE (DEG)

    1       9.794E+02    8.093E+00   1.000E+00    7.443E+00    0.000E+00
    2       1.959E+03    2.483E-01   3.068E-02    1.691E+01    9.469E+00
    3       2.938E+03    2.917E-01   3.604E-02   -4.998E+01   -5.742E+01
    4       3.918E+03    6.764E-02   8.358E-03   -1.021E+00   -8.464E+00
    5       4.897E+03    1.125E-01   1.390E-02   -1.121E+01   -1.866E+01
    6       5.877E+03    4.797E-02   5.927E-03   -2.027E+01   -2.772E+01
    7       6.856E+03    6.630E-02   8.193E-03    3.834E+01    3.090E+01
    8       7.835E+03    6.487E-02   8.016E-03    8.205E+00    7.612E-01
    9       8.815E+03    8.325E-03   1.029E-03    1.134E+02    1.059E+02

    TOTAL HARMONIC DISTORTION =    5.168163E+00 PERCENT
```

We repeat this analysis for the output voltage waveform appearing at the op amp output using the command:

```
.FOUR 979.43Hz V(4)
```

and obtain the results:

```
FOURIER COMPONENTS OF TRANSIENT RESPONSE V(4)

DC COMPONENT =  -1.196819E-02

HARMONIC    FREQUENCY    FOURIER     NORMALIZED    PHASE       NORMALIZED
   NO         (HZ)      COMPONENT    COMPONENT    (DEG)       PHASE (DEG)

   1        9.794E+02   8.748E+00    1.000E+00    7.390E+00    0.000E+00
   2        1.959E+03   2.595E-01    2.967E-02    1.659E+01    9.195E+00
   3        2.938E+03   3.824E-01    4.371E-02   -2.690E+01   -3.429E+01
   4        3.918E+03   8.218E-02    9.394E-03    1.496E+00   -5.895E+00
   5        4.897E+03   1.669E-01    1.908E-02    4.072E+00   -3.318E+00
   6        5.877E+03   6.224E-02    7.115E-03   -1.151E+01   -1.890E+01
   7        6.856E+03   9.173E-02    1.049E-02    4.158E+01    3.419E+01
   8        7.835E+03   7.950E-02    9.088E-03    1.081E+01    3.416E+00
   9        8.815E+03   1.452E-02    1.660E-03    9.456E+01    8.717E+01

    TOTAL HARMONIC DISTORTION =   5.906799E+00 PERCENT
```

When we compare the results, we find that the voltage waveform appearing at the op amp output contains slightly more harmonic distortion than the voltage output marked V_o as indicated by the THD measures. But, in all fairness, the difference is very small, and we may prefer to obtain the output from the low-impedance op amp output terminal instead of the high-impedance node presently marked as output.

12.1.2 An Active Filter–Tuned Oscillator

Figure 12.5 shows an active filter–tuned oscillator. It consists of a high-Q bandpass filter connected in a positive-feedback loop around a hard limiter. The filter circuit portion of this oscillator was designed to have a center frequency of 1 kHz and an adjustable Q determined by R_1. Using Spice we investigate the spectral purity of the output voltage waveform (V_1) by computing its harmonic content for different values of filter Q.

We begin our analysis with a filter circuit in the feedback loop which has a moderate Q of 5 when $R_1 = 50$ kΩ. Figure 12.6 shows the Spice description for this circuit. The op amps are modeled after the commercial 741 op amp circuit, and the diodes are modeled after the 1N4148 type. A transient analysis is requested with the circuit initially at rest, so the initial voltages across C_1 and C_2 are set to 0 V, as indicated by the key word $IC = 0$ V for C_1 and C_2. The circuit response must have reached steady state in order to correctly compute the harmonic content of the output waveform. To

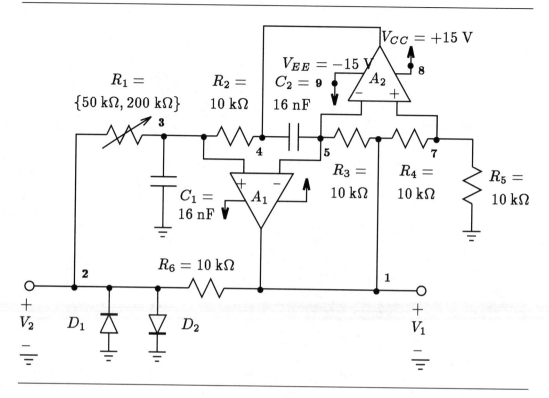

Figure 12.5   Active filter–tuned oscillator with the Q of the filter adjustable by R_1.

```
The Active Filter-Tuned Oscillator

** Circuit Description **

* op amp subcircuit

***** place uA741 op amp subcircuit here (see Fig. 12.1) *****

** Main Circuit **
* power supplies
Vcc 8 0 DC +15V
Vee 9 0 DC -15V
* high-Q filter circuit
Xopamp1 3 5 8 9 1 uA741
Xopamp2 7 5 8 9 4 uA741
R1 2 3 50k
C1 3 0 16nF IC=0V
R2 3 4 10k
C2 4 5 16nF IC=0V
R3 5 1 10k
R4 1 7 10k
R5 7 0 10k
* diode limiter circuit
D1 0 2 D1N4148
D2 2 0 D1N4148
R6 1 2 10k
* model statement
.model D1N4148 D (Is=0.1p Rs=16 CJO=2p Tt=12n Bv=100 Ibv=0.1p)
** Analysis Requests **
.OPTIONS itl5=0
.TRAN 20us 50ms 45ms 20us UIC
.FOUR 1kHz V(1)
** Output Requests **
.PLOT TRAN V(1) V(2)
.probe V(1) V(2)
.end
```

Figure 12.6 Spice input deck for computing the transient behavior of the active filter–tuned oscillator circuit shown in Fig. 12.5.

find a lower bound on this time requirement we assume that the filter circuit within the feedback loop of the oscillator has an underdamped response characterized by a pair of complex poles described by ω_o and Q. As a result, its impulse response can be described by a damped sinusoidal of the following form:

$$Ke^{-(\omega_o/2Q)t}\cos\left(\omega_o\sqrt{1-\frac{1}{4Q^2}}\,t+\phi\right) \qquad (12.1)$$

Clearly for the transient response to die out to less than 0.1% of its initial value, the exponential term in Eq. (12.1) must reduce in value to something less than 0.001. If T_S is the time required to settle to steady state, it can be found from

$$e^{-(\omega_o/2Q)T_S} \leq 0.001 \qquad (12.2)$$

with rearrangement:

$$T_S \geq -\frac{2Q}{\omega_o} \ln(0.001) \qquad (12.3)$$

Thus, for a filter Q of 5 we can expect the transient response to last for about 11 ms. Because the filter circuit is inside a positive-feedback loop, the effective Q of the oscillator circuit will be much larger than the Q of the filter circuit. From Eq. (12.3) we can expect that the time for the oscillator circuit to reach its steady state will be much longer than 11 ms.

Through several iterations with Spice we found that the oscillator circuit in Fig. 12.5 for a filter Q of 5 required 50 ms to reach its steady state. To get good plotting coverage of the output waveform, 10 points per period of oscillation were taken. This analysis revealed that the frequency of oscillation was precisely 1 kHz.

The output request asks only for plots of the output voltage and of the voltage across the limiter. Also, we requested that PROBE store only the results of these two voltages, because transient analysis produces an enormous amount of data. Last, a .FOUR command was added to the Spice input file to compute the harmonic content in the output voltage waveform:

```
.FOUR 1kHz V(1)
```

Figure 12.7 shows results of the transient analysis. The top graph displays the output voltage waveform of the oscillator, and the lower graph displays the voltage signal appearing across the diode limiter. The output signal appears undistorted, unlike that of the voltage appearing across the diode limiter. The frequency of oscillation of either waveform is very close to 1 kHz. The results of the Fourier series calculation reveal the following harmonic content in the output voltage waveform:

```
FOURIER COMPONENTS OF TRANSIENT RESPONSE V(1)

DC COMPONENT =   -6.624248E-03
```

HARMONIC NO	FREQUENCY (HZ)	FOURIER COMPONENT	NORMALIZED COMPONENT	PHASE (DEG)	NORMALIZED PHASE (DEG)
1	1.000E+03	1.270E+00	1.000E+00	4.450E+01	0.000E+00
2	2.000E+03	1.351E-02	1.063E-02	2.849E+01	-1.601E+01
3	3.000E+03	2.765E-02	2.177E-02	3.920E+01	-5.298E+00
4	4.000E+03	5.878E-03	4.627E-03	2.221E+01	-2.229E+01
5	5.000E+03	4.676E-03	3.681E-03	6.410E+01	1.960E+01
6	6.000E+03	3.573E-03	2.813E-03	2.499E+01	-1.952E+01
7	7.000E+03	2.726E-03	2.146E-03	2.529E+01	-1.921E+01
8	8.000E+03	2.635E-03	2.074E-03	2.553E+01	-1.897E+01
9	9.000E+03	2.277E-03	1.793E-03	3.218E+01	-1.232E+01

```
TOTAL HARMONIC DISTORTION =   2.533441E+00 PERCENT
```

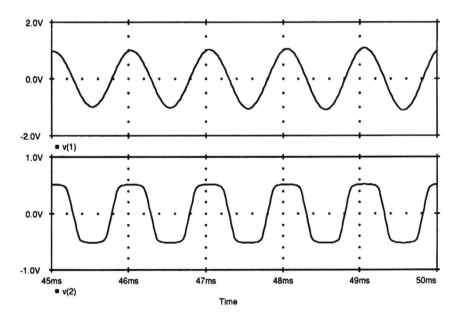

Figure 12.7   Steady-state transient response of the active filter–tuned oscillator shown in Fig. 12.5 for Q = 5.

It is evident that the total harmonic distortion (THD) is about 2.5%. Recall that for the Wien bridge oscillator of the previous section, the THD was about two times as large at 5%.

To repeat the preceding analysis for a filter Q of 20 (i.e., R_1 = 200 kΩ), we should increase the time for the transient analysis. According to Eq. (12.3), if the filter Q is increased by a factor of 4, the transient time also increases by the same factor, so the transient analysis time goes to 200 ms while maintaining the same time step. On doing so, we obtain the following Fourier series components for the output voltage waveform:

```
FOURIER COMPONENTS OF TRANSIENT RESPONSE V(1)

DC COMPONENT =    1.321303E-02
```

HARMONIC NO	FREQUENCY (HZ)	FOURIER COMPONENT	NORMALIZED COMPONENT	PHASE (DEG)	NORMALIZED PHASE (DEG)
1	1.000E+03	1.268E+00	1.000E+00	-6.697E+01	0.000E+00
2	2.000E+03	9.201E-03	7.254E-03	-5.044E+01	1.653E+01
3	3.000E+03	5.859E-03	4.619E-03	2.491E+01	9.188E+01
4	4.000E+03	2.674E-03	2.108E-03	-2.493E+01	4.204E+01
5	5.000E+03	2.754E-03	2.171E-03	-3.491E+01	3.206E+01

6	6.000E+03	1.652E-03	1.303E-03	-1.777E+01	4.920E+01
7	7.000E+03	1.227E-03	9.673E-04	-1.296E+01	5.401E+01
8	8.000E+03	1.162E-03	9.163E-04	-1.031E+01	5.666E+01
9	9.000E+03	9.852E-04	7.767E-04	-9.472E+00	5.750E+01

```
TOTAL HARMONIC DISTORTION =   9.337391E-01 PERCENT
```

As expected, the THD in the output waveform has decreased due to the increased selectivity of the filter circuit portion of the oscillator. In fact, we see that the THD decreased by almost the same factor as the increase of filter Q.

12.2

Multivibrator Circuits

Multivibrators are a class of circuits that can be used to produce the square waveforms needed for some applications. Multivibrators are classified as bistable circuits, astable circuits, or monostable circuits. Figure 12.8 presents one of each, and we analyze each with Spice, assuming that the op amps are modeled after the 741 type.

12.2.1   A Bistable Circuit

In Fig. 12.8(a) the bistable op amp circuit has a positive-feedback loop with two stable states. Any given state of the circuit depends on the input level and the previous state of the circuit. As a result, this is a regenerative circuit that shows hysteresis that can be deduced from its DC transfer characteristic. We might be tempted to perform a DC sweep of the input voltage V_i over the range supported by the two power supplies, but two issues arise. First, producing the complete transfer characteristic of the bistable circuit requires that the input voltage level be swept from the negative supply level to the positive supply level then returned to the negative supply. This requires two Spice runs, one for each sweep direction. Second, Spice has problems with convergence at the point where the circuit changes state, and it usually fails to complete the solution.

A better technique for obtaining the transfer characteristics is to apply a low-frequency, triangular waveform whose level varies between the two power supply levels. In this way we mimic the action taken when measuring the transfer characteristics of a circuit in the laboratory. The frequency of the input signal is kept low to minimize the effects of op amp dynamics on the circuit transfer characteristic.

Figure 12.9 shows the input file for the bistable circuit in Fig. 12.8(a). The input triangular waveform is specified as a piecewise linear transient source beginning at -15 V, rising to $+15$ V in the first second, then decreasing back to -15 V in the next second. A transient analysis is requested to compute the output voltage over a 2 s interval. The top graph of Fig. 12.10 displays the resulting input and output transient waveforms. The information contained in this figure is then translated into the transfer characteristic shown in the bottom graph by plotting V(3) versus V(1). From this plot

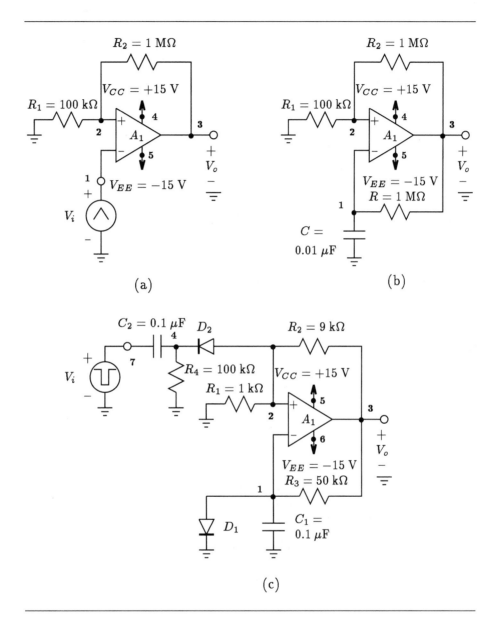

Figure 12.8 A collection of multivibrator circuits: (a) bistable circuit, (b) astable circuit, and (c) monostable circuit.

```
The Bistable Circuit

** Circuit Description **

* op amp subcircuit

***** place uA741 op amp subcircuit here (see Fig. 12.1) *****

** Main Circuit **
* power supplies
Vcc 4 0 DC +15V
Vee 5 0 DC -15V
* input triangular waveform
Vi 1 0 PWL ( 0,-15V 1s,+15V 2s,-15V )
* positive feedback op amp circuit
Xopamp1 2 1 4 5 3 uA741
R1 2 0 100k
R2 2 3 1Meg
** Analysis Requests **
.OPTIONS itl5=0
.TRAN 10ms 2s 0ms 10ms
** Output Requests **
.PLOT TRAN V(1) V(3)
.probe
.end
```

Figure 12.9 Spice input deck for computing the transfer characteristic of the bistable circuit shown in Fig. 12.8(a).

we find that the lower threshold limit (V_{T_L}) equals -1.3266 V, the upper threshold limit (V_{T_H}) equals $+1.3121$ V, and the width of the hysteresis is thus 2.639 V.

It is interesting to compare these threshold levels with those predicted by the theory from Sedra and Smith in their Section 12.4. For instance, we see from Fig. 12.10 that the saturation level of the 741 op amp circuit is ±14.6 V. Since $\beta = R_1/(R_1 + R_2)$, we substitute the appropriate values and find that $\beta = 0.091$. Thus, $V_{T_H} = -V_{T_L} = 1.33$ V, which is reasonably close to the threshold levels obtained through Spice simulation.

12.2.2   Generation of a Square Wave Using an Astable Multivibrator

A square-wave generator can be realized using a bistable multivibrator with an *RC* circuit in its feedback loop, as illustrated in Fig. 12.8(b). Such a circuit has no stable

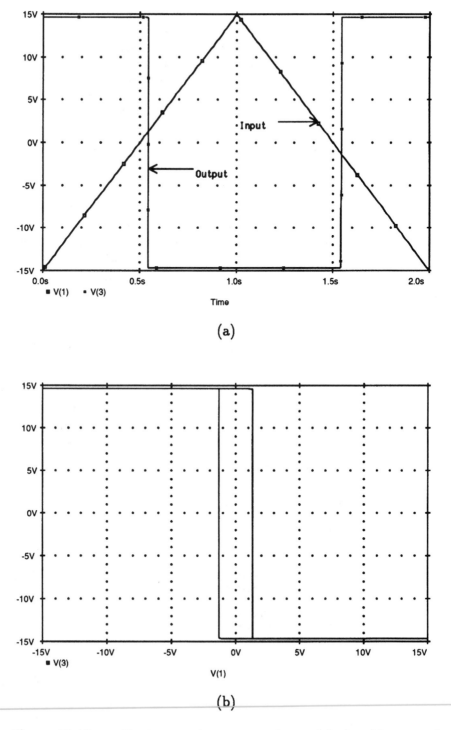

Figure 12.10 (a) The input and output waveforms of the bistable circuit shown in Fig. 12.8(a) as a function of time. (b) Transfer characteristic of the same bistable circuit.

states and is referred to as an astable multivibrator. The components of the bistable circuit are the same as in the previous section, so its expected behavior is known. Because the upper and lower saturation limits of the bistable portion of this circuit are equal (i.e., $L_+ = -L_- = 14.6$ V) and because $\beta = 0.091$ and $\tau = RC = 10$ ms, the expected period of oscillation is then computed by

$$T = 2\tau \ln\left(\frac{1 + \beta}{1 - \beta}\right) \tag{12.4}$$

to give 3.65 ms.

To confirm that this circuit indeed oscillates with a period of 3.65 ms, we create the input description of this circuit shown in Fig. 12.11 and submit it to Spice. The output waveform is shown plotted in the top graph of Fig. 12.12. The bottom graph displays the almost-triangular voltage waveform appearing at the positive input terminal to the

---

```
An Astable Multivibrator: Square-Wave Generator

** Circuit Description **

* op amp subcircuit

***** place uA741 op amp subcircuit here (see Fig. 12.1) *****

** Main Circuit **
* power supplies
Vcc 4 0 DC +15V
Vee 5 0 DC -15V
* multivibrator circuit
Xopamp1 2 1 4 5 3 uA741
R1 2 0 100k
R2 2 3 1Meg
R 1 3 1Meg
C 1 0 0.01uF IC=0V
** Analysis Requests **
.OPTIONS itl5=0
.TRAN 500us 50ms 0ms 500us
** Output Requests **
.PLOT TRAN V(3) V(1)
.probe
.end
```

Figure 12.11   Spice input deck for computing the transient behavior of the square-wave generator shown in Fig. 12.8(b).

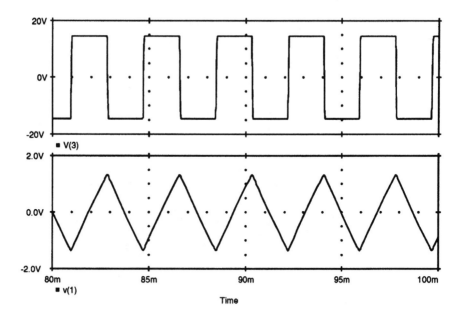

Figure 12.12 Several waveforms associated with the square-wave generator shown in Fig. 12.8(b). Top curve: output voltage waveform; bottom curve: voltage at input negative terminal of op amp.

op amp. The period of oscillation for each of these waveforms is found to be 3.74 ms, which agrees fairly well with that predicted by theory.

12.2.3   *The Monostable Multivibrator*

A monostable circuit is often used to create a pulse of a fixed duration that is triggered by another pulse of arbitrary duration. In Fig. 12.8(c) we present a monostable circuit that has been designed to generate a pulse of 755 μs duration when it receives a negative-going trigger pulse whose level varies between V_{CC} and ground, according to the formula:

$$T \cong C_1 R_3 \ln\left(\frac{0.7/L - 1}{\beta - 1}\right) \tag{12.5}$$

We use the input deck in Fig. 12.13 to compute the response of the monostable circuit when triggered with a negative-going pulse that is long enough that the op amp has time to change state. Because the slew rate of a 741 op amp is nominally 0.63 V/μs and, for a complete change of state, the op amp output voltage changes about 30 V, the pulse width required is about 50 μs.

```
The Monostable Multivibrator

** Circuit Description **

* op amp subcircuit

***** place uA741 op amp subcircuit here (see Fig. 12.1) *****

** Main Circuit **
* power supplies
Vcc 5 0 DC +15V
Vee 6 0 DC -15V
* input trigger signal + circuit
Vtrig 7 0 PWL ( 0,+15V 10us,+15V 10.01us,0V 60us,0V 60.01us,+15V
+ 10ms, +15V 10.00001ms,0V 10.090ms,0V 10.09001ms,+15V 1s,+15V )
C2 7 4 0.1uF
R4 4 0 100k
D2 2 4 D1N4148
* monostable multivibrator circuit
Xopamp1 2 1 5 6 3 uA741
R1 2 0 1k
R2 2 3 9k
R3 1 3 50k
C1 1 0 0.1uF
D1 1 0 D1N4148
* model statements
.model D1N4148 D (Is=0.1p Rs=16 CJO=2p Tt=12n Bv=100 Ibv=0.1p)
** Analysis Requests **
.IC V(3)=+15V
.TRAN 1ms 2ms 0ms
** Output Requests **
.PLOT TRAN V(7) V(3) V(1)
.probe
.end
```

Figure 12.13 Spice input deck for computing several time waveforms of the monostable op amp circuit shown in Fig. 12.8(c).

The simulation results are shown in Fig. 12.14, where the top graph displays the trigger pulse and the middle graph illustrates the output-generated pulse. The duration of the output pulse is about 750 μs, as measured from the points where the signal crosses the 0 V axis. This result agrees reasonably well with the 755 μs estimated using Eq. (12.5). The time necessary for the op amp output to change state is found to be 67 μs. The bottom graph displays the voltage appearing at the negative input

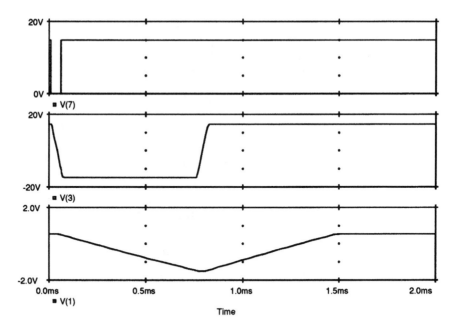

Figure 12.14   Several waveforms associated with the monostable op amp circuit shown in Fig. 12.8(c). Top curve: input trigger waveform; middle curve: output voltage waveform; bottom curve: voltage at input negative terminal of op amp.

terminal of the op amp. It takes this waveform 1441.1 μs to recover from the trigger pulse, so that the maximum frequency of operation for this circuit is 693 Hz.

12.3

Precision Rectifier Circuits

Here we investigate several precision rectifier circuits used in instrumentation. Our investigations include Spice analysis of the precision half-wave rectifier, peak detector, and clamping circuits shown in Fig. 12.15. Our analysis purposely keeps the input signal frequency low in order to minimize the dynamic effects (i.e., slew rate) of the op amp. With higher frequencies, these effects will begin to play a significant role in circuit operation. Readers are encouraged to investigate this themselves.

12.3.1   A Half-Wave Rectifier Circuit

Figure 12.15(a) shows a precision half-wave rectifier circuit. The active components include a single 741 op amp and two 1N4148-type diodes. Using a DC sweep of the input voltage level ranging between the two power supply levels, we compute the input–output transfer characteristic.

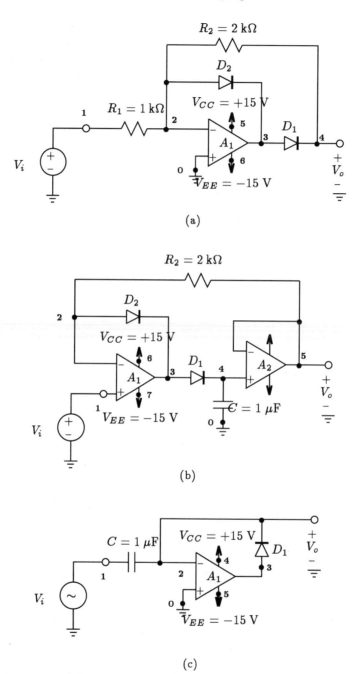

Figure 12.15 Precision rectifier circuits: (a) half-wave rectifier circuit, (b) peak detector circuit, and (c) clamping circuit.

```
Precision Rectifier Circuit

** Circuit Description **

* op amp subcircuit

***** place uA741 op amp subcircuit here (see Fig. 12.1) *****

** Main Circuit **
* power supplies
Vcc 5 0 DC +15V
Vee 6 0 DC -15V
* input signal source
Vi 1 0 DC 0V SIN( 0V 1V 1kHz )
* limiter circuit
Xopamp 0 2 5 6 3 uA741
R1 1 2 1k
R2 2 4 2k
D1 2 3 D1N4148
D2 3 4 D1N4148
* model statements
.model D1N4148  D (Is=0.1p Rs=6 CJO=1p Tt=3n Bv=100 Ibv=0.1p)
** Analysis Requests **
.DC Vi -15V +15V 0.5V
** Output Requests **
.PLOT DC V(4)
.probe
.end
```

Figure 12.16 Spice input deck for computing the DC transfer characteristic of the precision rectifier circuit shown in Fig. 12.15(a).

The Spice input file describing this circuit is given in Fig. 12.16 and includes a DC sweep analysis request. The resulting transfer characteristic is plotted in Fig. 12.17. The curve in Fig. 12.17(a) illustrates the output voltage as a function of the input voltage when the latter is varied between the positive and negative supplies. For negative input signals, the output signal is twice the size of the input signal with a 180° phase shift. For positive input signals, no signal appears at the output, so that half-wave rectification is performed. Figure 12.17(b) provides an expanded view of the transfer characteristic for signals in the range $-50 \mu V$ to $200 \mu V$. Here we see that the transfer characteristic does not go through the origin but has an output offset voltage of 211.9 μV. Also, this curve shows that for positive input voltages larger than 100 μV, the output voltage does not go to zero but levels off at 20.9 μV.

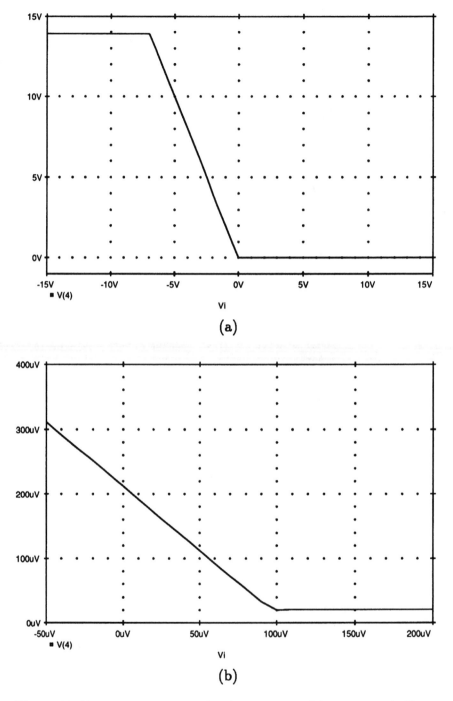

Figure 12.17 Input–output transfer characteristic of the precision half-wave rectifier circuit shown in Fig. 12.15(a) (a) Input voltage varied between the positive and negative supply rails. (b) Input voltage varied between −50 μV and 200 μV.

12.3.2   A Buffered Peak Detector

Figure 12.15(b) shows the buffered precision peak detector. This circuit is designed to hold the peak value of some input signal over long time periods. To demonstrate the behavior of this circuit, we apply an input triangular signal with increasing peak values and observe the output signal. The Spice circuit description is in Fig. 12.18 with the piecewise linear description of the input signal. A transient analysis is requested to compute the output signal over an 11 ms interval of the input signal.

The input and output signals for this experiment as calculated by Spice are shown in Fig. 12.19. This circuit does indeed hold the output level constant at the most recent peak level of the input signal.

```
A Buffered Precision Peak Detector

** Circuit Description **

* op amp subcircuit

***** place uA741 op amp subcircuit here (see Fig. 12.1) *****

** Main Circuit **
* power supplies
Vcc 6 0 DC +15V
Vee 7 0 DC -15V
* input signal source
Vi 1 0 DC 0V PWL ( 0s,-15V 1ms,-7.5V 2ms,-15V 3ms,-15V 4ms,0V 5ms,-15V
+            6ms,-15V 7ms,+7.5V 8ms,-15V 9ms,-15V 10ms,+10V 11ms,-15V )
* limiter circuit
Xopamp1 1 2 6 7 3 uA741
Xopamp2 4 5 6 7 5 uA741
D1 3 4 D1N4148
D2 2 3 D1N4148
R 2 5 1k
C 4 0 1uF
* model statements
.model D1N4148 D (Is=0.1p Rs=16 CJO=2p Tt=12n Bv=100 Ibv=0.1p)
** Analysis Requests **
.OPTIONS itl5=0
.TRAN 0.5ms 11ms 0s 0.5ms
** Output Requests **
.PLOT TRAN V(1) V(3)
.probe
.end
```

Figure 12.18   Spice input deck for computing the transient behavior of the peak detector shown in Fig. 12.15(b).

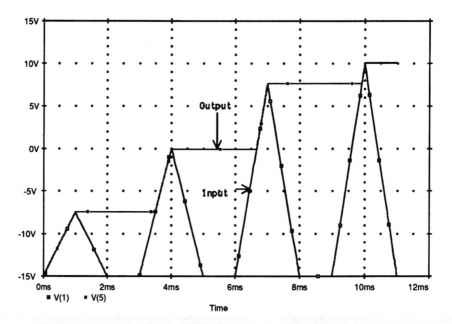

Figure 12.19   Input and output transient waveforms of the peak detector shown in Fig. 12.15(b).

12.3.3   A Clamping Circuit

As the final example, Fig. 12.15(c) shows the precision op amp clamping circuit whose function is to restore a DC level to some AC coupled input signal. We use a 5 V peak input sinusoidal signal of 1 kHz frequency with no DC offset. In the Spice input file in Fig. 12.20 a transient analysis is requested, and the input and output signals are to be observed. Figure 12.21 shows the analysis results. The output signal has been offset by the peak of the input signal and now varies between 0 V and +10 V. It takes less than one cycle of the input waveform to reach steady state.

12.4

Spice Tips

- Various electronic manufacturers are making Spice models of their components available in the form of subcircuits that can be included directly in a Spice deck.
- The Fourier analysis command (.FOUR) of Spice requires a very good estimate of the frequency of the signal to be analyzed. A fractional error in this frequency estimate can produce noticeable change in the harmonic content of the output waveform.

```
A Precision Clamping Circuit

** Circuit Description **

* op amp subcircuit

***** place uA741 op amp subcircuit here (see Fig. 12.1) *****

** Main Circuit **
* power supplies
Vcc 4 0 DC +15V
Vee 5 0 DC -15V
* input signal source
Vi 1 0 SIN ( 0 5V 1kHz )
* limiter circuit
Xopamp1 0 2 4 5 3 uA741
D1 3 2 D1N4148
C 1 2 1uF IC=0V
* model statements
.model D1N4148 D (Is=0.1p Rs=16 CJO=2p Tt=12n Bv=100 Ibv=0.1p)
** Analysis Requests **
.OPTIONS itl5=0
.TRAN 50us 4ms 0s 50us UIC
** Output Requests **
.PLOT TRAN V(3) V(1)
.probe
.end
```

Figure 12.20 Spice input deck for computing the transient behavior of the clamping circuit shown in Fig. 12.15(c).

- A Fourier analysis should be performed only on a periodic waveform that has reached steady state. The transient in the output waveform must be allowed to decrease to a negligible level. As a rule of thumb, for a circuit dominated by a pair of complex conjugate poles described by ω_o and Q, the transient portion of the waveform will require $-(2Q/\omega_o)\ln(0.001)$ seconds for it to decrease to less than 0.1% of its initial value.

- Initializing the reactive elements of a circuit with values that are close to their steady-state current or voltage values can reduce the time required for a circuit to reach its steady state.

- The transfer characteristics of regenerative circuits with hysteresis are best computed with Spice by applying a low-frequency triangular waveform input whose level varies between the two power supply levels.

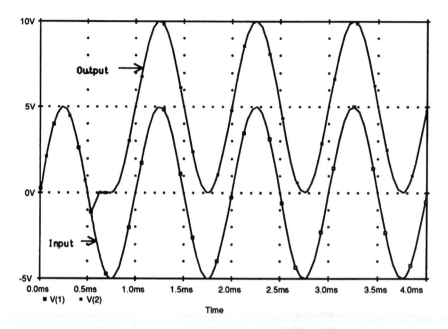

Figure 12.21   Input and output transient waveforms of the clamping circuit shown in Fig. 12.15(c).

12.5

Bibliography

Linear Circuits: Operational Amplifier Macromodels, Data Manual, Texas Instruments, Dallas, TX, 1990.

12.6

Problems

12.1 Compare the amplitude of oscillation of the output voltage generated by the Wien bridge oscillator shown in Fig. 12.2 with and without the diode limiter in the op amp feedback path.

12.2 Using Spice determine the transfer characteristic of the comparator circuit shown in Fig. P12.2. Then connect a DC source $V_B = +1$ V to the virtual ground of the op amp through a resistor of $R_B = 10$ kΩ, and observe that the transfer characteristic is shifted along the v_i-axis to the point $v_i = -(R_1/R_B)V_B$. Use the subcircuit description for the 741 op amp provided in Fig. 12.1 as the macromodel for the op amp in the comparator circuit. Assume that the two diodes have parameters $I_S = 10^{-14}$ A and $n = 1$.

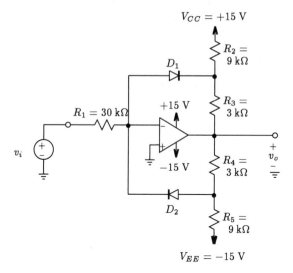

Figure P12.2

12.3 For the circuit in Fig. P12.3 determine the loop transmission $L(j\omega)$ and the frequency for zero loop phase. Verify that the circuit indeed oscillates at this frequency. Assume that the op amp is of the 741 type.

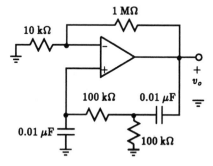

Figure P12.3

12.4 For the circuit in Fig. P12.4, plot its input–output transfer characteristic v_o–v_i using Spice. Assume that the op amp is modeled after the 741 op amp and the two diodes after the commercial diode 1N4148. What is the maximum diode current?

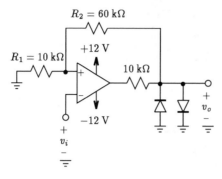

Figure P12.4

12.5 Consider the circuit of Fig. P12.4 with R_1 eliminated and R_2 short-circuited. Plot the input–output transfer characteristic. Model the op amp after the 741 op amp and the two diodes after the commercial diode 1N4148.

12.6 Find the frequency of oscillation of the circuit in Fig. 12.8(b) for the case $R_1 = 10\ \text{k}\Omega$, $R_2 = 16\ \text{k}\Omega$, $C = 10\ \text{nF}$, and $R = 62\ \text{k}\Omega$. Assume that the op amp is of the 741 type.

12.7 An oscillator circuit that generates two sine-wave signals that are in quadrature (i.e., 90° apart) is shown in Fig. P12.7. It oscillates at a frequency of 5 kHz. Using Spice,

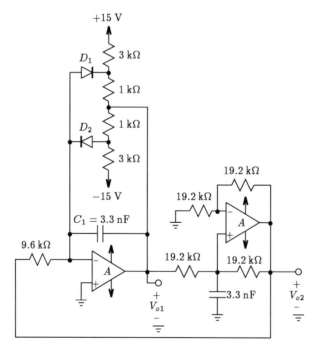

Figure P12.7

confirm that the two output signals are indeed in quadrature and that the frequency of oscillation is 5 kHz. Also, using the Fourier analysis capability of Spice, determine the total harmonic distortion present in the two output waveforms.

12.8 Two superdiode circuits, such as the one shown in Fig. P12.8, are connected to a common load resistor of 20 kΩ. They have the same input signal but have their diodes reversed: one with cathode to load, the other with anode to load. For a sine-wave input of 10 V peak to peak and 1 kHz frequency, use Spice to plot the output voltage waveform and the current supplied by each superdiode circuit. Assume that the op amp and the diode are modeled after the 741 and the 1N4148, respectively.

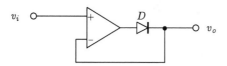

Figure P12.8

12.9 The circuit shown in Fig. P12.9 is used as a voltmeter that is intended to function at frequencies of 10 Hz and above. It consists of precision half-wave rectifier followed by a first-order lowpass filter. Apply a 100 mV$_{rms}$ sinewave input signal of 10 kHz and observe the voltage signal appearing at the output. Does the output signal correspond to the average value of the input signal? Repeat with a triangular waveform and a square wave as input. Model the op amp as a high-gain VCVS with a diode clamping circuit to limit the range of output voltage to ± 12 V. Assume that the diode has parameters $I_S = 10^{-14}$ A and $n = 1.6$.

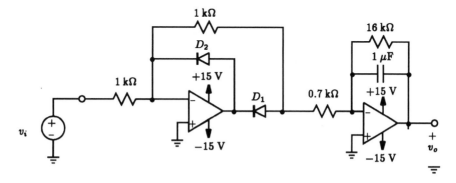

Figure P12.9

12.10 Using Spice, plot the transfer characteristic of the circuit in Fig. P12.10. Assume that the op amp is modeled after the 741 op amp and the diode after the 1N4148.

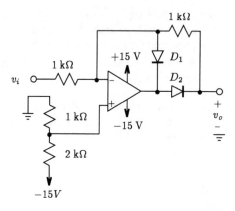

Figure P12.10

13

MOS Digital Circuits

In this chapter we investigate mainly the static and dynamic behavior of several basic inverter circuits found in several types of simple MOS digital circuits, particularly digital circuits constructed in NMOS, CMOS, or GaAs integrated circuit technologies.

13.1

NMOS Inverter with Enhancement Load

Figure 13.1 displays an NMOS inverter circuit with an enhancement load and with the substrate connected to the ground. Because the source-to-body voltage (V_{SB}) of M_1 is zero and that of M_2 is equal to V_O, the threshold voltage of M_2 is no longer equal to the

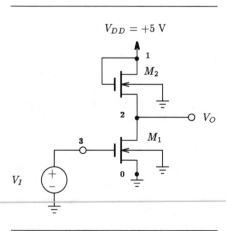

Figure 13.1 Enhancement-load inverter with substrate connected to ground.

threshold voltage of M_1. For straightforward hand analysis, it is common to neglect the transistor body effect and assume that each transistor has the same threshold voltage. However, as we shall see from the following Spice circuit simulations, neglecting the body effect results in substantial error in the large-signal input–output transfer characteristic. This is especially true for input signal levels near ground potential.

Figure 13.2 is the input description of the inverter circuit of Fig. 13.1. The parameters for each transistor of the inverter are assumed as follows: $V_{t0} = 0.7$ V, $W_1 = 9$ μm, $L_1 = 3$ μm, $W_2 = 3$ μm, $L_2 = 9$ μm, $\mu_n C_{OX} = 40$ μA/V^2, $2\phi_f = 0.6$ V. Each transistor is also assumed to have a body effect coefficient (γ) of 1.1 V$^{1/2}$.

Figure 13.3 shows the simulation results and the input–output transfer characteristic when the body effect is eliminated (i.e., another Spice run was done with gamma = 0 in the transistor model statement in Fig. 13.2. At low input levels the two curves differ significantly, whereas for higher input levels ($V_I > 2.5$ V) the two results are quite close.

It is interesting to compare the noise margins of these two situations. If we use Probe on the transfer characteristic that includes the body effect, from Fig. 13.3 we see that $V_{IL} = 0.7$ V, $V_{OL} = 0.24$ V, and $V_{OH} = 3.05$ V. To find V_{IH} it is necessary to determine the input level where the transfer characteristic has a slope of -1 in the vicinity of a logic low level. We can easily use the derivative procedure command available in Probe. Taking the derivatives of the output voltages (V_O) with respect to the input voltages (V_I) for both cases and plotting them versus the input voltage levels, we obtain the graphic results shown in Fig. 13.4. The input level in the vicinity of a logic low level corresponding to a slope of -1 is 1.78, making $V_{IH} = 1.78$ V. The two noise margins for the enhancement-load inverter with body effect included are

$$NM_H = V_{OH} - V_{IH} = 3.05 - 1.78 = 1.27 \text{ V}$$
$$NM_L = V_{IL} - V_{OL} = 0.7 - 0.24 = 0.46 \text{ V}$$

Carrying out the same procedure for the characteristic of the inverter excluding the body effect produces these two noise margins:

$$NM_H = V_{OH} - V_{IH} = 4.3 - 1.96 = 2.34 \text{ V}$$
$$NM_L = V_{IL} - V_{OL} = 0.7 - 0.26 = 0.44 \text{ V}$$

We see that the body effect has decreased the high noise margin (NM_H) from 2.34 V to 1.27 V. The low noise margin (NM_L) remains relatively unchanged.

We can complete our investigation of the static behavior of the enhancement-load inverter circuit by computing its average static power dissipation, assuming that the inverter spends equal time in each of its two states. We include the body effect in this analysis.

Returning to the Spice deck for the inverter in Fig. 13.2, we set the input to the inverter to logic low, say $V_I = V_{IL} = 0.7$ V, to force the inverter into its high state. We replace the DC sweep command with an .OP analysis command to compute the power dissipated by the inverter. The results of this analysis indicate that the power dissipated is very low at 5.56×10^{-11} W.

Repeating the above with the input at $V_I = 3.05$ V to correspond to an output voltage of V_{OL} for some preceding stage, we find that the static power dissipated by

```
An Enhancement-Load Inverter

** Circuit Description **
* DC supplies
Vdd 1 0 DC +5V
* input digital signal
Vi 3 0 DC 0V
* MOSFET circuit
M1 2 3 0 0 nmos_enhancement_mosfet L=3um W=9um
M2 1 1 2 0 nmos_enhancement_mosfet L=9um W=3um
* MOSFET model statement (by default level 1)
.model nmos_enhancement_mosfet nmos (kp=40u Vto=0.7V phi=0.6V gamma=1.1)
** Analysis Requests **
.DC Vi 0V +5V 100mV
** Output Requests **
.Plot DC V(2)
.probe
.end
```

Figure 13.2   Spice input file for calculating the input–output voltage transfer characteristic of the enhancement-load inverter circuit in Fig. 13.1.

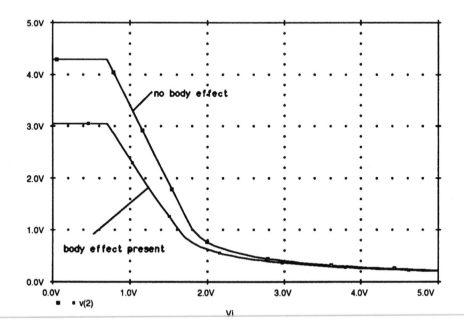

Figure 13.3   Input-output voltage transfer characteristic of the enhancement-load inverter circuit shown in Fig. 13.1 with and without the body effect present.

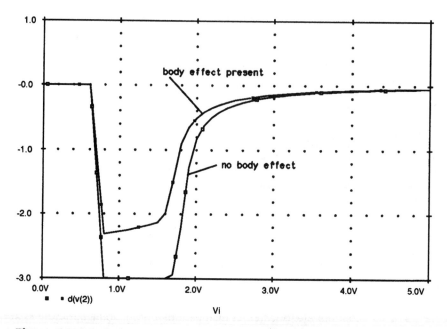

Figure 13.4 dV_O/dV_I as a function of the input voltage level V_I with and without the body effect present. These curves are used to determine the minimum inverter input voltage V_{IH}.

the inverter in its low state has increased substantially to a level of 462 μW. Thus, the average static power dissipated by the inverter with enhancement load is 231 μW.

13.1.1 Dynamic Operation

We next consider the dynamic operation of the enhancement-load NMOS inverter. To facilitate our investigation, we shall cascade two inverter circuits as shown in Fig. 13.5. The output of each inverter is assumed to be loaded with a 0.1 pF capacitor to represent the capacitance associated with the interconnect wiring between the two gates. A 0 V to 5 V pulse of 100 ns is applied to the input of the first inverter circuit. To obtain realistic results, we shall model the NMOS transistors after a commercial 3 μm CMOS process. These model parameters are included in the Spice deck in Fig. 13.6. The static parameters remain the same as previously, but it should be noted that the addition of the other parameters will alter the DC behavior of this circuit somewhat. A transient analysis is requested over a 200 ns interval using a 0.5 ns time step. The voltage at the output of each inverter will be plotted.

The transient response of the cascade of inverters is presented in Fig. 13.7. The top graph displays the input voltage signal applied to the first inverter and the bottom graph displays the voltage waveform at the output of each inverter circuit. The rise time of each inverter is much slower than its fall time. To better estimate the rise and fall times of the inverter, as well as the propagation delay, Fig. 13.8 provides an expanded view of the high-to-low and low-to-high output transitions of the first inverter.

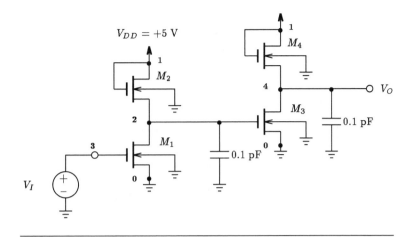

Figure 13.5 Cascade of two enhancement-load NMOS inverter circuits.

```
A Cascade of Enhancement-Load Inverters (dynamic effects included)

** Circuit Description **
* DC supplies
Vdd 1 0 DC +5V
* input digital signal
Vi 3 0 PWL (0,0V 10ns,0V 10.1ns,5V 110ns,5V 110.1ns,0V 210ns,0V)
* first MOSFET inverter circuit
M1 2 3 0 0 MN L=3um W=9um
M2 1 1 2 0 MN L=9um W=3um
Cl1 2 0 0.1pF
* second MOSFET inverter circuit
M3 4 2 0 0 MN L=3um W=9um
M4 1 1 4 0 MN L=9um W=3um
Cl2 4 0 0.1pF
* BNR 3um transistor model statements (level 3)
.MODEL MN nmos (level=3 vto=.7 kp=4.e-05 gamma=1.1 phi=.6
+ lambda=.01 rd=40 rs=40  pb=.7 cgso=3.e-10 cgdo=3.e-10
+ cgbo=5.e-10 rsh=25 cj=.00044  mj=.5 cjsw=4.e-10 mjsw=.3
+ js=1.e-05 tox=5.e-08 nsub=1.7e+16  nss=0 nfs=0 tpg=1 xj=6.e-07
+ ld=3.5e-07 uo=775  vmax=100000  theta=.11  eta=.05 kappa=1)
** Analysis Requests **
.TRAN 0.5ns 210ns 0ns 0.5ns
** Output Requests **
.Plot TRAN V(2) V(3) V(4)
.probe
.end
```

Figure 13.6 Spice input file for computing the transient response for the NMOS inverter cascade shown in Fig. 13.5.

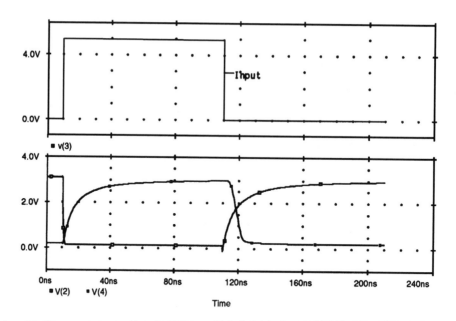

Figure 13.7 Transient response of a cascade of two enhancement-load NMOS inverter circuits. The top graph displays the input voltage signal, and the bottom graph displays the two output voltage signals appearing at each inverter output.

Using Probe on the graph of Fig. 13.8(a), we found that the 90%-to-10% fall time t_{THL} is 1.063 ns, based on a logic low level V_{OL} of 0.128 V and a logic high value V_{OH} of 3.12 V. Also, the high-to-low input-to-output delay t_{PHL} was found to be 0.524 ns. Repeating this for the low-to-high output voltage transition shown in Figure 13.8(b), we find that the 10%-to-90% rise time t_{TLH} is 45.9 ns, and the low-to-high input-to-output delay t_{PLH} is 7.71 ns. Averaging these two input-to-output delays, we find the propagation time delay t_P for the NMOS enhancement-load inverter with a 0.1 pF load to be 4.12 ns.

It is interesting to note that the voltage waveform at the output of the second inverter is somewhat different from that at the output of the first inverter, giving rise to different transient attributes for the second inverter. Specifically, we find that the rise time t_r is 38.6 ns and the fall time t_f is 7.95 ns. We also find that $t_{PLH} = 5.81$ ns and $t_{PHL} = 1.28$ ns, which when combined will give a second inverter propagation delay t_P of 3.54 ns. The second inverter has different transient attributes from the first one because the signals at the input to each inverter are different. The first inverter has an almost instantaneous voltage change at its input with a rise and fall time of about 0.1 ns, whereas the second inverter sees an input voltage signal having unsymmetrical rise and fall times of 38.6 ns and 7.95 ns, respectively.

It would be instructive to check the formulas derived by Sedra and Smith in their Section 13.2 for estimating the low-to-high and high-to-low transition times t_{PLH} and t_{PHL}. According to their analysis,

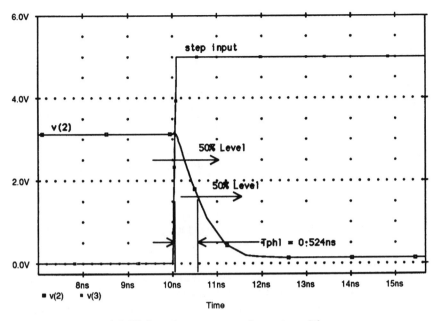

(a) High-to-low output voltage transition.

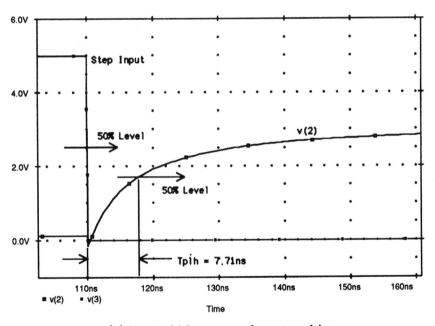

(b) Low-to-high output voltage transition.

Figure 13.8 Expanded view of the (a) high-to-low and (b) low-to-high output voltage transitions of the first inverter of the circuit shown in Fig. 13.5 when subjected to a step input.

$$t_{PHL} = \frac{C[V_{OH} - \frac{1}{2}(V_{OH} + V_{OL})]}{I_{HL}} \qquad (13.1)$$

and

$$t_{PLH} = \frac{C[\frac{1}{2}(V_{OH} + V_{OL}) - V_{OL}]}{I_{LH}} \qquad (13.2)$$

where I_{HL} and I_{LH} represent the average discharge and charge currents. These two currents are obtained from the static characteristic of each device. In terms of the development in Sedra and Smith, with the input V_I held high at $+5$ V (the same logic high level used to compute the transient response of the first inverter shown in Fig. 13.7) and with a simplified set of device parameters obtained from the level 3 NMOS model of the transistor given in the Spice deck of Fig. 13.6 (i.e., $k_p = 40 \, \mu A/V^2$, $V_t = 0.7$ V, $\gamma = 1.1 \, V^{1/2}$, and $\lambda = 0.01 \, V^{-1}$), we find that $I_{HL} = 861.2 \, \mu A$ and $I_{LH} = 120.6 \, \mu A$.

Under the assumption that the effective capacitance seen by the inverter is dominated[†] by the interconnect capacitance of 0.1 pF, we can compute the transition times t_{PHL} and t_{PLH} according to Eqs. (13.1) and (13.2) and obtain

$$t_{PHL} = 0.174 \text{ ns} \qquad t_{PLH} = 1.24 \text{ ns}$$

The times can be averaged to obtain an estimate of the propagation delay time at 0.71 ns. The values obtained directly from Spice were $t_{PHL} = 0.524$ ns and $t_{PLH} = 7.71$ ns, so we see that the preceding estimates are off by quite a large amount (for t_{PLH}, by a factor of 6).

The large discrepancy is caused by the assumption that the static parameters of the MOSFET are those seen listed in the level 3 MOSFET model included in Fig. 13.6. This is unfortunately not true. The value of many of the parameters in a level 3 Spice model of a MOSFET have no physical basis; they are simply empirical parameters that a computer optimization algorithm has made to fit the measured device characteristics. Using a subset of the MOSFET device parameters can result in significant error, as in this particular case. If we use Spice with the level 3 MOSFET model, we will find the average discharge and charge currents, I_{HL} and I_{LH}, to be much less than those computed previously at 432 μA and 36 μA, respectively. Substituting these two results into Eqs. (13.1) and (13.2), we then obtain

$$t_{PHL} = 0.347 \text{ ns} \qquad t_{PLH} = 4.16 \text{ ns}$$

These results are much closer to those read directly from the Spice graph in Fig. 13.8 (i.e., $t_{PHL} = 0.524$ ns and $t_{PLH} = 7.71$ ns). Based on the accuracy of the preceding two results, we can conclude that Eqs. (13.1) and (13.2) are accurate to only 50% of the Spice computed values.

A similar result is obtained when the transition times for the second inverter are computed. Using Eqs. (13.1) and (13.2) and the average discharge and charge currents

---

[†]This is a reasonable assumption here because the parasitic capacitances associated with each MOSFET of the inverter are on the order of femtofarads. This was confirmed through a small-signal analysis of each transistor at each critical point of the inverter transfer characteristic.

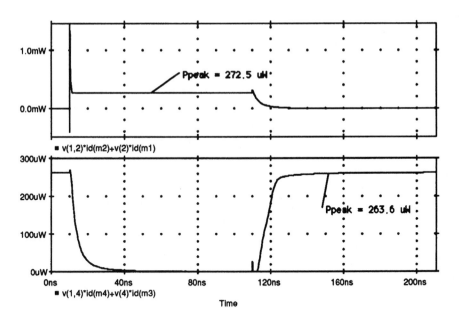

Figure 13.9 Instantaneous power dissipated by the first and second enhancement-load NMOS inverters in the cascade shown in Fig. 13.5. The top graph displays the power dissipated by the first inverter, and the bottom graph displays the power dissipated by the second inverter.

of 198 μA and 36 μA, respectively, we find that $t_{PHL} = 0.75$ ns and $t_{PLH} = 4.15$ ns. This in turn indicates a propagation delay of 2.45 ns. Comparing these results with those directly obtained from Spice (i.e., $t_{PHL} = 1.28$ ns and $t_{PLH} = 5.81$ ns), we once again see that these estimates are within 50% of those found from the transient waveform computed by Spice.

Finally, the instantaneous power dissipated by each inverter circuit of Fig. 13.5 is computed by Spice and shown in Fig. 13.9. The top graph depicts the power dissipated by the first inverter and the bottom graph that of the second inverter. From the top graph during the first transition we see that a sudden power glitch occurs as the inverter moves from the logic high to the logic low state. The power peaks at about 1.46 mW then settles into a steady state at 272.5 μW. Thus, we can state that the inverter dissipates a static power of 272.5 μW when in the logic low state. During the next transition when the inverter is driven into the logic high state, we see that the power dissipated by this gate reduces to nearly zero. An infinitesimal leakage current of the reverse-biased pn junctions accounts for a very small power dissipation of 31 pW. The steady-state power dissipated by the second inverter in the logic low state is quite similar to that dissipated by the first inverter, a constant peak power of 263.6 μW. Also, when in the high state the power dissipated by this gate reduces to nearly zero. The difference in power dissipated by the two inverter circuits is due to the difference in the levels of their respective input signals.

13.2

NMOS Inverter with Depletion Load

Replacing the enhancement-load MOSFET in the inverter circuit of Fig. 13.1 by a depletion MOSFET will result in an inverter circuit with a sharper voltage transfer characteristic and thus higher noise margins. In the following discussion we compare the voltage transfer characteristic of the enhancement-load inverter circuit with the corresponding characteristic of the depletion-load inverter circuit in Fig. 13.10. For a fair comparison we assume that the geometries are the same for the two inverters' corresponding devices. The device parameters of both the enhancement- and depletion-type MOSFET will be assumed to be the same as in first part of the previous example except that the threshold voltage of the depletion mode device will be set at -3 V.

Figure 13.11 is the Spice input file describing the depletion-load inverter circuit of Fig. 13.10. A DC sweep of the input voltage is requested between 0 V and 5 V. A plot of the output voltage is then requested. To compare the results of this analysis with those previously generated for the enhancement-load inverter, we have concatenated the two Spice listings in Figs. 13.2 and 13.11 into one input file before submitting the file to Spice.

The results of the two Spice analyses are shown in Fig. 13.12. Clearly, the depletion-load inverter has a steeper transition region than the enhancement-load inverter and therefore approaches the ideal inverter characteristic more closely. To quantify this we shall compare their respective noise margins.

From the curve for the depletion-load inverter shown in Fig. 13.12 we can directly obtain the output voltage levels V_{OL} and V_{OH} as 0.112 V and 5 V, respectively. To obtain the input threshold levels V_{IL} and V_{IH}, we must determine the input levels (V_I) that correspond to $d\,V_O/d\,V_I = -1$. With the procedure described in the previous section and the aid of Probe we find that $V_{IL} = 0.845$ V and $V_{IH} = 1.67$ V. Thus,

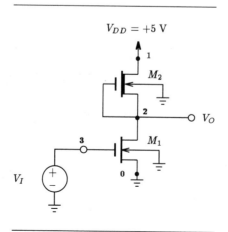

Figure 13.10   Depletion-load inverter with substrate connected to ground.

```
A Depletion-Load Inverter

** Circuit Description **
* DC supplies
Vdd 1 0 DC +5V
* input digital signal
Vi 3 0 DC 0V
* MOSFET circuit
M1 2 3 0 0 nmos_enhancement_mosfet L=3um W=9um
M2 1 2 2 0 nmos_depletion_mosfet L=9um W=3um
* MOSFET model statements (level 1 by default)
.model nmos_enhancement_mosfet nmos (kp=40u Vto=0.7V phi=0.6V gamma=1.1)
.model nmos_depletion_mosfet   nmos (kp=40u Vto=-3V  phi=0.6V gamma=1.1)
** Analysis Requests **
.DC Vi 0V +5V 20mV
** Output Requests **
.PLOT DC V(2)
.probe
.end
```

Figure 13.11   Spice input file for calculating the input–output voltage transfer characteristic of the depletion-load inverter circuit shown in Fig. 13.10.

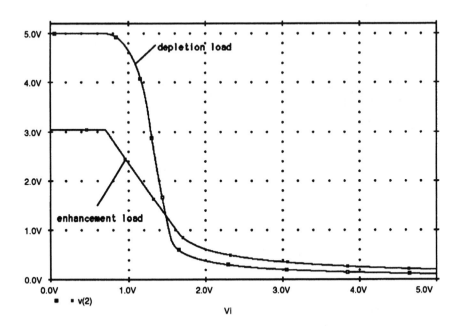

Figure 13.12   Comparison of the voltage transfer characteristic of a depletion-load inverter and of an equally sized enhancement-load inverter. The depletion-load inverter has a steeper transition region and therefore more closely approaches the ideal inverter characteristic.

the noise margins of the depletion-load inverter are $NM_L = 0.733$ V and $NM_H = 3.33$ V. Both of these values are larger than the corresponding values obtained for the enhancement-load inverter of the previous section, which supports the claim that the depletion-load inverter behaves more ideally than the corresponding enhancement-load inverter.

13.3

The CMOS Inverter

For both custom and semicustom VLSI (very-large-scale integration), CMOS is generally preferred over all other available technologies. Unlike NMOS technologies, CMOS digital circuits use both n-channel and p-channel enhancement-type MOS-FETS. Figure 13.13 is an example of a CMOS inverter, and we use Spice to compute its voltage transfer characteristic. The n-channel and p-channel MOSFETs are assumed to be modeled after the same commercial 3 μm CMOS technology used in previous examples in this chapter. Because the mobility μ for the n-channel device is about three times greater than that for the corresponding p-channel device, we make the p-channel device three times as wide as the n-channel device. Moreover, we make the n-channel device the size of the smallest transistor possible in this technology, 3 μm by 3 μm.

Figure 13.14 shows the Spice input file for the CMOS inverter. A sweep of the input voltage level is performed between ground and V_{DD}. Figure 13.15 graphs the resulting voltage transfer characteristic. We see that the output voltage of this gate extends completely between the limits of the two power supply levels. Thus, $V_{OL} = 0$ V and $V_{OH} = +5$ V. Using the derivative feature of Probe, we find the input logic levels, V_{IL} and V_{IH}, as 2.03 and 2.825 V, respectively. In the case of V_{IL} we find that the simulated result correlates very closely with that predicted by the simple formula derived in Section 13.5 of Sedra and Smith, $V_{OL} = \frac{1}{8}(3V_{DD} + 2V_{tn}) = 2.05$ V. Similarly, the result predicted by $V_{IH} = \frac{1}{8}(5V_{DD} - 2V_{tp}) = 2.93$ V is quite close to that obtained by simulation. Combining these results we find that the high and low noise margins are $NM_H = 2.175$ V and $MN_L = 2.03$ V, respectively. These noise margins are not equal because M_1 and M_2 are not exactly matched.

It is an important attribute of CMOS logic that the transistors do not conduct any current when settled into a particular logic state. To appreciate this, we plot in Fig. 13.16 the current that flows in the CMOS inverter circuit as a function of the input voltage level. These results indicate that insignificant current flows through the inverter when the input is either smaller than 0.7 V or greater than 4.2 V. We also see that a peak current of 43.8 μA flows through the inverter when the input signal level reaches the midpoint of the voltage supply.

13.3.1 Dynamic Operation

Next we consider the dynamic operation of a cascade of two CMOS inverters shown in Fig. 13.17. The output of each inverter is assumed to be loaded with a 0.1 pF capacitor to represent the capacitance associated with the interconnect wiring between the two

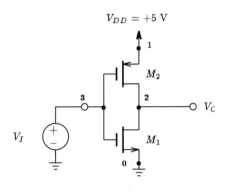

Figure 13.13   CMOS inverter.

```
The CMOS Inverter

** Circuit Description **
* DC supplies
Vdd 1 0 DC +5V
* input digital signal
Vi   3 0 DC +5V
* MOSFET inverter circuit
M1 2 3 0 0 MN L=3um W=3um
M2 2 3 1 1 MP L=3um W=9um
* BNR 3um transistor model statements (level 3)
.MODEL MN nmos level=3 vto=.7 kp=4.e-05 gamma=1.1 phi=.6
+ lambda=.01 rd=40 rs=40  pb=.7 cgso=3.e-10 cgdo=3.e-10
+ cgbo=5.e-10 rsh=25 cj=.00044  mj=.5 cjsw=4.e-10 mjsw=.3
+ js=1.e-05 tox=5.e-08 nsub=1.7e+16  nss=0 nfs=0 tpg=1 xj=6.e-07
+ ld=3.5e-07 uo=775  vmax=100000  theta=.11  eta=.05 kappa=1
.MODEL MP pmos level=3 vto=-.8 kp=1.2e-05 gamma=.6 phi=.6
+ lambda=.03 rd=100 rs=100  pb=.6 cgso=2.5e-10 cgdo=2.5e-10
+ cgbo=5.e-10 rsh=80 cj=.00015  mj=.6 cjsw=4.e-10 mjsw=.6
+ js=1.e-05 tox=5.e-08 nsub=5.e+15  nss=0 nfs=0 tpg=1 xj=5.e-07
+ ld=2.5e-07 uo=250  vmax=70000  theta=.13  eta=.3 kappa=1
** Analysis Requests **
.DC Vi 0 5 50mV
** Output Requests **
.PLOT DC V(2) Id(M1)
.probe
.end
```

Figure 13.14   Spice input file for calculating the input–output voltage transfer characteristic of the CMOS inverter shown in Fig.13.13.

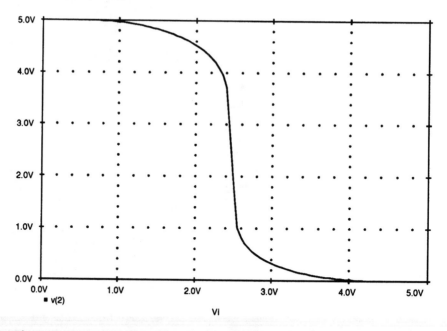

Figure 13.15 Input–output voltage transfer characteristic of the CMOS inverter shown in Fig. 13.13.

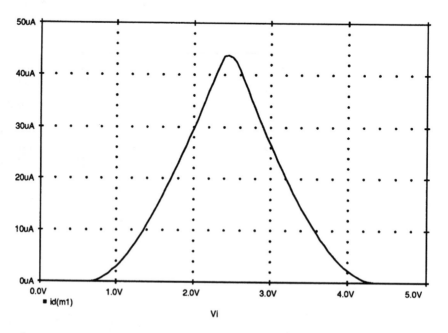

Figure 13.16 Current in the CMOS inverter versus the input voltage.

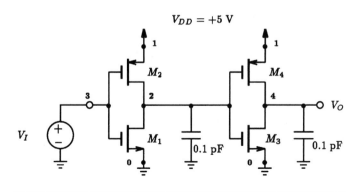

Figure 13.17   Cascade of two CMOS inverter circuits.

gates. The Spice deck for this circuit is provided in Fig. 13.18. A 0 V to 5 V pulse of 20 ns duration is applied to the input of the first inverter circuit. A transient analysis is requested over a 50 ns interval using a 0.1 ns time step. The voltage at the output of each inverter will be plotted.

The transient response of the cascade of CMOS inverters as calculated by Spice is presented in Fig. 13.19. The top graph displays the input voltage signal applied to the first inverter, and the bottom graph displays the voltage waveform that appears at the output of each inverter circuit. Let us consider the voltage signal that appears at the output of the first inverter. We see that both the rise and fall times of the output voltage signal are quite similar in duration. With the aid of Probe we find that the 90%-to-10% fall time t_{THL} is 3.82 ns and the high-to-low input-to-output transition delay t_{PHL} is 2.44 ns. Repeating this for the low-to-high output voltage transition, we find that the 10%-to-90% rise time t_{TLH} is 4.17 ns. The low-to-high input-to-output delay t_{PLH} is 2.11 ns. Averaging the two input-to-output delays for the CMOS inverter with a 0.1 pF load, we obtain the propagation time delay t_P of 2.28 ns.

The voltage waveform at the output of the second inverter appears similar to that at the output of the first inverter. Using Probe we find that $t_{THL} = 3.73$ ns, $t_{TLH} = 3.92$ ns, $t_{PHL} = 2.7$ ns, and $t_{PLH} = 2.59$ ns. The propagation delay t_P is then 2.64 ns. These timing attributes are quite close to those found for the first inverter.

For hand analysis (see Section 13.5 of Sedra and Smith) the expected transitional delays, t_{PHL} and t_{PLH}, are found from the following two expressions:

$$t_{PHL} \approx \frac{0.8C}{\frac{1}{2}\mu_n C_{OX}(W/L)\mid_n V_{DD}} \tag{13.3}$$

and

$$t_{PLH} \approx \frac{0.8C}{\frac{1}{2}\mu_p C_{OX}(W/L)\mid_p V_{DD}} \tag{13.4}$$

```
A Cascade of CMOS Inverters (dynamic effects included)

** Circuit Description **
* DC supplies
Vdd 1 0 DC +5V
* input digital signal
Vi 3 0 DC 1.237 PWL (0,0V 10ns,0V 10.1ns,5V 30ns,5V 30.1ns,0V 50ns,0V)
* first CMOS inverter
M1 2 3 0 0 MN L=3um W=3um
M2 2 3 1 1 MP L=3um W=9um
Cl1 2 0 0.1pF
* second CMOS inverter
M3 4 2 0 0 MN L=3um W=3um
M4 4 2 1 1 MP L=3um W=9um
Cl2 4 0 0.1pF
* BNR 3um transistor model statements (level 3)
.MODEL MN nmos level=3 vto=.7 kp=4.e-05 gamma=1.1 phi=.6
+ lambda=.01 rd=40 rs=40  pb=.7 cgso=3.e-10 cgdo=3.e-10
+ cgbo=5.e-10 rsh=25 cj=.00044  mj=.5 cjsw=4.e-10 mjsw=.3
+ js=1.e-05 tox=5.e-08 nsub=1.7e+16 nss=0 nfs=0 tpg=1 xj=6.e-07
+ ld=3.5e-07 uo=775  vmax=100000  theta=.11  eta=.05 kappa=1
.MODEL MP pmos level=3 vto=-.8 kp=1.2e-05 gamma=.6 phi=.6
+ lambda=.03 rd=100 rs=100  pb=.6 cgso=2.5e-10 cgdo=2.5e-10
+ cgbo=5.e-10 rsh=80 cj=.00015  mj=.6 cjsw=4.e-10 mjsw=.6
+ js=1.e-05 tox=5.e-08 nsub=5.e+15  nss=0 nfs=0 tpg=1 xj=5.e-07
+ ld=2.5e-07 uo=250  vmax=70000  theta=.13  eta=.3 kappa=1
** Analysis Requests **
.TRAN 0.1ns 50ns 0ns 0.1ns
** Output Requests **
.Plot TRAN V(2) V(3) V(4)
.probe
.end
```

Figure 13.18   Spice input file for computing the transient response for the CMOS inverter cascade shown in Fig. 13.17.

Substituting the appropriate parameters from the Spice deck in Fig. 13.18, we find that $t_{PHL} = 0.8$ ns and $t_{PLH} = 0.88$ ns. The propagation delay t_P is then 0.84 ns. We find that the transition times predicted by the preceding equations are underestimated by a factor of about 3 when compared with the Spice results, an unreasonably large error. Investigation reveals that the theory used to derive Eqs. (13.3) and (13.4) assumed that the mobility of each MOSFET (μ_n and μ_p) was constant and independent of the voltages that appear at the terminals of the device. Experimental evidence indicates otherwise, and a more complicated relationship for the mobility factor has been included in the level 3 MOSFET model of Spice. Through additional simulations and some indirect calculations, it was found for an n-channel MOSFET over a 5 V range of $V_{GS} = V_{DS}$ that $\mu_n C_{OX}(k_p)$ varied between 10 μA/V^2 and 40 μA/V^2. Separately

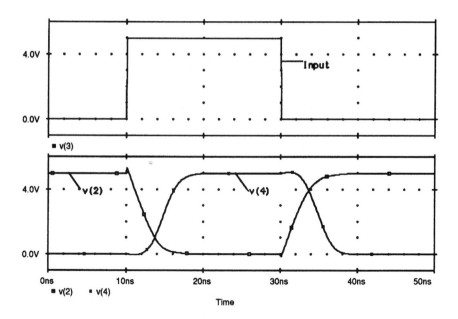

Figure 13.19 Transient response of a cascade of two CMOS inverter circuits. The top graph displays the input voltage signal, and the bottom graph displays the voltage signals appearing at output of the two inverters.

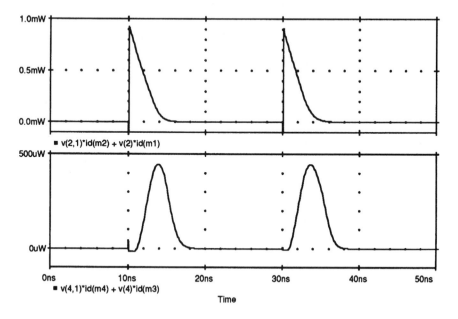

Figure 13.20 Instantaneous power dissipated by the first and second CMOS inverters in the cascade of Fig. 13.17. The top graph displays the power dissipated by the first inverter, and the bottom graph displays the power dissipated by the second inverter.

substituting these two extreme values of $\mu_n C_{OX}$ back into Eq. (13.3) indicates that t_{PHL} can range from 0.8 ns to 3.2 ns. The Spice-derived result of 2.7 ns lies between the two limits predicted by theory.

Similarly, for the p-channel MOSFET we find that $\mu_p C_{OX}$ varies between 4 μA/V^2 and 12 μA/V^2. Thus we should expect that the Spice-derived transition time t_{PLH} should lie somewhere between 0.86 ns and 2.66 ns, which it does at $t_{PLH} = 2.59$ ns.

In Fig. 13.20 the top graph depicts the instantaneous power dissipated by the first inverter, and the bottom graph shows that of the second inverter. The shapes of the power waveforms are different because different voltage signals appear at the inputs.

13.4

A Gallium-Arsenide Inverter Circuit

As an emerging IC technology, GaAs shows promise for high-speed electronic design. Figure 13.21 shows such an inverter circuit, known as the direct-coupled FET (DCFL) inverter. Two inverter circuits are connected in cascade with a 30 fF lumped capacitor representing the distributed capacitance of the wiring that interconnects the two inverters. The lower two MESFETs of each inverter, B_1 and B_3, are enhancement MESFETs, and their corresponding loads, B_2 and B_4, are depletion MESFETs. We assume that the parameters describing these two types of MESFETs per micron width are as follows: $L = 1$ μm, $V_{tD} = -1$ V, $V_{tE} = 0.2$ V, $\beta = 0.1$ mA/V^2, and $\lambda = 0.1$ V^{-1}. Moreover, in order to model the behavior of the Schottky diode that makes up the gate-to-source region of each MESFET, we shall set $n = 1.1$ and $I_S = 5 \times 10^{-16}$ A. The capacitive effects of each transistor will be modeled using Spice parameters

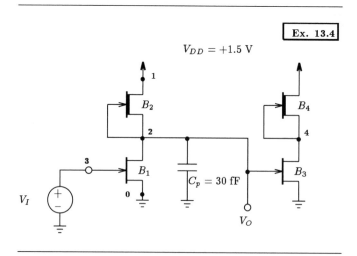

Figure 13.21   Cascade of two DCFL GaAs inverter circuits with the first gate loaded with 30 fF.

$C_{GS} = 1.2 \times 10^{-15}$ F and $C_{GD} = 1.2 \times 10^{-15}$ F. Let the width of the input MESFET of each inverter be 50 μm, and let the width of the load MESFETs be 6 μm.

Using Spice we compute the voltage transfer characteristic of the first inverter with the second inverter present. We also compute the transient response of the first inverter for a 2 ns duration pulse input.

With the Spice deck provided in Fig. 13.22, the input–output voltage transfer characteristic shown in Fig. 13.23 was calculated for the DCFL inverter. Using Probe we find that V_{OH} = 708 mV and V_{OL} = 247 mV; by taking the derivative of the output voltage with respect to the input voltage, we find that V_{IH} = 651 mV and V_{IL} = 491 mV. Therefore, the low and high noise margins are 245 mV and 56.7 mV, respectively. We see that the high noise margin appears quite low, a disadvantage of this type of logic gate. Nevertheless, these gates are presently receiving considerable attention in the development of GaAs digital circuits because they are simple and have low power dissipation.

Figure 13.24 shows the transient response of the DCFL inverter circuit for a 2 ns duration pulse having a voltage swing between ground and V_{OH}. The input pulse has

```
A DCFL GaAs Gate

** Circuit Description **
* DC supplies
Vdd 1 0 DC +1.5V
* input signal
Vi 3 0 PWL (0,0V 1ns,0V 1.1ns,700mV 3ns,700mV 3.1ns,0V 5ns,0V)
* amplifier circuit
B1 2 3 0 enchancement_n_mesfet 50
B2 1 2 2 depletion_n_mesfet 6
Cp1 2 0 30fF
B3 4 2 0 enchancement_n_mesfet 50
B4 1 4 4 depletion_n_mesfet 6
* MESFET model statements (level 1 by default)
.model enchancement_n_mesfet gasfet (beta=0.1m Vto=+0.2V lambda=0.1
+                    n=1.1 Is=5e-16 Cgd=1.2e-15 Cgs=1.2e-15 Rs=2500)
.model depletion_n_mesfet gasfet (beta=0.1m Vto=-1.0V lambda=0.1
+                    n=1.1 Is=5e-16 Cgd=1.2e-15 Cgs=1.2e-15 Rs=2500)
** Analysis Requests **
.DC Vi 0V 0.8V 10mV
.TRAN 0.1ns 5ns 0s 0.1ns
** Output Requests **
.PLOT DC V(2)
.PLOT TRAN V(2)
.probe
.end
```

Figure 13.22   Spice input file for calculating the input–output transfer characteristic and the pulse response of the DCFL inverter circuit shown in Fig. 13.21.

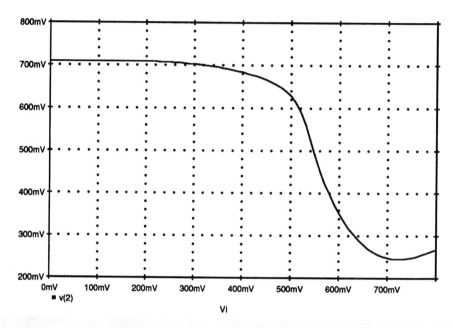

Figure 13.23 Input–output voltage transfer characteristic for the DCFL inverter circuit shown in Fig. 13.21.

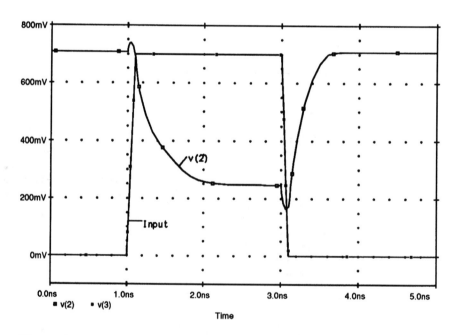

Figure 13.24 Transient response behavior of the DCFL inverter circuit shown in Fig. 13.21 subject to a 2 ns wide input pulse signal with 100 ps rise and fall times.

assigned rise and fall times of 100 ps. The output response has asymmetrical rise and fall times of 623.5 ps and 337 ps, respectively. The two propagation delays t_{PHL} and t_{PLH} are found to be 198 ps and 211 ps, respectively. The average propagation delay of this DCFL inverter gate is 204.5 ps.

13.5

Spice Tips

- The derivative $d\,V_O/d\,V_I$ can be computed using the derivative command of Probe. A plot of V_O versus V_I must first be computed using the DC sweep command.

- The behavior of transistors becomes more complex as their dimensions shrink, requiring greater reliance on Spice analysis because hand analysis with simple transistor models is no longer accurate.

13.6

Problems

13.1 An n-channel enhancement MOS device used in an inverter load operating from a 5 V supply has V_{t0} ranging from 1 to 1.5, γ ranging from 0.3 to 1 $V^{1/2}$, and $2\phi_f = 0.6$ V. What is the lowest value of V_{OH} that will be found using transistors of this kind?

13.2 With the aid of Spice determine the average propagation delay, the average power dissipation, and the delay-power product of an inverter for the following conditions:

K_n ($\mu A/V^2$)	K_p ($\mu A/V^2$)	V_t (V)	V_{DD} (V)	C_L (pF)
10	5	1	5	0.1
1	1	1	5	0.1
5	5	0.5	5	0.1
10	1	1	5	0.1
10	1	1	10	1.0
7.5	10	1	5	0.1

13.3 The circuit shown in Fig. P13.3 can be used in applications, such as output buffers and clock drivers, where capacitive loads require high current-drive capability. Let $V_{tE} = 1$ V, $V_{tD} = -3$ V, $V_{DD} = 5$ V, $\mu_n C_{OX} = 20 \ \mu A/V^2$, $W_1 = W_2 = 12 \ \mu m$, $L_1 = L_2 = L_3 = L_4 = 6 \ \mu m$, $W_3 = 120 \ \mu m$, and $W_4 = 240 \ \mu m$.
  (a) Find V_{AH}, V_{OH}, V_{AL}, and V_{OL} for v_i (high) $= 5$ V. Note that V_{AH} denotes the high level at node A and V_{AL} the low level.
  (b) Find the total static current with the output high and with the output low.
  (c) As v_o goes low, find the peak current available to discharge a load capacitance $C = 10$ pF. Also find the discharge current for v_o at the 50% output swing point, and thus find the average discharge current and t_{PHL}.
  (d) Repeat (c) for v_o going high, thus finding t_{PLH}.

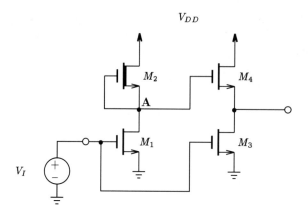

Figure P13.3

13.4 The circuit shown in Fig. P13.4, called a bootstrap driver, is intended to provide a large output voltage swing and a high output drive current with a reasonable standing current. Let $V_{DD} = 5$ V, $V_t = 1$ V, $\mu_n C_{OX} \vert_1 = 40\ \mu\text{A/V}^2$, $\mu_n C_{OX} \vert_2 = 10\ \mu\text{A/V}^2$, and $\mu_n C_{OX} \vert_3 = 1\ \mu\text{A/V}^2$, and assume $C = 10$ pF. With the aid of Spice, answer the following questions:

(a) Find the voltages at nodes X and Y when v_i has been high (at 4 V) for some time.

(b) When v_i goes low to 0 V and Q_1 turns off, to what voltages do nodes X and Y rise?

(c) Find the current available to charge a load capacitance at Y as soon as v_i goes low and Q_1 turns off.

(d) As v_i goes high again, what current is immediately available to discharge the load capacitance at Y?

(e) What is the current shortly thereafter, when $v_Y = 4$ V?

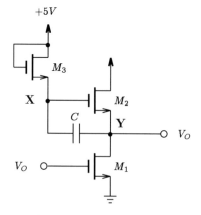

Figure P13.4

13.5 A particular CMOS inverter uses n- and p-channel devices of identical sizes. If $\mu_n C_{OX} = 2\mu_p C_{OX} = 10\ \mu\text{A/V}^2$, $|V_t| = 1$ V, and $V_{DD} = 5$ V, find V_{IL} and V_{IH} and hence the noise margins.

13.6 For a CMOS inverter having $V_{tn} = -V_{tp} = 1$ V, $V_{DD} = 5$ V, $\mu_n C_{OX} = 2\mu_p C_{OX} = 20$ μA/V^2, $(W/L)_n = 10$ μm/5 μm, and $(W/L)_p = 20$ μm/5 μm, use Spice to determine the maximum current that the inverter can sink with the output not exceeding 0.5 V. Also find the largest current that can be sourced with the output remaining within 0.5 V of V_{DD}.

13.7 For an inverter with $\mu_n C_{OX} = 2\mu_p C_{OX} = 50$ μA/V^2, $V_{tn} = -V_{tp} = 0.8$ V, $V_{DD} = 5$ V, $(W/L)_n = 4$ μm/2 μm, and $(W/L)_p = 8$ μm/2 μm, use Spice to find the peak current drawn from a 5 V supply during switching.

13.8 If the inverter specified in Problem 13.7 is loaded with a 0.2 pF capacitance, use Spice to determine the dynamic power dissipation when the inverter is switched at a frequency of 20 MHz. What is the average current drawn from the power supply?

13.9 A CMOS inverter having $V_{tn} = -V_{tp} = 2$ V, $V_{DD} = 5$ V, and $\mu_n C_{OX} = 2\mu_p C_{OX} = 50$ μA/V^2 is loaded with a 0.1 pF capacitance. If the inverter is to be clocked at 100 MHz, determine the values of (W/L) that will result in a delay-power product no larger than 0.1 pJ. Verify that this is indeed the case using Spice.

13.10 Consider a DCFL gate fabricated in GaAs technology for which $L = 1$ μm, $V_{tD} = -1$ V, $V_{tE} = +0.2$ V, and $\lambda = 0.1$ V^{-1}, and let $V_{DD} = 1.5$ V. Determine the widths of the two MESFETs that yields a $NM_H = 0.2$ V. Verify your choice using Spice. What is the resulting value of NM_L?

14

Bipolar Digital Circuits

In this final chapter we investigate the two dominant bipolar digital circuit families in use today: transistor–transistor logic (TTL) and emitter-coupled logic (ECL). The chapter concludes with a brief discussion of a simple digital circuit combining bipolar and MOS technology into a single technology referred to as BiCMOS.

14.1

Transistor–Transistor Logic (TTL)

The basic circuit for a TTL NAND gate shown in Fig. 14.1(a) consists of four *npn* transistors Q_1 to Q_4, one diode D_1, and several resistors. The input transistor Q_1 consists of a single BJT with two emitters and is referred to as a multi-emitter transistor. This particular multi-emitter transistor can be viewed as two *npn* transistors connected in parallel as shown in Fig. 14.1(b). If Q_1 has a single emitter, the NAND gate shown in Fig. 14.1(a) reduces to the basic TTL inverter circuit.

With the aid of Spice let us compute the voltage transfer characteristic of this gate from the *A* input to the NAND gate output. We shall sweep the DC voltage level at the *A* input between ground and the 5 V supply. The *B* input will be connected to $V_{CC} = 5$ V throughout this simulation. In this arrangement one emitter of the input transistor is reverse biased, and the NAND gate behaves exactly like an inverter circuit. The fan-out associated with this NAND gate will be set to 1. It is necessary to load the output of the NAND gate so that realistic results can be obtained. Fig. 14.1(c) shows the circuit used in this simulation, and its Spice deck is provided in Fig. 14.2.

Notice that a subcircuit definition has been provided for the NAND gate, an approach that is especially helpful when working with a large digital circuit consisting of many similar logic gates. Also, it is assumed that each resistor has a linear temperature coefficient of $TC = +1200$ ppm/°C. This will realistically model resistor behavior at different temperatures so that we can investigate circuit behavior versus temperature.

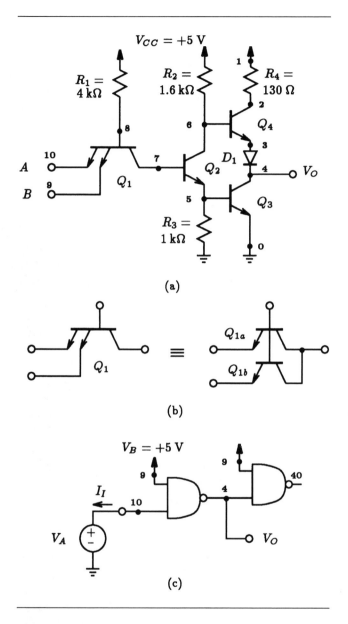

Figure 14.1 (a) Two-input TTL NAND gate. (b) Multi-emitter bipolar transistor and its equivalent representation for Spice simulation. (c) Cascade of two NAND gates.

```
Voltage Characteristics of a Two-Input TTL NAND Gate (Vb=logic 1)

* TTL two-input NAND gate
.subckt    NAND    10 9 4 1
* connections:       | | | |
*               inputA | | |
*                 inputB | |
*                   output |
*                      Vcc
*
Q1a  7 8 10  npn_transistor
Q1b  7 8 9   npn_transistor
Q2   6 7 5   npn_transistor
Q3   4 5 0   npn_transistor
Q4   2 6 3   npn_transistor
QD1  3 3 4   npn_transistor
R1 1 8 4k    TC=1200u
R2 1 6 1.6k  TC=1200u
R3 5 0 1k    TC=1200u
R4 1 2 130   TC=1200u
* BJT model statement
.model npn_transistor npn (Is=1.81e-15 Bf=50 Br=0.02 Va=100
+                         Tf=0.1ns Cje=1pF Cjc=1.5pF)
.ends NAND


** Main Circuit **
* DC supplies
Vcc 1 0 DC +5V
* input digital signals
Va 10 0 DC  0V
Vb  9 0 DC +5V
* first NAND gate: inputB is held high
Xnand_gate1 10 9 4  1 NAND
* second NAND gate: inputB is held high
Xnand_gate2  4 9 40 1 NAND
** Analysis Requests **
.DC Va 0V 5V 40mV
** Output Requests **
.Plot DC V(4) I(Va)
.probe
.end
```

Figure 14.2 Spice input deck for calculating the voltage transfer characteristic of the two-input NAND gate shown in Fig. 14.1 with one input connected to logic high (+5 V).

Figure 14.3 shows that the resulting voltage transfer characteristic at room temperature consists of four main segments. With an input between 0 V and approximately 0.4 V, the output of the NAND gate is held constant at a level of 3.85 V. Between inputs of 0.4 V and 1.2 V, the output voltage decreases at a linear rate of about -1.59 V/V. For input between 1.2 V and 1.4 V, the output voltage deceases to about 0.111 V. Above 1.4 V, the output remains a constant 0.111 V. From this information we can determine that $V_{OL} = 0.111$ V, $V_{OH} = 3.85$ V, $V_{IL} = 0.423$ V, and $V_{IH} = 1.4$ V. Thus, the low and high noise margins for this particular NAND gate with one input connected to V_{CC} are 0.312 V and 2.45 V, respectively.

Commercial TTL gates are specified to operate over a temperature range of 0 to 70° C. Using the Spice deck in Fig. 14.2, we recalculate the voltage transfer characteristic of the above NAND gate at three different temperatures: 0°, 27°, and 70° C. By adding the appropriate temperature statement into each of the three Spice decks, such as

```
.TEMP 0C
```

for setting the temperature of the circuit at 0° C, we obtain the three different voltage transfer characteristics in Fig. 14.4. From these results we summarize the variation in the logic level parameters at the extreme limits of the previously specified temperature range:

Parameter	Variation (70° C/0° C)
V_{IL}	0.366 V/0.493 V
V_{IH}	1.24 V/1.48 V
V_{OL}	0.126 V/0.101 V
V_{OH}	4.03 V/3.75 V
NM_L	0.240 V/0.392 V
NM_H	2.79 V/2.27 V

For this TTL gate, as the temperature increases the low noise margins NM_L decrease and the high noise margins NM_H increase.

In Fig. 14.5 we have plotted the input current characteristic of the NAND gate as the input voltage is varied between the limits of the power supplies. This current is denoted by I_I in Fig. 14.1(c). Three different curves are provided corresponding to the three different circuit temperatures. If we consider the results obtained at a circuit temperature of 27° C, we see that the current supplied by the input of the NAND gate decreases from 1.09 mA at 0 V to 0.722 mA at 1.36 V. When the input voltage goes above 1.36 V, the current decreases rapidly and changes direction, thus flowing into the gate. The magnitude of this current is quite small at about 7.4 μA. The current level remains constant for input voltages above 1.62 V. We can conclude from these results that the maximum input-low current I_{IL} is 1.09 mA and the maximum input-high current I_{IH} is 7.4 μA.

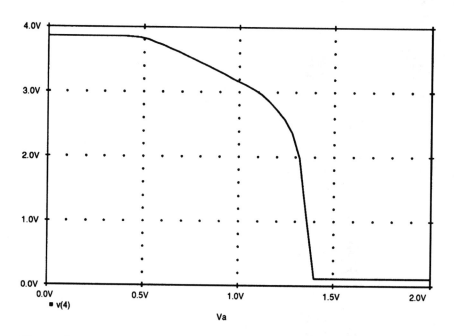

Figure 14.3 Voltage transfer characteristic of a two-input TTL NAND gate with one input connected to V_{CC}. The temperature of the circuit is assumed by default to be 27° C.

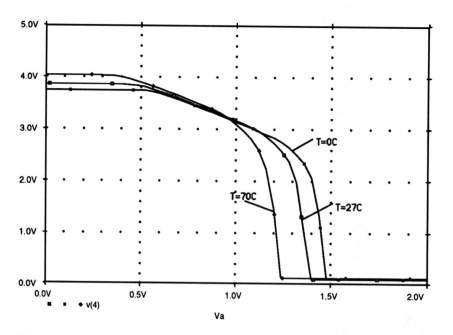

Figure 14.4 Temperature dependence of the voltage transfer characteristic of the TTL NAND gate shown in Fig. 14.1 when the B input is held at $V_{CC} = 5$ V.

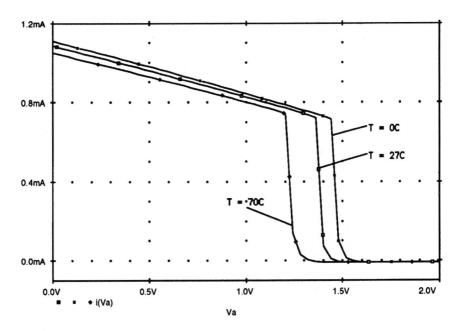

Figure 14.5 Input current characteristic of a single input of a TTL NAND gate as a function of the input voltage level.

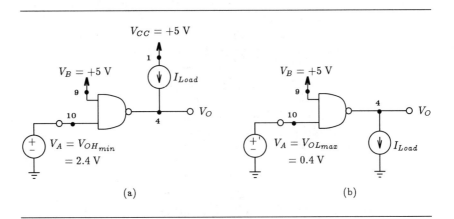

Figure 14.6 Circuit setup for generating the output voltage characteristics of the TTL NAND gate under different load conditions: (a) low state and (b) high state.

At other temperatures current input characteristics are very similar, although the point where the input current changes direction decreases with increasing temperature. Over the 0° to 70° C temperature range, we can conclude that the maximum input-low current I_{IL} is 1.11 mA and occurs when the temperature of the circuit is at its lowest. The input-high current level I_{IH} remains low with a maximum level of 7.4 μA over the two temperature extremes.

The low and high output characteristics of the NAND gate can be determined by varying the load. Consider the two circuit setups shown in Fig. 14.6. The circuit configuration in part (a) is used to determine the low-state output voltage characteristic. The input to this gate is set equal to the manufacturer's specified minimum output voltage for a TTL gate established in the high state (i.e., $V_A = V_{OH_{min}} = 2.4$ V). This is considered to be the worst input possible for establishing the output in the low state. The circuit shown in Fig. 14.6(b) is used to determine the high-state output voltage characteristic. The input to this gate is set equal to the manufacturer's specified maximum output voltage for a TTL gate set in the low state (i.e., $V_A = V_{OL_{max}} = 0.4$ V). This is also a worst-case input situation.

Let us consider the circuit setup in Fig. 14.6(a). Here the NAND gate is placed in the low state. With the Spice deck in Fig. 14.7 we have requested that the load current source I_{Load} be swept between 0 and 150 mA and that the output voltage be plotted. This will be performed over three different temperatures: 0° C, 27° C, and 70° C. We repeat this same experiment but with the NAND gate in the high state. The range of load current is to be reduced to somewhere between 0 and 20 mA. We leave to the reader the several minor modifications that must be made to the Spice deck in Fig. 14.7.

In the results in Fig. 14.8 the top graph displays the low output voltage characteristics for different load currents at three different temperatures, and the bottom graph displays the high output voltage characteristics. Let us first consider the low-state output voltage characteristics in Fig. 14.8(a) at room temperature. With no load (i.e., $I_{Load} = 0$ mA) the output voltage is the offset voltage of about 100 mV. With increasing load current, the output voltage rises at a rate of about 1 mV/mA for an equivalent output resistance of 1 Ω. With load currents exceeding 120 mA, the rate at which the output voltage changes increases dramatically to a near-vertical slope. We note that at 27° C, the low-state output voltage exceeds the manufacturer's maximum output voltage $V_{OL_{max}}$ of 0.4 V at a current of about 120 mA. Similar behavior is seen to occur at different temperatures. At 0° C the output resistance decreases to 0.85 Ω, and at 70° C it increases to 1.15 Ω.

In the high state the output voltage characteristics are shown in Fig. 14.8(b). With no load current V_{OH} is about 4.4 V at 27° C but quickly reduces to 3.85 V with a very slight load current. For load currents increasing to about 5 mA, the output voltage changes by about 500 mV, thus the output resistance is 46 Ω. Beyond this current limit we see that the output voltage changes at a new rate, with an output resistance of 118 Ω. At a load current of 13.3 mA the output voltage reduces to the manufacturer's specified minimum output voltage $V_{OH_{min}}$ of 2.4 V. Thus, the fan-out of this particular gate is limited to loads less than this current. From the other curves in Fig. 14.8(b) we see that for large current values the output voltage decreases as the circuit temperature rises, suggesting that the fan-out of this gate decreases with increasing temperature.

To conclude our study of TTL logic gates, we perform a transient analysis of the cascade of two NAND gates (connected as simple inverters) subject to a simple pulse input. The Spice deck in Fig. 14.2 is modified by replacing the DC source statement V_A with the following piecewise linear source statement:

```
Va 10 0 PWL( 0,0V 10ns,0V 20ns,5V 50ns,5V 60ns,0V 100ns,0V )
```

```
Fan-out Behavior of a NAND gate

* TTL Two-input NAND Gate
.subckt    NAND    10 9 4 1
* connections:     | | | |
*              inputA | | |
*                 inputB | |
*                    output |
*                       Vcc
*
Q1a 7 8 10   npn_transistor
Q1b 7 8 9    npn_transistor
Q2  6 7 5    npn_transistor
Q3  4 5 0    npn_transistor
Q4  2 6 3    npn_transistor
QD1 3 3 4    npn_transistor
R1 1 8 4k     TC=1200u
R2 1 6 1.6k TC=1200u
R3 5 0 1k     TC=1200u
R4 1 2 130   TC=1200u
* BJT model statement
.model npn_transistor npn (Is=1.81e-15 Bf=50 Br=0.02 Va=100
+                          Tf=0.1ns Cje=1pF Cjc=1.5pF)
.ends NAND


** Main Circuit **
* DC supplies
Vcc 1 0 DC +5V
* input digital signals
Va 10 0 DC  +2.4; Va=Vohmin
Vb  9 0 DC  +5V
* first NAND gate: inputB is held high
Xnand_gate1 10 9 4  1 NAND
* current source load condition
Iload 1 4 0A
** Analysis Requests **
.Temp 27C
.DC Iload 0A 150mA 1mA
** Output Requests **
.Plot DC V(4)
.probe
.end
```

Figure 14.7 Spice input deck for calculating the low-state output voltage characteristic of the NAND gate circuit setup shown in Fig. 14.6(a) at a circuit temperature of 27° C.

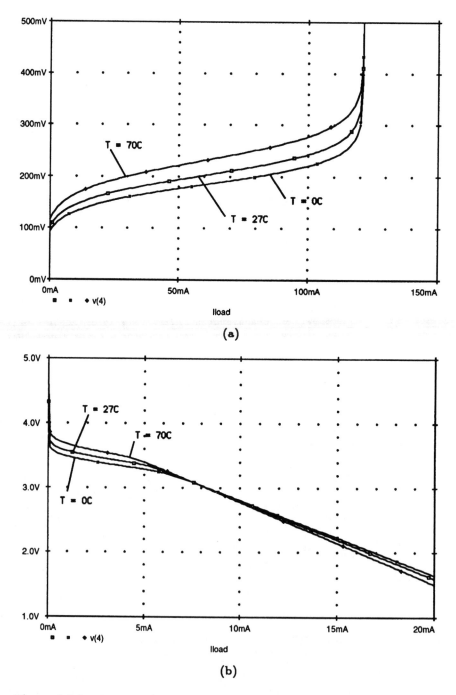

Figure 14.8 Output voltage characteristics of the TTL NAND gate in the two logic states under different load conditions: (a) low state and (b) high state.

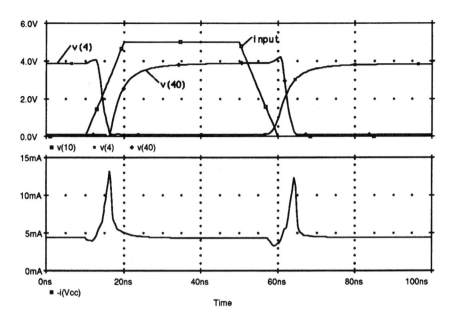

Figure 14.9 Transient response of a cascade of two TTL NAND gates (one input to each NAND gate is connected to logic high). The top graph displays the input voltage signal and the voltage waveforms that appear at the output of the two NAND gates. The bottom graph displays the current waveform associated with the power supply that powers each NAND gate.

This source statement creates a logic pulse with a 5 V amplitude and a duration of 40 ns. The rise and fall times are both equal to 10 ns. A transient analysis is then requested over 100 ns:

```
.TRAN 0.5ns 100ns 0s 0.5ns
```

Figure 14.9 shows several results of the Spice analysis. In the top graph of this figure the output voltage of the first and second NAND gates and the input voltage pulse are plotted. Using Probe we found the transition delay times of the first inverter to be $t_{PHL} = 0.62$ ns and $t_{PLH} = 6.64$ ns. Thus, the propagation time t_P for this TTL gate is 3.63 ns. Similar results are found for the second inverter. The lower graph in Fig. 14.9 displays the current supplied to the two NAND gates. Notice the current spike with a peak of 13.24 mA that occurs when the two NAND gates change states. During nonswitching times we see that the current drawn by the two NAND gates is a constant 4.41 mA.

14.2

Emitter-Coupled Logic (ECL)

Emitter-coupled logic, or ECL, is a nonsaturating family of bipolar digital circuits. Nonsaturating logic provides higher speed than digital circuits that incorporate transistors that saturate during normal operation, as the TTL logic family does. In the following discussion we use Spice to explore both the static and dynamic behavior of a simple ECL OR/NOR gate.

The basic schematic of an ECL OR/NOR gate is shown in Fig. 14.10. Unlike any of the previous logic families in this text, ECL logic has two outputs: an OR output and its complement, the NOR output. Both are connected to a -2 V level through two 50 Ω terminating resistors. The remaining transistors are biased using a -5.2 V supply. Applying Spice to the circuit of Fig. 14.10, we shall calculate the voltage transfer characteristics of both the OR and NOR outputs of the ECL gate as a function of the input voltage V_A. We shall keep the B input to the ECL gate at a logic low level (i.e., $V_B = -1.77$ V) throughout these calculations. We assume the following model parameters for each bipolar transistor: $I_S = 0.26$ fA, $\beta_F = 100$, $\beta_R = 1$, $\tau_F = 0.1$ ns, $C_{je} = 1$ pF, $C_{jc} = 1.5$ pF, and $V_A = 100$ V. The diodes of this ECL gate are realized by shorting the base and collector of an *npn* transistor together.

In Fig. 14.11 the Spice input file for the ECL requests that the level of the input voltage source V_A be varied between -2 V and 0 V. This represents the maximum voltage that can appear at the output of the ECL gate. The voltages at the OR and NOR outputs and at the base of Q_R (denoted V_R in Fig. 14.10) are all to be plotted.

Figure 14.12 shows the results for inputs between -2 and 0 V. Within the linear region of the gate the two transfer characteristics are symmetrical about an input voltage of -1.32 V. With inputs larger than -0.42 V the NOR transfer characteristic no

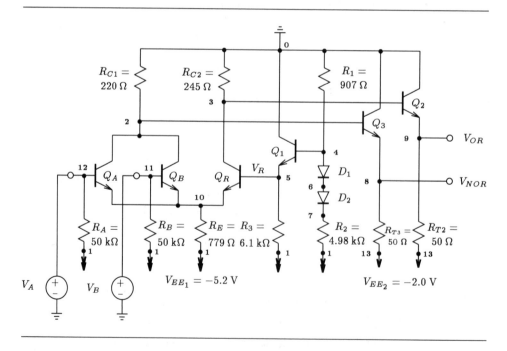

Figure 14.10 Two-input ECL OR gate with a complementary NOR output.

```
Two-Input ECL OR Gate with Complementary NOR Output

** Circuit Description **
* DC supplies
Vee1 1  0 DC -5.2V
Vee2 13 0 DC -2.0V
* input digital signals
Va 12 0 DC 0V
Vb 11 0 DC -1.77V
* ECL Gate
Qa 2 12 10 npn_transistor
Qb 2 11 10 npn_transistor
Qr 3  5 10 npn_transistor
Q2 0  3  9 npn_transistor
Q3 0  2  8 npn_transistor
Ra 12 1 50k TC=1200u
Rb 11 1 50k TC=1200u
Re 10 1 779 TC=1200u
Rc1 0 2 220 TC=1200u
Rc2 0 3 245 TC=1200u
Rt2 9 13 50 TC=1200u
Rt3 8 13 50 TC=1200u
* temperature-compensated voltage reference circuit
Q1 0  4 5 npn_transistor
QD1 4 4 6 npn_transistor
QD2 6 6 7 npn_transistor
R1  0 4 907    TC=1200u
R2  7 1 4.98k TC=1200u
R3  5 1 6.1k  TC=1200u
* BJT model statement
.model npn_transistor npn (Is=0.26fA Bf=100 Br=1
+                          Tf=0.1ns Cje=1pF Cjc=1.5pF Va=100)
** Analysis Requests **
.TEMP 27C
.DC Va -2V 0V 10mV
** Output Requests **
.Plot DC V(8) V(9) V(5)
.probe
.end
```

Figure 14.11 Spice input deck for calculating the input–output voltage transfer characteristics of the OR and NOR outputs of the ECL gate shown in Fig. 14.10.

longer behaves in the usual manner but begins to increase. It is therefore imperative that the input voltage to the ECL gate be limited to less than -0.42 V.

From the transfer curve for the OR output we use Probe to find that $V_{OL} = -1.77$ V, $V_{OH} = -0.884$ V, $V_{IL} = -1.41$ V, and $V_{IH} = -1.22$ V. Thus, the low and high noise margins are $NM_H = 0.339$ V and $NM_L = 0.358$ V. From the transfer curve for

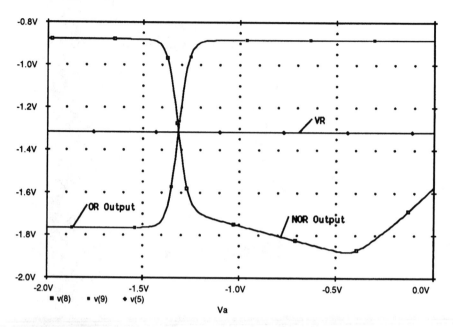

Figure 14.12 Input–output voltage transfer characteristics of the OR and NOR outputs of the ECL gate shown in Fig. 14.10.

the NOR output we find that $V_{OL} = -1.78$ V, $V_{OH} = -0.879$ V, $V_{IL} = -1.41$ V, and $V_{IH} = -1.22$ V. Thus, the corresponding low and high noise margins are $NM_H = 0.345$ V and $NM_L = 0.377$ V. In both cases, the NOR and OR outputs have noise margins that nearly equal one another (a maximum difference of 32 mV).

An important part of ECL logic is the voltage reference circuit. In Fig. 14.10 this consists of transistor Q_1, diodes D_1 and D_2, and resistors R_1, R_2, and R_3. This circuit is responsible for setting the DC voltage at the base of transistor Q_R. At a room temperature of 27° C, a Spice analysis shows the voltage at the base of Q_R to be -1.32 V, which is about equidistant from the high and low output voltage levels of the NOR and OR outputs. At different temperatures the level of this voltage will change. A well-designed ECL gate aims for the output voltage levels to remain at equal distance from the voltage generated by the reference voltage circuit V_R over a wide range of temperatures. We will demonstrate that the voltage reference circuit incorporated into the circuit in Fig. 14.10 accomplishes this goal by comparing the voltage transfer characteristics of an ECL OR/NOR gate with a temperature-compensated bias network in place (Fig. 14.10) and a corresponding gate with only a DC voltage source representing the bias network (Fig. 14.13). Our comparison will be based on computing the OR and NOR voltage characteristics of the gate at temperature extremes of 0° C and 70° C.

Figure 14.14 shows the results of the analysis. The graphs in Fig. 14.14(a) show the OR and NOR voltage transfer characteristics for the ECL gate with a temperature-compensated bias network and with the reference voltage V_R appearing at the base of Q_R. The graphs in Fig. 14.14(b) are the corresponding results without temperature compensation. We can see from Fig. 14.14(a) that regardless of the temperature of

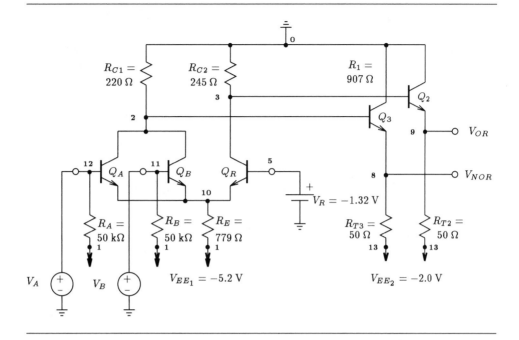

Figure 14.13 Removing the voltage reference bias circuit of the ECL OR/NOR gate shown in Fig. 14.10 and replacing it with a temperature-independent voltage source V_R.

the circuit the transfer characteristics are symmetrical about the reference voltage V_R. In contrast, the uncompensated results in Fig. 14.14(b) show that as the temperature changes the voltage characteristics are less symmetrical about V_R. To quantify our results, Table 14.1 tabulates the output voltages levels V_{OL} and V_{OH}, their corresponding average $V_{avg} = (V_{OL} + V_{OH})/2$, the reference level V_R, and $V_{avg} - V_R$. The closer V_{avg} is to V_R, the smaller $V_{avg} - V_R$ is and the greater the symmetry of the output voltage characteristic. We can see that over a 70° C temperature change V_{avg} for either the OR or NOR output deviates a maximum of 23 mV from V_R, but in the uncompensated ECL circuit V_{avg} deviates as much as 38 mV, thus indicating the advantages of incorporating a temperature-compensated bias circuit in an ECL gate.

Finally we investigate the dynamic operation of ECL gates. Let us consider the following situation: Two ECL gates are required to communicate over a distance of 1.5 m. A coaxial cable having a characteristic impedance (Z_O) of 50 Ω is used to connect them. Based on the manufacturer's data, signals propagate along this cable at about half the speed of light, or 15 cm/ns. Thus, we can expect a signal injected at one end of the coaxial cable to take 10 ns to reach the other end. If we assume that the transmission is lossless (i.e., ignore the skin effect), we can model the behavior of the coaxial cable using the **transmission line** element statement of Spice.

The general syntax of a lossless transmission line statement is provided in Fig. 14.15. The input port terminals of this transmission line are denoted $nA+$ and

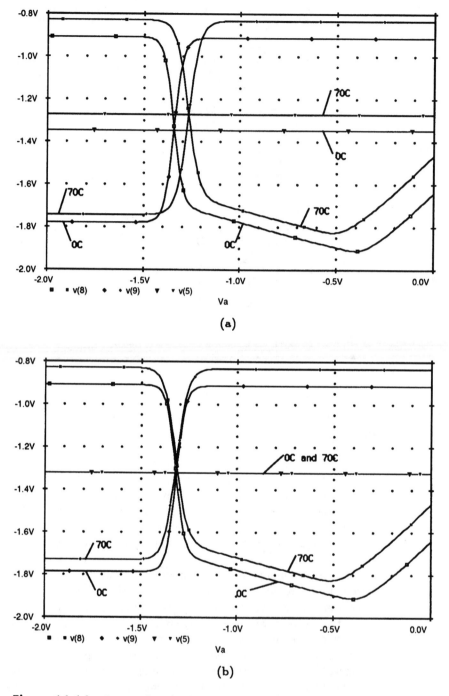

Figure 14.14 Comparing the input–output voltage transfer characteristics of the OR and NOR outputs of the ECL gate shown in Fig. 14.10 with (a) a temperature-compensated bias network and (b) a temperature-independent voltage source.

Table 14.1 Characteristics of an ECL gate with and without temperature compensation at two different temperatures.

Temperature	Parameter	Temperature Compensated		Not Temperature Compensated	
		OR	NOR	OR	NOR
0° C	V_{OL}	−1.779 V	−1.799 V	−1.786 V	−1.799 V
	V_{OH}	−0.9142 V	−0.9092 V	−0.9142 V	−0.9092 V
	$V_{avg} = \dfrac{V_{OL} + V_{OH}}{2}$	−1.3466 V	−1.3541 V	−1.3501 V	−1.3541 V
	V_R	−1.345 V	−1.345 V	−1.32 V	−1.32 V
	$\lvert V_{avg} - V_R \rvert$	1.6 mV	9.1 mV	30.1 mV	34.1 mV
70° C	V_{OL}	−1.742 V	−1.759 V	−1.729 V	−1.759 V
	V_{OH}	−0.8338 V	−0.8285 V	−0.8338 V	−0.8285 V
	$V_{avg} = \dfrac{V_{OL} + V_{OH}}{2}$	−1.288 V	−1.294 V	−1.2814 V	−1.294 V
	V_R	−1.271 V	−1.271 V	−1.32 V	−1.32 V
	$\lvert V_{avg} - V_R \rvert$	17 mV	23 mV	38 mV	26.2 mV

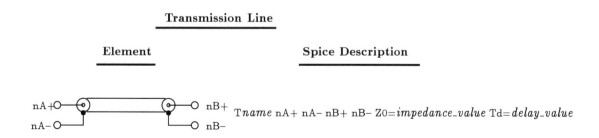

Transmission Line

Element	Spice Description

T*name* nA+ nA− nB+ nB− Z0=*impedance_value* Td=*delay_value*

Figure 14.15 Spice element description for a lossless transmission line.

$nA-$, and the output port terminals are denoted $nB+$ and $nB-$. The characteristic impedance Z_O of the transmission line and its propagation delay time T_d are specified by the two fields marked Z0 = *impedance_value* and Td = *delay_value*, respectively. Initial conditions at the input and output ports can also be specified, but these are purposely left out of the diagram of Fig. 14.15. Further details on the transmission line statement can be obtained from any Spice reference guide, such as [Vladimirescu et al., 1981].

Figure 14.16 shows the two ECL gates connected with 1.5 m of coaxial cable. It is assumed that the outputs of each gate are open collectors, thus each must be terminated with 50 Ω. Notice that the output of the ECL gate that is driving the coaxial cable is terminated by a 50 Ω resistor at the input to the second gate. This is

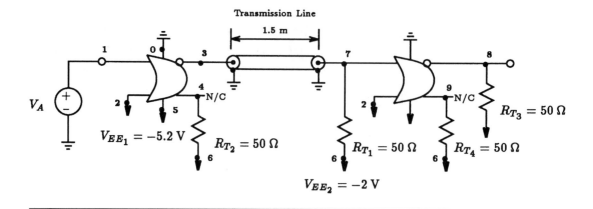

Figure 14.16   Circuit arrangement depicting a cascade of two ECL gates interconnected by 1.5 m of coaxial cable.

necessary to minimize line reflections. With the aid of Spice let us simulate this situation and determine the time needed for a logic change at one end (i.e., at the input to the first gate) to appear at the output of the second gate. Figure 14.17 shows the input deck for this circuit. A voltage step beginning at –1.77 V and rising to –0.884 V in 1 ns is input to the first gate, and a transient analysis over a 30 ns interval is requested. The inputs and outputs for each gate are to be plotted.

Figure 14.18 illustrates the results. The output change of the first ECL gate appears at the input to the second gate after 10 ns, which causes a new logic state to appear at the output of the second gate 1.06 ns later. Thus, we see that the overall propagation delay from the input to the first gate to the output of the second gate is 12.15 ns.

To complete our transient analysis, consider the effect of replacing this 50 Ω coaxial cable with one that has a characteristic impedance of 300 Ω. With the terminating resistance R_{T_1} remaining at 50 Ω the transmission line is no longer terminated properly. We should then expect signal reflections to appear along the transmission line. Let us resimulate this situation and determine what effect it has on the overall propagation delay. We can use the same Spice deck as before, except that we replace the transmission line element statement by

```
Tinterconnect 3 0 7 0 Z0=300 Td=10ns
```

We shall also modify the .TRAN command so that the transient analysis is performed over 400 ns, which was found sufficient for our purposes.

Figure 14.19 shows the results of this analysis. These results are very different from the results when the transmission line was terminated properly. Most notable is the voltage signal that appears at the input to the second gate (V(7)), which seems to change in small increments beginning at −0.879 V and decreasing to −1.16 V, followed by another small step change to −1.35 V, and again to −1.48 V, and so on, until it reaches the logic low level of −1.77 V. Probe finds the 90%-to-10% fall time associated with this decreasing signal to be 102 ns. Such a slow falling signal at the

```
* ECL two-input OR/NOR gate
.subckt OR/NOR_ECL    12 11 8 9 1
* connections:            |  |  |  |  |
*                  inputA |  |  |  |  |
*                    inputB |  |  |  |
*                    NORoutput |  |  |
*                        ORoutput |  |
*                            Vee1 |
*
* ECL gate
Qa 2 12 10 npn_transistor
Qb 2 11 10 npn_transistor
Qr 3  5 10 npn_transistor
Q2 0  3  9 npn_transistor
Q3 0  2  8 npn_transistor
Ra 12 1 50k TC=1200u
Rb 11 1 50k TC=1200u
Re 10 1 779 TC=1200u
Rc1 0 2 220 TC=1200u
Rc2 0 3 245 TC=1200u
* temperature-compensated voltage reference circuit
Q1 0  4 5 npn_transistor
QD1 4 4 6 npn_transistor
QD2 6 6 7 npn_transistor
R1  0 4 907   TC=1200u
R2  7 1 4.98k TC=1200u
R3  5 1 6.1k  TC=1200u
* BJT model statement
.model npn_transistor npn (Is=0.26fA Bf=100 Br=1
+                          Tf=0.1ns Cje=1pF Cjc=1.5pF Va=100)
.ends OR/NOR_ECL

** Main Circuit **
* DC supplies
Vee1 5 0 DC -5.2V
Vee2 6 0 DC -2.0V
* input digital signals
Va 1 0 PWL (0,-1.77V 2ns,-1.77V 3ns,-0.884V 30ns,-0.884V)
Vb 2 0 DC -1.77V
* first OR/NOR gate: inputB is held low
Xnor_gate1 1 2 3 4 5 OR/NOR_ECL
* second OR/NOR gate: inputB is held low
Xnor_gate2 7 2 8 9 5 OR/NOR_ECL
* transmission line interconnect + terminations
Tinterconnect 3 0 7 0 Z0=50 Td=10ns
Rt1 7 6 50
Rt2 4 6 50
Rt3 8 6 50
Rt4 9 6 50
** Analysis Requests **
.TRAN 0.05ns 30ns 0s 0.05ns
** Output Requests **
.Plot TRAN V(1) V(3) V(7) V(8)
.probe
.end
```

Figure 14.17 Spice input deck for calculating the transient response of a cascade of two ECL gates interconnected by 1.5 m of coaxial cable as shown in Fig. 14.16.

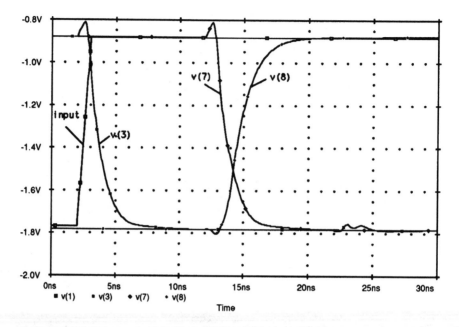

Figure 14.18   Transient response of a cascade of two ECL gates interconnected by a 1.5 m coaxial cable having a characteristic impedance of 50 Ω.

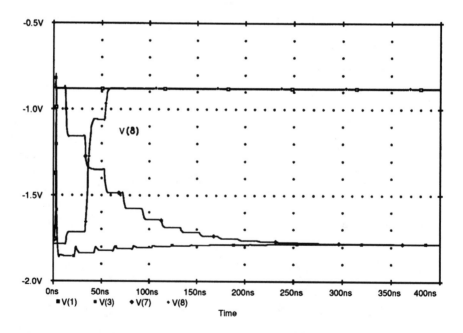

Figure 14.19   Transient response of a cascade of two ECL gates interconnected by a 1.5 m coaxial cable having a characteristic impedance of 300 Ω.

input to the second gate means this gate changes state 33.8 ns after the initiation of the level change at the input of the first gate—a rather long propagation delay. Moreover, the rise time of the second gate is also very slow at 20.7 ns. Our conclusion is that two ECL gates connected by a transmission line not properly terminated can give rise to unpredictably long propagation delays.

14.3

BiCMOS Digital Circuits

We conclude our study of digital circuits with a look at a new technology called BiCMOS that combines the low power, high packing density, and high input imped- ance of CMOS with the faster operation and higher drive capability of bipolar technol- ogy. We compare the transient behavior of a two-input NAND gate implemented with 5 μm[†] BiCMOS and CMOS technology. The advantages of a BiCMOS technology over a CMOS technology should become evident.

Figure 14.20 shows the circuit schematics of the two NAND gates. The BiC- MOS NAND gate shown in Fig. 14.20(a) consists of both n-channel and p-channel MOSFETs, *npn* bipolar transistors, and two 20 kΩ resistors, R_1 and R_2, that provide the gate with rail-to-rail voltage swing once the bipolar transistors turn off. These two resistors typically would be implemented using MOSFETs, but to keep the ideas sim- ple we shall stick to resistors. The logic gate shown in Fig. 14.20(b) is the familiar two-input CMOS NAND gate. Both circuits of Fig. 14.20 are loaded at the output with a 5 pF capacitor.

When we compare the BiCMOS NAND gate with its corresponding CMOS implementation in Fig. 14.20, we see that the BiCMOS circuit is made up of a CMOS NAND gate with a bipolar output stage. It is precisely the introduction of these bipolar transistors that increases the drive capability of the BiCMOS gate. For direct comparison we have Spice compute the output voltage and output current waveform of both types of gates.

Figure 14.21 provides the input file for BiCMOS NAND gate. The *A* input is driven by a voltage signal generator whose output produces a 5 V symmetrical square- wave signal of 500 kHz frequency. The *B* input is to be maintained at 5 V so that we can observe the effect of a single change at the input on the output. Initially, the *A* input to the NAND gate is logic low so that the output will begin in the logic high state. We may want to begin the transient analysis with the load capacitor initialized to 5 V, but we must be aware that the internal nodes of the gate will not be properly initialized. It is important that we allow at least one cycle of the square-wave input to complete before we conclude that we have a typical output waveform. Therefore, we shall request that the transient analysis be performed over two periods of the input signal. We will then focus our attention on the high-to-low output transition of the gate.

Figure 14.22 shows the analysis results. The top graph depicts the input and out- put voltage waveforms of the BiCMOS gate circuit between 1.8 μs and 2.3 μs. We can

---

[†] BiCMOS technologies tend to be in the submicron region. We use a 5 μm technology simply because its models are readily available to us.

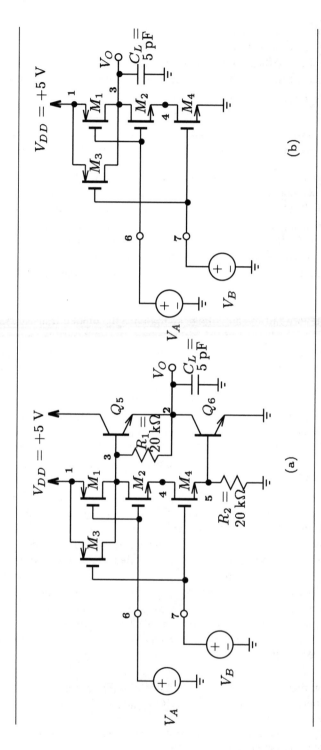

Figure 14.20 Logic NAND gate: (a) BiCMOS implementation and (b) CMOS implementation.

```
A BiCMOS Two-Input NAND Gate

** Circuit Description **
* DC supplies
Vdd 1 0 DC +5V
* input digital signals
Va 6 0 PULSE (0V 5V 0 10ns 10ns 1us 2us)
Vb 7 0 DC +5V
* CMOS input stage
M1 3 6 1 1 pmos_transistor L=5u W=30u
M2 3 6 4 0 nmos_transistor L=5u W=15u
M3 3 7 1 1 pmos_transistor L=5u W=30u
M4 4 7 5 0 nmos_transistor L=5u W=15u
* bipolar output stage
Q5 1 3 2 npn_transistor
Q6 2 5 0 npn_transistor
* high-swing resistors
R1 5 0 20k
R2 2 3 20k
* load capacitance
CL 2 100 5pF IC=+5V
* capacitor discharge current meter
VCL 100 0 0
* MOS and BJT model statements
.MODEL nmos_transistor nmos ( level=2 vto=1 nsub=1e16 tox=8.5e-8 uo=750
+ cgso=4e-10 cgdo=4e-10 cgbo=2e-10 uexp=0.14 ucrit=5e4 utra=0 vmax=5e4 rsh=15
+ cj=4e-4 mj=2 pb=0.7 cjsw=8e-10 mjsw=2 js=1e-6 xj=1u ld=0.7u )
.MODEL pmos_transistor pmos ( level=2 vto=-1 nsub=2e15 tox=8.5e-8 uo=250
+ cgso=4e-10 cgdo=4e-10 cgbo=2e-10 uexp=0.03 ucrit=1e4 utra=0 vmax=3e4 rsh=75
+ cj=1.8e-4 mj=2 pb=0.7 cjsw=6e-10 mjsw=2 js=1e-6 xj=0.9u ld=0.6u )
.model npn_transistor npn (Is=10fA Bf=100 Br=1 Tf=0.1ns Cje=1pF Cjc=1.5pF Va=100)
** Analysis Requests **
.TRAN 100ns 2.3us 1.8us 100ns UIC
** Output Requests **
.PLOT TRAN V(2) V(6) I(VC1)
.probe
.end
```

Figure 14.21   Spice input file used to calculate the discharging current waveform of the BiCMOS NAND gate shown in Fig. 14.20(a) when a square wave is applied to its A input. The B input is held logical high throughout this example. A zero-valued voltage source V_{CL} is placed in series with the load capacitor to monitor the capacitor discharge current.

deduce from these two waveforms that the high-to-low transitional delay t_{PHL} is about 25.2 ns. The waveform in the bottom graph represents the current that flows into the gate from the 5 pF load capacitor. The discharge current waveform simply rises and falls in accordance with the output voltage waveform. Using Probe we find that the peak value is 1.75 mA.

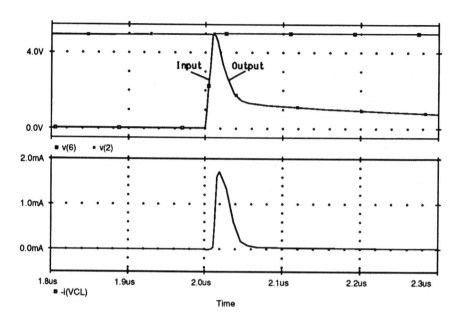

Figure 14.22 Voltage and current waveforms associated with the BiCMOS NAND gate shown in Fig. 14.20(a). The top graph illustrates the input and output voltage of the gate circuit. The bottom graph depicts the load capacitor C_L discharge current waveform.

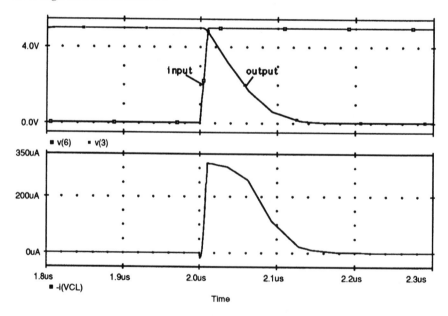

Figure 14.23 Voltage and current waveforms associated with the CMOS NAND gate shown in Fig. 14.20(b). The top graph illustrates the input and output voltage of the gate circuit. The bottom graph depicts the load capacitor C_L discharge current waveform.

The voltage and current waveforms associated with the CMOS NAND gate are shown in Figure 14.23. In constrast, here we see that the output voltage has a much longer high-to-low transitional delay t_{PHL} of 43.5 ns, an increase to 1.6 times the delay associated with the BiCMOS gate. In addition the discharge current waveform shown in the bottom graph does not simply increase to a peak level and fall back down, as was the case for the BiCMOS gate. The discharge current reaches a level of 317 μA and stays there for about 30 ns. In other words, the current available to discharge the load capacitor saturates. As a result the output voltage of the gate begins to slew and the delay time increases significantly. This shows clearly in the output waveform in the top graph of Fig. 14.23. Similar results also occur during the charging of the load capacitor.

In conclusion, the BiCMOS gate provides extra drive capability, preventing the gate from slew-rate limiting. As a result it has a shorter propagation delay than a corresponding CMOS gate.

14.4

Bibliography

A. Vladimirescu, K. Zhang, A. R. Newton, D. O. Pederson, and A. Sangiovanni-Vincentelli, *SPICE Version 2G6 User's Guide,* Dept. of Electrical Engineering and Computer Sciences, University of California, Berkeley, 1981.

14.5

Problems

14.1 A BJT for which $\beta_F = 100$ and $I_S = 10$ fA operates with a constant base current of 1 mA but with the collector open. Using Spice, determine the saturation voltage $V_{CE\text{sat}}$ for this transistor.

14.2 Determine the logic function implemented by the circuit shown in Fig. P14.2 by cycling through the four possible inputs using Spice.

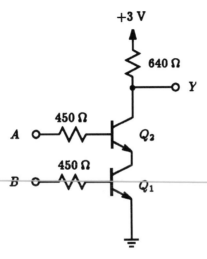

Figure P14.2

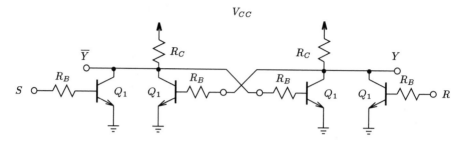

Figure P14.3

14.3 Consider the circuit in Fig. P14.3. If $V_{CC} = 5$ V and $R_C = R_B = 1$ kΩ, with the aid of Spice determine the voltage levels at Y and $\overline{Y}$ when the set and reset inputs are inactive but following an interval in which S was high while R was low. Assume that the transistors have device parameters $I_S = 10$ fA, $\beta_F = 100$, $\beta_R = 1$, $V_A = 50$ V, $\tau_F = 0.1$ ns, $C_{je} = 1$ pF, and $C_{jc} = 1.5$ pF.

14.4 For the diode–transistor logic (DTL) gate shown in Fig. P14.4, use Spice to determine the logic levels V_{OH}, V_{OL}, V_{IL}, and V_{IH}. Then calculate the corresponding noise margins. Assume that the transistor has device parameters $I_S = 10$ fA, $\beta_F = 100$, $\beta_R = 1$, $V_A = 50$ V, $\tau_F = 0.1$ ns, $C_{je} = 1$ pF, and $C_{jc} = 1.5$ pF. Further assume a base–collector shorted transistor for the pn junction diodes.

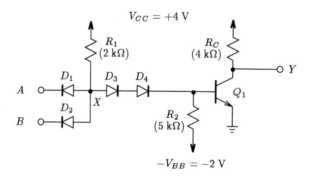

Figure P14.4

14.5 Repeat Problem 14.4 for a fan-out of 5.

14.6 For the DTL gate of Fig. P14.4, with the aid of Spice determine the total current in each supply and also the gate power dissipation for the two cases v_Y high and v_Y low. Then find the average power dissipation in the DTL gate. Use the same transistor parameters given in Problem 14.4.

14.7 A variant of the TTL gate shown in Fig. 14.1 in which all resistances are tripled is being considered. For both inputs high, use Spice to determine all node voltages and branch currents. Assume that the transistors have Spice parameters $I_S = 1.81$ fA, $\beta_F = 50$, $\beta_R = 0.1$, and $V_A = 100$ V.

14.8 For the BJT circuit shown in Fig. P14.8, use Spice to determine the following:
(a) What logic function is performed?
(b) Determine V_{OL} and V_{OH}.

(c) Find V_{IL} and V_{IH} at the A input when both the B and C inputs are held at ground potential.

(d) What are the noise margins?

(e) Determine the average power dissipated by this gate when both inputs B and C are held at ground.

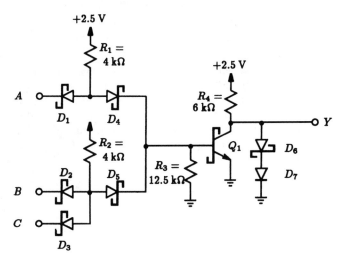

Figure P14.8

Assume that the transistors have device parameters $I_S = 10$ fA, $\beta_F = 100$, $\beta_R = 1$, $V_A = 50$ V, $\tau_F = 0.1$ ns, $C_{je} = 1$ pF, and $C_{jc} = 1.5$ pF. Further assume a base–collector shorted transistor for all pn junction diodes. For the Schottky diodes assume $I_S = 1$ pA, $n = 1$, $C_{j0} = 0.2$ pF, $\phi_o = 0.7$ V, and $m = 0.5$.

14.9 For the ECL circuit shown in Fig. P14.9, with the aid of Spice determine the following:

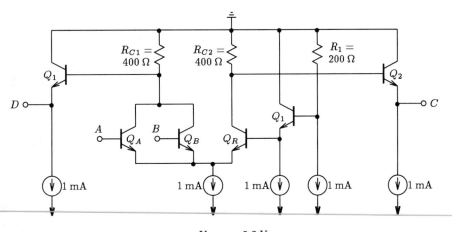

Figure P14.9

(a) Find V_{OH} and V_{OL}.

(b) With the input at B sufficiently negative for Q_B to be cut off, what voltage at A causes a current of 0.5 mA to flow in Q_R?

(c) Repeat (b) for a current in Q_R of 0.99 mA.

(d) Repeat (b) for a current in Q_R of 0.01 mA.

(e) Use the results of (c) and (d) to specify V_{IL} and V_{IH}.

(f) Find NM_H and NM_L.

Assume that the transistors have Spice parameters $I_S = 10$ fA, $\beta_F = 50$, $\beta_R = 0.1$, and $V_A = 75$ V.

14.10 For the BiCMOS and CMOS NAND gates shown in Fig. 14.20, with the aid of Spice compute the propagation delay time for the following capacitive loads: 10 fF, 100 fF, 1 pF, and 10 pF. The propagation delay time is defined as the time between the midway points of the input and output voltage waveforms. Sketch the delay time versus load capacitance for both gates. What can you conclude from these results? Use the same transistor models and dimensions as in Section 14.3.

A

Device Model Parameters

Each built-in semiconductor device of Spice is modeled with a sophisticated set of mathematical equations that describe the static and dynamic terminal behavior of the device. To allow users to customize a particular device to their applications, a set of parameters can be specified on a .MODEL statement according to the following Spice syntax:

```
.MODEL   model_name   type   (parameter_list   . . . )
```

Here the field labeled *model_name* is the model name, and the field labeled *type* is one of the following eight device types:

Type	Description
D	Diode model
NPN	*npn* BJT model
PNP	*pnp* BJT model
NJF	N-channel JFET model
PJF	P-channel JFET model
NMOS	N-channel MOSFET model
PMOS	P-channel MOSFET model
GASFET	N-channel MESFET model

The next field, labeled *parameter_list,* describes a set of parameters that characterize the semiconductor device. A discussion of these parameters is the focus of this appendix.

In the following discussion we shall outline all the parameters that make up the various models of Spice. This includes a description of the model parameters for semiconductor diodes, bipolar transistors (BJTs), and junction and metal-oxide-semiconductor field-effect transistors (JFETs and MOSFETs). Finally, we conclude with a description of the model parameters that make up the metal-semiconductor field-effect transistor (MESFET) found in PSpice.

Table A.1   Semiconductor diode model parameters.

Spice Name	Model Parameter	Units	Default	Example
IS	Saturation current	A	1×10^{-14}	1.0×10^{-14}
RS	Ohmic resistance	Ω	0	10
N	Emission coefficient	—	1	1.0
TT	Transit time	s	0	0.1 ns
CJO	Zero-bias junction capacitance	F	0	2 pF
VJ	Junction potential	V	1	0.6
M	Grading coefficient	—	0.5	0.5
EG	Activation energy	eV	1.11	1.11 Si
				0.69 Sbd
				0.67 Ge
XT1	Saturation-current temperature coefficient	—	3.0	3.0 pn-junction
				2.0 Sbd
KF	Flicker noise coefficient	—	0	
AF	Flicker noise exponent	—	1	
FC	Coefficient for forward-bias depletion capacitance formula	—	0.5	
BV	Reverse-bias breakdown voltage	V	∞	40.0
IBV	Reverse-bias breakdown current	A	1×10^{-10}	1.0×10^{-10}

A.1

Diode Model

The DC characteristics of the diode are determined by the parameters IS and N. A resistor RS accounts for the series ohmic resistance of the diode. Charge storage effects are modeled by a transit time TT and a nonlinear depletion-layer capacitance. The parameters that affect the depletion layer capacitance are CJO, VJ, M, and FC. The temperature dependence of the saturation current is defined by the parameters EG and XTI. The flicker noise behavior of the diode is defined by the parameters KF and AF. Reverse breakdown is modeled by an exponential increase in the reverse diode current and is determined by the parameters BV and IBV.

The parameters used to model a semiconductor diode in Spice are listed in Table A.1.

A.2

BJT Model

The bipolar junction trasistor model in Spice is an adaptation of the integral charge control model of Gummel and Poon. The model will automatically simplify to the Ebers–Moll model when certain parameters are not specified.

The forward static current gain characteristic of the BJT is defined by the parameters IS, BF, NF, ISE, IKF, and NE. The corresponding reverse current gain characteristic of the BJT is defined by the parameters IS, BR, NR, ISC, IKR, and NC. The output conductances of the forward and reverse regions of the transistor are determined by VAF and VAR, respectively. Resistors RB, RC, and RE represent an ohmic resistance in series with each terminal of the BJT. The current dependence of RB can be modeled by the parameters IRB and RBM. Base charge storage is modeled by forward and reverse transit times, TF and TR. The nonlinear depletion-layer capacitances are determined by CJE, VJE, and MJE for the base–emitter junction, by CJC, VJC, and MJC for the base–collector junction, and by CJS, VJS, and MJS for the collector–substrate junction. The temperature dependence of the saturation current IS is determined by parameters EG and XTI. Additionally, base current temperature dependence is modeled by parameter XTB. The flicker noise behavior of the diode is defined by the parameters KF and AF.

The BJT parameters used in the modified Gummel–Poon model are listed in Table A.2. There are 40 parameters associated with this model.

Table A.2   BJT model parameters.

Spice Name	Model Parameter	Units	Default	Example
IS	Transport saturation current	A	1×10^{-16}	1.0×10^{-15}
BF	Ideal maximum forward beta	—	100	100
NF	Forward current emission coefficient	—	1	1.0
VAF	Forward Early voltage	V	∞	200
IKF	Corner for forward beta high-current roll-off	A	∞	0.01
ISE	B–E leakage saturation current	A	0	1.0×10^{-13}
NE	B–E leakage emission coefficient	—	1.5	2.0
BR	Ideal maximum reverse beta	—	1	0.1
NR	Reverse current emission coefficient	—	1	1.0
VAR	Reverse Early voltage	V	∞	200
IKR	Corner for reverse beta high-current roll-off	A	∞	0.01
ISC	B–C leakage saturation current	A	0	1.0×10^{-13}
NC	B–C leakeage emission coefficient	—	2	1.5
RB	Base ohmic resistance	Ω	0	100
IRB	Current where base resistance falls halfway to its minimum value	A	∞	0.1
RBM	Minimum base resistance at high currents	Ω	RB	10
RE	Emitter resistance	Ω	0	1
RC	Collector resistance	Ω	0	10
CJE	B–E zero-bias depletion capacitance	F	0	2 pF

Continued

Table A.2   (*Continued*)

Spice Name	Model Parameter	Units	Default	Example
VJE	B–E built-in potential	V	0.75	0.6
MJE	B–E junction exponential factor	—	0.33	0.33
TF	Ideal forward tansit time	s	0	0.1 ns
XTF	Coefficient for bias dependence of TF	—	0.75	0.6
VTF	Voltage describing VBC dependence on TF	V	∞	
ITF	High-current parameter for effect on TF	A	0	
PTF	Excess phase at freq = $1/(2\pi TF)$	degree	0	
CJC	B–C zero-bias depletion capacitance	F	0	2 pF
VJC	B–C built-in potential	V	0.75	0.5
MJC	B–C junction exponential factor	—	0.33	0.5
XCJC	Fraction of B–C depletion capacitance connected to internal base node	—	1	
TR	Ideal reverse transit time	s	0	10 ns
CJS	Zero-bias collector–substrate capacitance	F	0	2 pF
VJS	Substrate junction built-in potential	V	0.75	
MJS	Substrate junction expontial factor	—	0	0.5
XTB	Forward and reverse beta temperature exponent	—	0	
EG	Energy gap for temperature effect on IS	eV	1.11	
XTI	Temperature exponent for effect on IS	—	3	
KF	Flicker-noise coefficient	—	0	
AF	Flicker-noise exponent	—	1	
FC	Coefficient for forward-bias depletion capacitance formula	—	0.5	

A.3

JFET Model

The JFET model in Spice is derived from the FET model of Shichman and Hodges. The DC characteristics are defined by the parameters VTO, BETA, LAMBDA, and IS. Two ohmic resistances RD and RS are included in series with the drain and source terminals of the JFET. Charge storage is modeled by a nonlinear depletion-layer capacitance for both gate junctions using parameters CGS, CGD, PB, and FC. The flicker noise behavior of the JFET is defined by the parameters KF and AF.

The JFET model parameters are listed in Table A.3.

Table A.3   JFET model parameters.

Spice Name	Model Parameter	Units	Default	Example
VTO	Threshold voltage	V	−2.0	−2.0
BETA	Transconductance parameter	A/V^2	1×10^{-4}	1×10^{-3}
LAMBDA	Channel length modulation parameter	1/V	0	1×10^{-4}
RD	Drain ohmic resistance	Ω	0	100
RS	Source ohmic resistance	Ω	0	100
CGS	Zero-bias G–S junction capacitance	F	0	5 pF
CGD	Zero-bias G–D junction capacitance	F	0	1 pF
PB	Gate junction potential	V	1	0.6
IS	Gate junction saturation current	A	1×10^{-14}	1×10^{-14}
KF	Flicker noise coefficient	—	0	
AF	Flicker noise exponent	—	1	
FC	Coefficient for forward-bias depletion capacitance formula	—	0.5	

A.4

MOSFET Model

Spice provides three MOSFET device models that have different large-signal *i–v* characteristics. The variable LEVEL specifies the model that is to be used to represent a particular MOSFET:

LEVEL = 1 $\longrightarrow$ Shichman–Hodges

LEVEL = 2 $\longrightarrow$ MOS2, an analytical model (as described in [A. Vladimirescu et al., 1981])

LEVEL = 3 $\longrightarrow$ MOS3, a semi-empirical model (see [A. Vladimirescu et al., 1981])

The DC characteristics of the MOSFET are defined by the device parameters VTO, KP, LAMBDA, PHI, and GAMMA. These parameters are computed by Spice if process parameters (NSUB, TOX,...) are given and user-specified values are not given instead. Two ohmic resistances RD and RS are included in series with the drain and source terminals of the MOSFET. Charge storage is modeled by a nonlinear thin-oxide capacitance, several nonlinear depletion-layer capacitances, and overlap capacitances. There are two built-in models of the charge storage effects associated with the thin-oxide. The flag/coefficient XQC determines which of the two models will be used; a voltage-dependent or a charge-controlled capacitance model [A. Vladimirescu et al., 1981]. Other parameters of the MOSFET model that determine the charge storage effects are CBD, CBS, CJ, CJSW, MJ, MJSW, PB, and FC. The overlap capacitances are set by the parameters CGSO, CGDO, and CGBO. The flicker noise behavior of the diode is defined by the parameters KF and AF.

The MOSFET parameters used for the three different MOSFET models in Spice are listed in Table A.4. There are 42 parameters associated with the three models of the MOSFET.

Table A.4   MOSFET model parameters.

Spice Name	Model Parameter	Units	Default	Example
LEVEL	Model index (e.g., 1, 2, or 3)	—	1	
VTO	Zero-bias threshold voltage	V	0	1.0
KP	Transconductance parameter	A/V^2	2.0×10^{-5}	3.1×10^{-5}
GAMMA	Bulk threshold parameter	V$^{1/2}$	0	0.37
PHI	Surface potential	V	0.6	0.65
LAMBDA	Channel-length modulation (level 1 and 2 only)	1/V	0	0.02
RD	Drain ohmic resistance	Ω	0	1.0
RS	Source ohmic resistance	Ω	0	1.0
CBD	Zero-bias B–D junction capacitance	F	0	20 fF
CBS	Zero-bias B–S junction capacitance	F	0	20 fF
IS	Bulk junction saturation current	A	1.0×10^{-14}	1.0×10^{-15}
PB	Bulk junction potential	V	0.8	0.87
CGSO	Gate–source overlap capacitance per meter channel width	F/m	0	4.0×10^{-11}
CGDO	Gate–drain overlap capacitance per meter channel width	F/m	0	4.0×10^{-11}
CGBO	Gate–bulk overlap capacitance per meter channel length	F/m	0	2.0×10^{-10}
RSH	Drain and source diffusion sheet resistance	Ω/sq.	0	10.0
CJ	Zero-bias bulk junction bottom capacitance per square meter of junction area	F/m^2	0	2.0×10^{-4}
MJ	Bulk junction bottom grading coefficient	—	0.5	0.5
CJSW	Zero-bias bulk junction sidewall capacitance per meter of junction perimeter	F/m	0	2.0×10^{-9}
MJSW	Bulk junction sidewall coefficient	—	0.33	
JS	Bulk junction saturation current per square meter of junction area	A/m^2	1.0×10^{-8}	
TOX	Oxide thickness	m	1.0×10^{-7}	1.0×10^{-7}
NSUB	Substrate doping	1/cm^3	0	4.0×10^{15}
NSS	Surface state density	1/cm^2	0	1.0×10^{10}
NFS	Fast surface state density	1/cm^2	2×10^{-5}	1.0×10^{10}
TPG	Type of gate material: +1 opposite to substrate −1 same as substrate 0 Al gate	—	1	
XJ	Metallurgical junction depth	m	0	1.0 μm
LD	Lateral diffusion	m	0	0.8 μm
UO	Surface mobility	cm^2/(V · s)	600	700
UCRIT	Critical field for mobility degradation (level 2 only)	V/cm	1×10^4	1.0×10^4

Table A.4   *(Continued)*

Spice Name	Model Parameter	Units	Default	Example
UEXP	Critical field exponent in mobility degradation (level 2 only)	—	0	0.1
UTRA	Transverse field coefficient (mobility) (deleted for level 2)	—	0	0.3
VMAX	Maximum drifty velocity of carriers	m/s	0	5.0×10^4
NEFF	Total channel charge (fixed and mobile) coefficient (level 2 only)	—	1	5.0
XQC	Thin-oxide capacitance model flag and coefficient of channel charge share attributed to drain (0–0.5)	—	1	0.4
KF	Flicker-noise coefficient	—	0	1.0×10^{-26}
AF	Flicker-noise exponent	—	1	1.2
FC	Coefficient for forward-bias depletion capacitance formula	—	0.5	
DELTA	Width effect on threshold voltage (level 2 and level 3)	—	0	1.0
THETA	Mobility modulation (level 3 only)	1/V	0	0.1
ETA	Static feedback (level 3 only)	—	0	1.0
KAPPA	Saturation field factor (level 3 only)	—	0.2	0.5

A.5

MESFET Model

The MESFET model that we describe here is that provided in the PSpice program by MicroSim Corporation. Spice version 2G6 does not have a device model for the MESFET.

PSpice provides three MESFET device models that have different large-signal *i–v* characteristics. The variable LEVEL specifies the model that is to be used to represent a particular MESFET:

$$\text{LEVEL} = 1 \longrightarrow \text{Curtice}$$
$$\text{LEVEL} = 2 \longrightarrow \text{Raytheon}$$
$$\text{LEVEL} = 3 \longrightarrow \text{TriQuint}$$

The MESFET is modeled as an intrinsic JFET with resistances RD, RS, and RG in series with the drain, source, and gate, respectively. The DC characteristics are defined by the parameters VTO, BETA, ALPHA, LAMBDA, IS, N, and M. Charge storage is modeled by a nonlinear depletion-layer capacitance for both gate junctions using parameters CGS, CGD, PB, and FC. A capacitance between drain and source

Table A.5 MESFET model parameters.

PSpice Name	Model Parameter	Units	Default	Example
LEVEL	Model index (e.g., 1, 2, or 3)	—	1	2
VTO	Threshold voltage	V	−2.5	−2.0
ALPHA	Saturation voltage parameter	V^{-1}	2.0	1.0
BETA	Transconductance parameter	A/V^2	0.1	1×10^{-2}
B	Doping tail extending parameter (level = 2 only)	V^{-1}	0.3	
LAMBDA	Channel length modulation parameter	V^{-1}	0	
GAMMA	Static feedback parameter (level = 3 only)	—	0	
DELTA	Output feedback parameter (level = 3 only)	$(A \cdot V)^{-1}$	0	
Q	Power-law parameter (level = 3 only)	—	2	
TAU	Conduction current delay time	s	0	
RG	Gate ohmic resistance	Ω	0	
RD	Drain ohmic resistance	Ω	0	100
RS	Source ohmic resistance	Ω	0	100
IS	Gate junction saturation current	A	1×10^{-14}	1×10^{-14}
N	Gate junction emission coefficient	—	1	
M	Gate junction grading coefficient	—	0.5	
VBI	Gate junction potenital	V	1	
CGS	Zero-bias G–S junction capacitance	F	0	1 pF
CGD	Zero-bias G–D junction capacitance	F	0	1 pF
CDS	D–S capacitance	F	0	1 pF
FC	Coefficient for forward-bias depletion capacitance formula	—	0.5	
EG	Bandgap voltage	eV	1.11	
XTI	IS temperature exponent	—	0	
VTOTC	VTO temperature coefficient	V/°C	0	
BETATCE	BETA exponential temperature coefficient	%/°C	0	
TRG1	RG temperature coefficient (linear)	$°C^{-1}$	0	
TRD1	RD temperature coefficient (linear)	$°C^{-1}$	0	
TRS1	RS temperature coefficient (linear)	$°C^{-1}$	0	
KF	Flicker-noise coefficient	—	0	
AF	Flicker-noise exponent	—	1	

CDS can also be declared. The flicker noise behavior of the diode is defined by the parameters KF and AF. Effects of temperature can also be modeled using parameters EG, XTI, VTOTC, BETATCE, TRG1, TRD1, and TRS1.

The MESFET parameters used to describe the three different PSpice models are listed in Table A.5.

A.6

Bibliography

A. Vladimirescu, K. Zhang, A. R. Newton, D. O. Pederson, and A. Sangiovanni-Vincentelli, *SPICE Version 2G6 User's Guide,* Dept. of Electrical Engineering and Computer Sciences, University of California, Berkeley, 1981.

P. Antognetti and G. Massobrio, *Semiconductor Device Modeling with SPICE,* New York: McGraw-Hill, 1988.

A. Vladimirescu and S. Liu, *The Simulation of MOS Integrated Circuits Using SPICE2,* Memorandum No. M80/7, Electronic Research Laboratory, University of California, Berkeley, Feb. 1980.

PSpice Users' Manual, MicroSim Corporation, Irvine, CA, Jan. 1991.

B

Spice Options

Spice allows the user to reset program control and specify user options for various simulation purposes. This is accomplished using an .OPTIONS statement in the Spice input file. The syntax of the .OPTIONS statement has the following form:

`.OPTIONS` `list_of_options`

Table B.1 lists the various options available in Spice version 2G6. Similar options are also available in PSpice; refer to the *PSpice User's Manual* for the exact details. Any combination of the options listed in Table B.1 may be included in the *list_of_options* in any order. There are two kinds of options: flags that initiate specific action and flags that reassign a value to a specific parameter. The variable *x* associated with these types of flags in this table represents some positive number.

Table B.1 Options available in Spice.

Options	Effect
ACCT	Causes accounting and run time statistics to be printed.
LIST	Causes the summary listing of the input data to be printed.
NOMOD	Suppresses the printout of the model parameters.
NOPAGE	Suppresses page ejects.
NODE	Causes the printing of the node table.
OPTS	Causes the options values to be printed.
GMIN=x	Sets the value of GMIN, the minimum conductance allowed by the program. The default value is 1.0×10^{-12}.
RELTOL=x	Resets the relative error tolerance of the program. The default value is 0.001.
ABSTOL=x	Resets the absolute current error tolerance of the program. The default value is 1 pA.
VNTOL=x	Resets the absolute voltage error tolerance of the program. The default value is 1 μV.

Table B.1   *(Continued)*

Options	Effect
TRTOL = x	Resets the transient error tolerance. The default value is 7.0. This parameter is an estimate of the factor by which Spice overestimates the actual truncation error.
CHGTOL = x	Resets the charge tolerance of the program. The default value is 1.0×10^{-14}.
PIVTOL = x	Resets the absolute minimum value for a matrix entry to be accepted as a pivot. The default is 1.0×10^{-13}.
PIVREL = x	Resets the relative ratio between the largest column entry and an acceptable pivot value. The default value is 1.0×10^{-3}.
NUMDGT = x	The number of significant digits printed for output variable values. The variable x must satisfy the relation $0 < x < 8$. The default is 4. Note: This option is independent of the error tolerance used by Spice (i.e., if the values of options RELTOL, ABSTOL, etc., are not changed, one may be printing numerical "noise" for NUMDGT > 4.
TNOM = x	Resets the nominal temperature. The default value is 27° C (300 K).
ITL1 = x	Resets the DC iteration limit. The default is 100.
ITL2 = x	Resets the DC transfer curve iteration limit. The default is 50.
ITL3 = x	Resets the lower transient analysis iteration limit. The default is 4.
ITL4 = x	Resets the transient analysis timepoint iteration limit. The default is 10.
ITL5 = x	Resets the transient analysis total iteration limit. The default is 5000. Set ITL5 = 0 to omit this test.
ITL6 = x	Resets the DC iteration limit at each step of the source-stepping method. The default is 0, which means not to use this method.
CPTIME = x	The maximum CPU time in seconds allowed for this job.
LIMTIM = x	Resets the amount of CPU time reserved by Spice for generating plots should a CPU time limit cause job termination. The default value is 2 s.
LIMPTS = x	Resets the total number of points that can be printed or plotted in a DC, AC, or transient analysis. The default value is 201.
LVLTIM = x	If x is 1, the iteration timestep control is used. If x is 2, the truncation-error timestep is used. The default value is 2. If METHOD = GEAR and MAXORD > 2, then LVLTIM is set to 2 by Spice.
METHOD = name	Sets the numerical integration method used by Spice. Possible names are GEAR or TRAPEZOIDAL. The default is trapezoidal.
MAXORD = x	Sets the maximum order for the integration method if Gear's variable-order method is used. The variable x must be between 2 and 6. The default is 2.
DEFL = x	Resets the value for MOS channel length. The default is 100.0 μm.
DEFW = x	Resets the value for MOS channel width. The default is 100.0 μm.
DEFAD = x	Resets the value for MOS drain diffusion area. The default is 0.0.
DEFAS = x	Resets the value for MOS source diffusion area. The default is 0.0.

Index